微机原理及应用

The Principle and Application of Microcomputer

丁艳 编著

国防工业出版社

·北京·

内 容 简 介

本书以16位微处理器为核心,全面讲述了微型计算机的基本组成、工作原理以及硬件接口技术。全书共8章,逐一讲述了计算机基础知识、微型计算机的基本组成及工作原理、16位微处理器8086/8088CPU的寻址方式、指令系统、汇编程序的设计、输入/输出接口、存储器系统和可编程定时/计数控制器等内容。本书是作者多年教学经验的结晶,内容编排合理,由浅入深,体系完整,重点突出并配有丰富的例题及详尽的注释。

本书采用全英文编写,并配有部分汉语注释,适合用作普通高等院校非计算机类各专业学生"微型计算机原理及应用"课程的教材,也可用作成人高等教育的培训教材及广大科技工作者的参考书。

图书在版编目(CIP)数据

微机原理及应用 = The Principle and Application of Microcomputer:英文 / 丁艳编著. —北京:国防工业出版社,2016.1
ISBN 978-7-118-10410-3

Ⅰ. ①微… Ⅱ. ①丁… Ⅲ. ①微型计算机-英文 Ⅳ. ①TP36

中国版本图书馆 CIP 数据核字(2015)第234161号

※

国防工业出版社出版发行
(北京市海淀区紫竹院南路23号 邮政编码100048)
天利华印刷装订有限公司印刷
新华书店经售

*

开本 787×1092 1/16 印张 21¼ 字数 490千字
2016年1月第1版第1次印刷 印数1—2500册 定价58.00元

(本书如有印装错误,我社负责调换)

国防书店:(010)88540777 发行邮购:(010)88540776
发行传真:(010)88540755 发行业务:(010)88540717

前　言

从第一台计算机诞生至今,微型计算机技术得到了迅速发展。为了满足高素质人才培养的需求,在高等院校中,很多专业都开设了微型计算机原理及应用的相关课程。

本书编写的主要目的是为高等院校非计算机专业"微型计算机原理及应用"课程提供教材,为了满足双语教学的需求,本书采用全英文编写。考虑到非计算机专业学时的限制、本门课程的定位以及学生对语言理解的精准度,本书仍然以 IBM PC/XT 或 AT 机型为背景机,以 16 位微处理器为核心,全面讲述微机系统的组成、工作原理、硬件接口技术和典型应用。在此基础上,帮助学生系统掌握汇编语言程序设计基本方法和微机硬件接口技术,建立微机系统的整体概念,以使其具有一定的微机软件及硬件初步开发、设计的能力。

全书分为 8 章。第 1 章介绍计算机中的数制、逻辑电路等基础知识。第 2 章介绍微型计算机的系统组成及基本组成电路,包括算术逻辑部件(ALU)、触发器、寄存器及存储器的基本原理及其符号等,同时以一个简化计算机为例剖析了微型计算机的工作原理。第 3 章以 8086/8088CPU 为例重点介绍 16 位微处理器的结构、工作模式及操作功能。第 4 章介绍 8086/8088CPU 的寻址方式及指令系统。第 5 章介绍微型计算机的汇编语言及汇编程序。第 6 章介绍微型计算机的程序设计步骤及基本的程序形式。第 7 章介绍微型计算机存储器分类及存储器扩展技术。第 8 章介绍输入/输出接口及其编程方法,并对典型的串行通信接口芯片、并行接口芯片及可编程定时/计数控制器进行介绍。

在内容的取舍及编排上,我们从非计算机专业的特点出发,在保证体系完整的前提下,力求深入浅出,循序渐进,适用实用,系统与原理并重,既有利于教师组织课程教学、实验教学,又便于学生自学。为了保证学生的学习效果,本书对部分难理解的知识点给出了汉语注释。

本书的编写过程中,我们参阅了大量的资料,在此谨向引为本书内容和作为本书参考资料的各位作者、译者表示由衷的感谢。在本书的成稿过程中,薛斌、王建林、杨慧蕾、王晓熔、李浩、袁瑜键、胡雪梅、李佳等同学做了大量的文字校对工作,在此向他们致以诚挚的谢意。

由于编者水平有限,本书中错误和不当之处在所难免,敬请各位读者和专家批评指正。

编者

CONTENTS

Chapter 1　Fundamentals of Computer 1
　1.1　Number System 1
　　1.1.1　Basic Number Systems 1
　　1.1.2　Conversions between Different Number Systems 3
　　1.1.3　Common Data Units 7
　1.2　Logic Algebra and Logic Gates 9
　　1.2.1　"OR" Operation and "OR" Gate 10
　　1.2.2　"AND" Operation and "AND" Gate 10
　　1.2.3　"NOT" Operation and the NOT-Inverter 11
　　1.2.4　Basic Rules of Boolean Algebra 12
　1.3　Binary Numbers and Binary Addition/Subtraction 13
　　1.3.1　Negative Binary Numbers 13
　　1.3.2　Binary Addition 14
　　1.3.3　Binary Subtraction 14
　　1.3.4　Adder Circuits 15
　　1.3.5　Switchable Inverter and Binary Addition/Subtraction Circuit 16
　1.4　Computer Data Formats 17
　　1.4.1　ACSII Code 17
　　1.4.2　BCD (Binary-Coded Decimal) Format 19
　1.5　Logic Circuit 19
　　1.5.1　Logic Gates 19
　　1.5.2　Flip-Flops 19
　　1.5.3　Registers 23
　　1.5.4　Tri-state Gate and BUS Structure 25
　Tips 29
　Exercise 30

Chapter 2　System Organization of Microcomputer 32
　2.1　The Basic System Components 32
　　2.1.1　CPU (Central Processing Unit) 33
　　2.1.2　The System Bus 33
　　2.1.3　The Memory Subsystem 35
　　2.1.4　The I/O Subsystem 38

  2.2 A Simple Computer ……………………………………………………………… 41
    2.2.1 Introduction ……………………………………………………………… 41
    2.2.2 Architecture of the Simple Computer ……………………………………… 41
    2.2.3 Instruction Set of the Simple Computer …………………………………… 45
    2.2.4 Encoding Instructions …………………………………………………… 46
    2.2.5 Organization of the Control Unit ………………………………………… 47
    2.2.6 Step-by-Step Instruction Execution ……………………………………… 49
  Tips ……………………………………………………………………………………… 51
  Exercise ………………………………………………………………………………… 52

Chapter 3 Intel 8086 Microprocessor ……………………………………………………… 54
  3.1 The History of Intel Microprocessor Family ………………………………………… 54
  3.2 8086 CPU Architecture ………………………………………………………………… 59
    3.2.1 Execution Unit and Bus Interface Unit …………………………………… 59
    3.2.2 Organization of Execution Unit …………………………………………… 61
    3.2.3 Organization of Bus Interface Unit (BIU) ………………………………… 64
  3.3 Internal Memory ……………………………………………………………………… 66
    3.3.1 Addressing Data in Memory ……………………………………………… 66
    3.3.2 Memory Segment ………………………………………………………… 67
    3.3.3 Segment Boundary ……………………………………………………… 68
    3.3.4 Segment Offset …………………………………………………………… 68
    3.3.5 About Stack Segment …………………………………………………… 71
  3.4 System Timing ………………………………………………………………………… 72
    3.4.1 The System Clock ………………………………………………………… 72
    3.4.2 Memory Access Time …………………………………………………… 73
    3.4.3 Wait States ……………………………………………………………… 73
    3.4.4 Bus Cycle ………………………………………………………………… 75
  3.5 8086 Pin Assignments and Working Modes ………………………………………… 75
    3.5.1 Pins and Their Function Descriptions …………………………………… 76
    3.5.2 Working Modes …………………………………………………………… 78
  3.6 Basic Operations of 8086/8088 ……………………………………………………… 88
    3.6.1 Reset Operation ………………………………………………………… 88
    3.6.2 Input and Output for 8086 Minimum Mode ……………………………… 89
    3.6.3 Bus Request and Bus Grant Timing in Minimum Mode ………………… 91
    3.6.4 Interrupt Operation ……………………………………………………… 92
    3.6.5 Interrupt Operations in Maximum Mode ………………………………… 97
  Tips ……………………………………………………………………………………… 99
  Exercise ………………………………………………………………………………… 100

Chapter 4 8086 Address Mode and Assembly Instructions …………………………… 103
  4.1 8086 Assembly Instruction Format ………………………………………………… 103

4.2　8086 Addressing Modes ……………………………………………… 104
　　4.2.1　Immediate Addressing ………………………………………… 105
　　4.2.2　Direct Addressing ……………………………………………… 105
　　4.2.3　Register Addressing …………………………………………… 106
　　4.2.4　Register Indirect Addressing ………………………………… 106
　　4.2.5　Register Relative Addressing ………………………………… 107
　　4.2.6　Base-plus-Index Addressing ………………………………… 108
　　4.2.7　Base Relative-plus-Index Addressing ……………………… 109
4.3　Data Movement Instructions ……………………………………… 112
　　4.3.1　MOV Instruction ……………………………………………… 112
　　4.3.2　PUSH and POP ………………………………………………… 114
　　4.3.3　XCHG Instruction …………………………………………… 115
　　4.3.4　XLAT Instruction …………………………………………… 115
　　4.3.5　LEA Instruction ……………………………………………… 117
　　4.3.6　LDS and LES ………………………………………………… 118
　　4.3.7　Flags Register Movement Instruction ……………………… 119
　　4.3.8　IN and OUT …………………………………………………… 120
4.4　Arithmetic Instructions and Logic Instructions ………………… 120
　　4.4.1　ADD and SUB Function ……………………………………… 121
　　4.4.2　INC and DEC Function ……………………………………… 124
　　4.4.3　NEG and CMP Function ……………………………………… 124
　　4.4.4　MUL and DIV Function ……………………………………… 126
　　4.4.5　Type Conversion Functions ………………………………… 127
　　4.4.6　BCD Conversion Functions ………………………………… 128
　　4.4.7　Boolean Operations ………………………………………… 132
　　4.4.8　Shifting and Rotation ………………………………………… 134
4.5　String Instructions ………………………………………………… 137
　　4.5.1　The Direction Flag …………………………………………… 138
　　4.5.2　String Data Transfers ………………………………………… 139
　　4.5.3　String Comparisons ………………………………………… 140
4.6　Program Control Instructions …………………………………… 144
　　4.6.1　Program Flow Control Instructions ………………………… 144
　　4.6.2　Machine Control and Miscellaneous Instructions ………… 154
4.7　The Symbolic Instruction Set …………………………………… 156
Tips ………………………………………………………………………… 158
Exercise …………………………………………………………………… 159

Chapter 5　Directives and Macro Processing …………………… 163
5.1　The Format of the Directives ……………………………………… 163
5.2　Operators and Expression ………………………………………… 165

5.3　Directives ······ 171
　5.3.1　Data Definition and Storage Allocation ······ 172
　5.3.2　EQU Directive ······ 174
　5.3.3　Segment Definition Directive ······ 175
　5.3.4　Assume Directive ······ 176
　5.3.5　PROC Directive ······ 177
　5.3.6　END Directive ······ 178
　5.3.7　ORG Directive ······ 178
　5.3.8　Structures ······ 180
　5.3.9　Records ······ 182
　5.3.10　The PAGE and TITLE Listing Directives ······ 185
　5.3.11　EXTRN/EXTERN Directive ······ 185
　5.3.12　GROUP Directive ······ 185
　5.3.13　INCLUDE Directive ······ 186
　5.3.14　LABEL Directive ······ 186
5.4　Macro Processing ······ 188
　5.4.1　Macro Definition ······ 188
　5.4.2　Macro Sequence and Procedure Calling ······ 191
　5.4.3　Macro Directives ······ 191
5.5　DOS Function Calls ······ 193
　5.5.1　The IBM PC BIOS ······ 193
　5.5.2　An Introduction to MS-DOS' Services ······ 194
　5.5.3　MS-DOS Calling Sequence ······ 195
　5.5.4　Frequently Used MS-DOS Functions ······ 195
5.6　Assembling, Linking and Executing a Program ······ 197
　5.6.1　The Assembler and Linker ······ 197
　5.6.2　Assembling a Source Program ······ 199
　5.6.3　Linking an Object Program ······ 199
　5.6.4　Executing a Program ······ 200
　5.6.5　Using the DEBUG Program ······ 200
　5.6.6　DEBUG Commands Exercise ······ 201
Tips ······ 204
Exercise ······ 205

Chapter 6　Programming with Assembly Language ······ 207
6.1　Design of Assembly Program ······ 207
6.2　Simple Procedures Designing ······ 208
6.3　Branch and Looping Procedures ······ 209
6.4　Procedure Call and Return ······ 216
6.5　Programming Examples ······ 220

Tips ……………………………………………………………………………………………… 227
Exercise ………………………………………………………………………………………… 228

Chapter 7 Memory System ……………………………………………………………… 229
7.1 Overview of the Memory ……………………………………………………………… 229
7.1.1 Non-Volatile Memory ……………………………………………………… 230
7.1.2 Volatile Memory …………………………………………………………… 232
7.1.3 Performance Index of Memory System ………………………………… 233
7.2 Memory Devices ………………………………………………………………………… 235
7.2.1 SRAM 6264 ………………………………………………………………… 235
7.2.2 SRAM 6116 ………………………………………………………………… 240
7.2.3 DRAM 2164A ……………………………………………………………… 242
7.2.4 EPROM 2764A ……………………………………………………………… 243
7.3 Memory Module Design ………………………………………………………………… 245
7.3.1 Memory Pin Connections ………………………………………………… 245
7.3.2 Memory Module Design …………………………………………………… 247
7.3.3 Memory Expansion Examples …………………………………………… 248
Tips ……………………………………………………………………………………………… 251
Exercise ………………………………………………………………………………………… 251

Chapter 8 I/O Interfaces ………………………………………………………………… 253
8.1 I/O Instructions ………………………………………………………………………… 254
8.2 I/O Interfacing Methods ……………………………………………………………… 256
8.3 Serial Interface and Serial Communication ………………………………………… 257
8.3.1 Serial Interface …………………………………………………………… 257
8.3.2 Basic Serial Transmission Lines ………………………………………… 258
8.3.3 Asynchronous and Synchronous Communication …………………… 258
8.4 8251A Programmable Communication Interface ………………………………… 261
8.4.1 The Architecture of the 8251A ………………………………………… 261
8.4.2 The Application of the 8251A ………………………………………… 266
8.4.3 The Initialization of the 8251A ………………………………………… 271
8.5 Parallel Communication Interface …………………………………………………… 275
8.6 8255A Programmable Peripheral Interface ………………………………………… 276
8.6.1 The Architecture of the 8255A ………………………………………… 276
8.6.2 The Function Description of the 8255A ……………………………… 280
8.6.3 The Communication Mode of 8255A …………………………………… 282
8.6.4 The Initialization and Programming of 8255A ……………………… 288
8.7 Programmable Timer and Event Counter ………………………………………… 289
8.8 Intel's 8253 Programmable Timer/Counter ……………………………………… 291
8.8.1 The Architecture of 8253 ………………………………………………… 292
8.8.2 The Operation Mode Definition of 8253 ……………………………… 296

IX

  8.8.3 Examples of 8253 Timer/Counter ……………………………………… 302
 Tips …………………………………………………………………………………… 305
 Exercise ……………………………………………………………………………… 306
Appendix A 8086/8088 Instruction Set Summary ……………………………… 308
Appendix B Vocabulary and Terms ……………………………………………… 322
Reference …………………………………………………………………………………… 330

Chapter 1 Fundamentals of Computer

1.1 Number System

Unlike human counting habits, computer memory does not store numbers in decimal (base 10). In order to make computers more reliable and easier to build, they are based on devices that can take on only two states, one of which is denoted by 0 and the other by 1. All information in computers is stored in a binary (base 2) format because it greatly simplifies the hardware.

This section is the basis of the follow-up study. It is helpful for those who are unfamiliar with number systems. Some basic concepts and conversions between decimal and binary, decimal and hexadecimal, and binary and hexadecimal are introduced.

1.1.1 Basic Number Systems

Nonnegative integers are normally represented by choosing a number **x**, called the **base**, and **x** different symbols, called **digits**, and then using a string of digits to indicate the number. Although a base of 10 is what is used in our everyday work, bases of 2 and 16 are commonly encountered when working with computers.

The number systems based of 10, 2, and 16, which are most common used, are called **decimal**, **binary**, and **hexadecimal** (base 16) **respectively**.

1. Decimal

When we began to learn to count, especially when we entered the primary school, we learned that a **decimal** (base 10) number was constructed with **10** digits: **0** through **9**. Note that the first digit in 10 numbering system is a 0 and the last one is a 9.

In a decimal number system, large numbers are constructed by using **positional notation**. For example, the number 19 is a compilation of two digits in a certain order. The digit 9 is the units position of this numeral and the 1 is the tens position. As shown in **Example 1-1**, the position to the left of the units position was the tens position, the position to the left of the tens position was the hundreds position, and so forth. It also uses the exponential value of each position to express. The units position has a weight of 10^0, or 1; the tens position has weight of 10^1, or 10; and the hundred position has a weight of 10^2, or 100.

【Example 1-1】

Number	3	2	8.	8	0	6
Power	10^2	10^1	10^0	10^{-1}	10^{-2}	10^{-3}
Weight	100	10	1	0.1	0.01	0.001
Numeric Value	300	20	8	0.8	0	0.006
Value	$300 + 20 + 8 + 0.8 + 0 + 0.006 = 328.806$					

2. Binary

Base 2 numbers are composed of **2** possible digits (**0 and 1**). Each digit of a number has a power of 2 associated with it based on its position in the number, just like **Example 1-2.**

【Example 1-2】

Number	1	1	1	0	0	1
Power	2^2	2^1	2^0	2^{-1}	2^{-2}	2^{-3}
Weight	4	2	1	0.5	0.25	0.125
Numeric Value	4	2	1	0	0	0.125
Value	$4 + 2 + 1 + 0 + 0 + 0.125 = 7.125$					

Example 1-2 shows a binary number. Its radix is 2. So its position to the left of the binary point is 2^0, or 1.

In fact, in any number systems, the position to the left of the radix (number base) point is always the units position and the position to the left of the units position is always the number base raised to the first power. As shown in **Example 1-1**, this is 10^1, or 10. In **Example 1-2**, it is 2^1, or 2.

3. Hexadecimal

Hexadecimal numbers use **base 16**. Hexadecimal (or hex for short) can be used as shorthand for binary numbers. Hex has **16** possible digits in which letters are used for the extra digits after 9. It means that the 16 hex digits are **0-9** then followed by **A** for 10, **B** for 11, **C** for 12, **D** for 13, **E** for 14 and **F** for 15. Each digit of a hex number has a power of 16 associated with it.

4. Other number systems

Besides the common number systems introduced above, the number system corresponding to the base of 8 which is called **octal** is also used in computers sometimes.

The real numbers often encountered in many high-level languages using the Intel family of microprocessors. A real number, or a **floating-point number** as it is often called, contains two parts: a mantissa significant, or fraction; and an exponent. **Figure 1-1** illustrates both the 4-byte and 8-byte forms of real numbers stored in any Intel system. Note that the 4-byte real number is called **single-precision** and the 8-byte form is called **double-precision**.

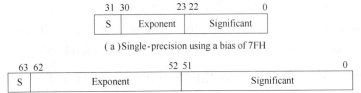

(a) Single-precision using a bias of 7FH

(b) double-precision using a bias of 3FFH

Figure 1-1   The floating-point numbers

1.1.2 Conversions between Different Number Systems

After understanding the basic concept of a number system, the conversions between different number base systems will be learned in this section.

1. Conversion to Decimal

To convert to decimal, we need to determine the weights of values of each position of the number, and then sum the weights to form the decimal equivalent. **Example 1-2** illustrates the conversion from binary number base to decimal, while **Example 1-3** gives the conversion from octal.

【Example 1-3】 Please convert 205.6_8 octal to decimal.

To accomplish this conversion, we need to write down the weight of each position of the number and then multiply them by the digit of each position.

Number	2	0	5.	6
Power	8^2	8^1	8^0	8^{-1}
Weight	64	8	1	0.125
Numeric Value	128	0	5	0.75
Value		128 + 0 + 5 + 0.75 = 133.75		

As shown in expression $2 \times 64 + 0 \times 8 + 5 \times 1 + 6 \times 0.125$, the value of 205.6_8 is 133.75 decimal.

Note that the weights of position which immediately to the right of the octal point is 1/8, or 0.125. The next position is 1/64. So the number in **Example 1-3** can also be written as the decimal number $133^6/_8$.

【Example 1-4】 Please complete the conversion from a binary number 01010.0101 to decimal.

As we did in prior example, first we should write down the weights and powers of each position, and then sum the numeric values together.

Number	0	1	0	1	0.	0	1	0	1
Power	2^4	2^3	2^2	2^1	2^0	2^{-1}	2^{-2}	2^{-3}	2^{-4}
Weight	16	8	4	2	1	0.5	0.25	0.125	0.0625
Numeric Value	0	8	0	2	0	0	0.25	0	0.0625
Value		0 + 8 + 0 + 2 + 0 + 0 + 0.25 + 0 + 0.0625 = 10.3125							

The value of the binary number 01010.0101 converted to decimal is 10.3125.

The fractional part of this number is 5/16. Note that 0101 is a 5 in binary code for the numerator and the rightmost one is in the 1/16 position for the denominator.

[Example 1-5] Please convert 2B.EH (H for hexadecimal) to decimal.

As we know, hexadecimal numbers are often used with computers 2B.EH is illustrated with its weights as follows.

Number	2	B.	E
Power	16^1	16^0	16^{-1}
Weight	16	1	0.0625
Numeric Value	32	11	0.875
Value	\multicolumn{3}{c}{32 + 11 + 0.875 = 43.875}	

The sum of its digits is 43.875, or $43^7/_8$. The whole number part is represented with 2×16 plus 11 (B) ×1. The fraction part is 14 (E) as a numerator and 16 (16^{-1}) as the denominator, or 14/16, which is reduced to 7/8.

2. Conversion from Decimal

Conversions from decimal to other number systems are more difficult to accomplish than that conversions to decimal. To convert the whole number portion of a number from decimal, we should divide them by the radix. To convert the fractional portion, we should multiply them by the radix.

1) Whole Number Conversion from Decimal

To convert a decimal whole number to another number system, we should divide it by the radix and save the remainders as significant digits of the result. An algorithm for this conversion is as follows:

(1) Divide the decimal number by the radix (number base).

(2) Save the remainder (first remainder is the least significant digit).

(3) Repeat steps 1 and 2 until the quotient is zero.

That is to say, if we want to convert decimal number to any other number systems, what we only need to do is dividing the decimal number by the corresponding base. **For example**, to convert a 15 decimal to binary, divide it by 2. To convert a decimal into base 8, divide it by 8. Conversion from decimal to hexadecimal is accomplished by dividing it by 16.

[Example 1-6] Please convert a 15 decimal to a binary.

$$
\begin{array}{rll}
2)\underline{15} & \text{remainder} = 1 & \uparrow \\
2)\underline{\ 7} & \text{remainder} = 1 & \uparrow \\
2)\underline{\ 3} & \text{remainder} = 1 & \uparrow \\
2)\underline{\ 1} & \text{remainder} = 1 & \uparrow \\
0 & & \text{result} = 1111
\end{array}
$$

Above Example shows the conversion process from a 15 decimal to binary. Dividing 15 by 2, the quotient is 7 with a remainder of 1. The first remainder 1 is the unit position of the conversion result. Then divide the 7 by 2 with a quotient of 3 and a remainder of 1. The 1 is the value of the two's (2^1) position. Continue the division until the quotient is a zero. The result is written as 1111_2, from the bottom to the top.

[Example 1-7] Please convert a 10 decimal to an octal.

$$8 \underline{)10} \quad \text{remainder} = 2 \quad \uparrow$$
$$8 \underline{)1} \quad \text{remainder} = 1 \quad \uparrow$$
$$0 \qquad\qquad\qquad\qquad \text{result} = 12$$

The base of an octal is 8, so divide 10 by 8 and follow the rules. As shown in **Example 1-7**, a 10 decimal is a 12 octal.

[Example 1-8] Please convert a 109 decimal to a hexadecimal.

Conversion from 109 decimal to hexadecimal is accomplished by dividing it by 16.

$$16 \underline{)109} \quad \text{remainder} = 13 \ (D) \quad \uparrow$$
$$16 \underline{)6} \quad \text{remainder} = 6 \quad \uparrow$$
$$0 \qquad\qquad\qquad\qquad \text{result} = 6D$$

The remainder will range from 0 through 15. Notice that any remainder of 10 to 15 is then converted to the letter A to F as the hexadecimal number. **Example 1-8** shows the decimal number 109 converted to a 6DH.

2) Conversion from a Decimal Fraction

Conversion from a decimal fraction to another number base is accomplished with multiplication by the radix. **For example**, to convert a decimal fraction to binary, multiply it by 2. After the multiplication, the whole number portion of the result is saved as a significant digit of the result, and the fractional remainder is again multiplied by the radix. When the fraction remainder is zero, multiplication ends. Note that some numbers are never ending. That is, a zero is never a remainder. An algorithm for conversion from a decimal fraction is as follows:

(1) Multiply the decimal fraction by the radix (number base).

(2) Save the whole number portion of the result (even if zero) as a digit. Note that the first result is written immediately to the right of the radix point.

(3) Repeat step 1 and 2, using the fractional part of step 2 until the fractional part of step 2 is zero.

[Example 1-9] Please convert a 0.125 decimal to binary.

Because the radix of binary is 2, the conversion is accomplished with multiplications by 2 and the multiplication continues until the fractional remainder is zero.

$$\begin{array}{r} 0.125 \\ \times \quad 2 \\ \hline 0.25 \end{array} \quad \text{digit is 0} \quad \downarrow$$

$$\begin{array}{r} \times \quad 2 \\ \hline 0.5 \end{array} \quad \text{digit is 0} \quad \downarrow$$

$$\begin{array}{r} \times \quad 2 \\ \hline 1.0 \end{array} \quad \text{digit is 1} \quad \downarrow$$

The result is written as 0.001 binary.

As illustrated in **Example 1-9**, the whole number portions are written as the binary fraction (0.00).

Likewise, the same technique is used to convert a decimal fraction into any other number bases. The examples given below show that how to convert a decimal fraction to an octal or a hexadecimal fraction respectively.

【**Example 1-10**】 Please convert a 0.125 decimal to octal.

$$\begin{array}{r} 0.125 \\ \times \quad 8 \\ \hline 1.0 \end{array} \quad \text{digit is 1} \quad \downarrow$$

Multiplying with an 8, we can get 0.125 decimal fraction is 0.1 octal.

【**Example 1-11**】 Please convert a 0.046875 decimal to hexadecimal.

Similarly, a decimal 0.046875 is converted to hexadecimal by multiplying it by the radix 16. As shown in **Example 1-11**, the conversion of 0.046875 is a 0.0CH.

$$\begin{array}{r} 0.046875 \\ \times \quad 16 \\ \hline 0.75 \end{array} \quad \text{digit is 0} \quad \downarrow$$

$$\begin{array}{r} \times \quad 16 \\ \hline 12.0 \end{array} \quad \text{digit is 12 (C)} \downarrow$$

The result is written as 0.0C hexadecimal.

3. Conversion between Binary and Hexadecimal

Conversion of a binary number to a hexadecimal number is a simply matter of putting the binary digits in groups of four. Similarly to convert a hex number to binary, we can simply convert each hex digit to a 4-bit binary number.

【**Example 1-12**】 Please convert a hexadecimal number 14D to a binary number.

Hexadecimal Number   1   4   D
Binary Number   0001   0100   1101

Note that the leading zeroes of the 4-bits are important. Converting from binary to hex is just as easy. One does the process in reverse. Convert each 4-bit segments of the binary to a hex. Start from the right end of the binary number to ensure conversion correct. **For example:**

【Example 1-13】 Please convert a binary number to a hexadecimal number.

Binary Number   0010 0000 0101 1010 0111 1110 0100
Hexadecimal Number   2   0   5   A   7   E   4

A 4-bit number is called a **nibble**. Thus each hex digit corresponds to a nibble. Two nibbles make a byte and so a 2-digit hex number can represent a byte. A byte's value ranges from 0 to 11111111 in binary, 0 to FF in hex and 0 to 255 in decimal.

Table 1-1 shows the decimal numbers 0 through 15 along with their equivalent binary and hexadecimal values

Table 1-1 Binary, decimal, and hexadecimal representation

Binary	Decimal	Hexadecimal	Binary	Decimal	Hexadecimal
0000	0	0	1000	8	8
0001	1	1	1001	9	9
0010	2	2	1010	10	A
0011	3	3	1011	11	B
0100	4	4	1100	12	C
0101	5	5	1101	13	D
0110	6	6	1110	14	E
0111	7	7	1111	15	F

1.1.3 Common Data Units

1. Byte-Sized Data

A group of nine **related** bits is called a **byte**. As shown in Figure 1-2, each byte consists of 8 bits for data and 1 bit for parity.

```
          0    0    0    0    0    0    0    0    1
        |←              Data bits              →| parity |
```

Figure 1-2 A byte with one bit parity

Byte-sized data is stored as unsigned or signed integers. **Figure 1-3** illustrates both the unsigned and signed forms of the byte-sized integer. The difference between those forms is the meaning of the leftmost bit position. In signed integer format, the leftmost bit represents the sign bit of the number. **That is to say**, an 80H represents a value of 128 as an unsigned number. While as a signed number, it represents a value of minus 128. The values of unsigned integers range from 00H to FFH (0-255). The values of signed integers range from -128 to 0 to +127.

Although negative signed numbers are represented in this way, they are stored in the two's complement form. The method of evaluating a signed number by using the weights of each bit position is much easier than the act of two's complementing a number to find its value. This is especially helpful to the programmers to design calculators.

Whenever a number is two's complemented, its sign changes from negative to positive or

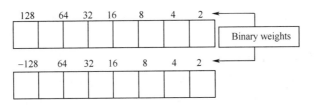

Figure 1-3 The unsigned and signed bytes

positive to negative. **For example**, the number 00001000 is a +8. Its negative value (−8) is found by two's complementing the +8. To form a two's complement: first, invert each bit of the number to get **one's complement**, then add 1 to the one's complement to form **two's complement**. Example 1-14 shows the process.

【Example 1-14】

$$
\begin{array}{r}
+8 = 0\,0\,0\,0\,1\,0\,0\,0 \\
1\,1\,1\,1\,0\,1\,1\,1 \ (\textbf{one's complement}) \\
+\qquad\qquad\qquad 1 \\
\hline
-8 \quad 1\,1\,1\,1\,1\,0\,0\,0 \ (\textbf{two's complement})
\end{array}
$$

2. Word-Sized Data

A word (16-bit) is formed with **two** bytes of data. The **least** significant byte is always stored in the **lowest-numbered** memory location, and the **most** significant byte is stored in the **highest**. This method of storing data is called the **little-endian** format. Another method, the big-endian format in which numbers are stored with the lowest location containing the most significant data is not used with the Intel family of microprocessors. But used with the Motorola family of microprocessors. **Figure 1-4** shows the process that storing the number 4350H in the memory location 3000H and 3001H. We can see that the lower-order byte is stored in the lowest-numbered memory location (3000H) and the high-order byte is stored in the highest-numbered memory location (3001H).

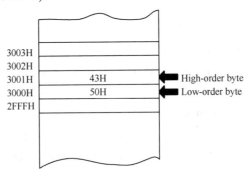

Figure 1-4 The storage format for a 16-bit word in two bytes memory

3. Doubleword-Sized Data

Doubleword-sized data requires **four** bytes of memory to store because it is a 32-bit number. Double-word data appear as a result of a multiplication or a dividend before a division. When a double word is stored in memory, its least significant byte is stored in the lowest-numbered memory location using the little-endian format. **Figure 1-5** shows the process of storing a double word 12345678H in the memory.

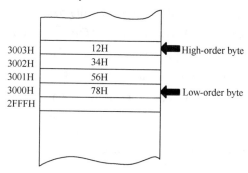

Figure 1-5　The storage format for a double word in four bytes memory

4. Other Related Bytes

（1）**Quadword**： An 8-byte (64-bit) data item.

（2）**Paragraph**： A 16-byte (128-bit) area.

（3）**Kilobyte（KB）**： The number 2^{10} equals 1,024, which happens to be the value K, for kilobyte. Thus 640K of memory is $640 \times 1,024 = 655,360$ bytes.

（4）**Megabyte（MB）**： The number 2^{20} equals 1,048,576 or 1 megabyte.

要点：

（1）数制是人们利用符号来计数的科学方法。数制可以有很多种,常用的有十进制、二进制、八进制和十六进制。

（2）数制的基和权:数制所使用的数码的个数称为基,数制每一位所具有的值称为权。

（3）十进制:基为"10",权为以 10 为底的幂——D。

二进制:基为"2",权为以 2 为底的幂——B。

八进制:基为"8",权为以 8 为底的幂——O。

十六进制:基为"16",权为以 16 为底的幂——H。

（4）十进制数转换成二进制代码的方法:对于十进制数整数部分采用除 2 取余法;对于十进制数小数部分采用乘 2 取整法。

（5）二进制数转换为十六进制数的方法:从小数点开始分别向左和向右把整数和小数部分每四位分段,每段分别转换为一位。若整数最高位的一组不足 4 位,则在其左边补零;若小数最低位的一组不足 4 位,则在其右边补零。

1.2　Logic Algebra and Logic Gates

Logic algebra is also called switch algebra. As same as common algebra it can be expressed as below:

$$Y = f(A, B, C, D)$$

It has two characters:

(1) The variables A, B, C, D should either be mark or space. The value of the variable doesn't relate to its amount. It only represents the two opposite properties of the variables. **For example**, the space (0) represents **False** or **Off** and the mark(1) represents **True** or **On**.

(2) Function f has only three basic fashions: "**OR**" operation, "**AND**" operation and "**NOT**" operation. Usually, we use " × " for "and", " + " for "or" and " − " for "not". The following passages will discuss the rules of these operations.

Logic Gate is a switching circuit (a network of interconnected switches). The basic digital gates (or "**Logic Gates**") are outlined in the following pages.

1.2.1 "OR" Operation and "OR" Gate

"OR" operation is also called logic "OR". It can be expressed as:
$$Y = A + B$$

As the name implies, if either input "A" or input "B" is HIGH, then the output will be HIGH, otherwise the output will remain LOW.

OR truth table:

A	B	Y
0	0	0
0	1	1
1	0	1
1	1	1

For "OR" operation, when the binary numbers A and B have more than one bit, **for example**:
$$A = A_1 A_2 A_3 \cdots A_n$$
$$B = B_1 B_2 B_3 \cdots B_n$$

The "OR" operation will be done based on each bit of the operands. The expression can be written as follows:
$$Y = A + B = (A_1 + B_1)(A_2 + B_2)(A_3 + B_3)\cdots(A_n + B_n)$$

The **OR Gate** produces 1 if one of its inputs is 1. The symbol of an OR Gate is usually drawn as that shown in **Figure 1-6**.

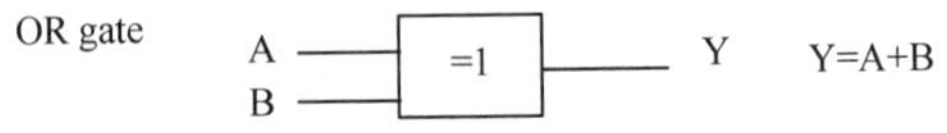

OR gate

Figure 1-6 OR gate

1.2.2 "AND" Operation and "AND" Gate

Similarly, "AND" operation is also called logic "AND". "AND" operation can be expressed as:
$$Y = A \times B$$

The output will remain LOW, if either input "A" or input "B" is LOW. The output will be High only when both A and B are High.

When the binary numbers A and B have more than one bit, **for example**:

$$A = A_1 A_2 A_3 \cdots A_n$$
$$B = B_1 B_2 B_3 \cdots B_n$$

The "AND" operation will be done based on each bit of the operands, the expression can be written as follows:

$$Y = A \times B = (A_1 \times B_1)(A_2 \times B_2)(A_3 \times B_3) \cdots (A_n \times B_n)$$

In digital logic, the "normal" state is a "LOW" (i.e. the voltage is OFF). The action of gates is usually described in terms of what must happen to produce a HIGH. In an AND gate, a HIGH is produced at the output only if the first AND second inputs are all High. Any other conditions would leave the output LOW.

An AND gate may be compared to two switches, A and B. If A **AND** B are closed then the output is HIGH. The output "Y" will remain LOW unless A and B are both HIGH. The possible combinations are summarized in a "Truth Table".

AND truth table

A	B	Y	
0	0	0	NOTE: "0" = Low
0	1	0	"1" = High
1	0	0	
1	1	1	

The AND gate produces 1 if both its input values are 1. The symbol of an AND gate is usually drawn as shown in **Figure 1-7**.

AND gate   A ——[&]—— Y   Y=AB

Figure 1-7   AND gate

Notice: an AND gate may have more than two inputs.

1.2.3 "NOT" Operation and the NOT-Inverter

The NOT operation can be expressed as:

$$Y = \overline{A}$$

For NOT operation, the output will NOT be the same as the input. It always gets the opposite value of the input. Hence the alternative name for this gate: "inverter".

NOT truth table

A	Y
0	1
1	0

When A has more than one bit, **for example**:
$$A = A_1 A_2 A_3 \cdots A_n$$
The result of logic "NOT" is:
$$Y = \overline{A_1} \overline{A_2} \overline{A_3} \cdots \overline{A_n}$$

It is obviously that the NOT gate (inverter) produces an output that is the opposite of its input value. The symbol of a NOT gate is usually drawn as shown in **Figure 1-8**.

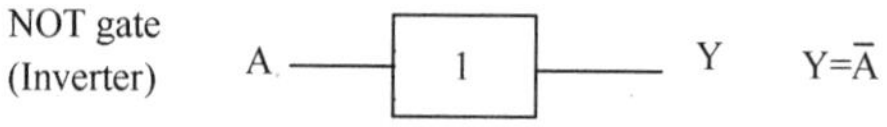

NOT gate (Inverter)　　A ──[1]── Y　　Y=$\overline{A}$

Figure 1-8　NOT gate

1.2.4 Basic Rules of Boolean Algebra

1. Identical Equation

$$A \times A = A \quad A \times 0 = 0 \quad A \times 1 = A$$
$$A + 0 = A \quad A + 1 = 1 \quad A + A = A$$
$$A + \overline{A} = 1 \quad A \times \overline{A} = 0 \quad \overline{\overline{A}} = A$$

2. General Rules

(1) Commutative:
$$A \times B = B \times A$$
$$A + B = B + A$$

(2) Associative:
$$(A \times B) \times C = A \times (B \times C)$$
$$(A + B) + C = A + (B + C)$$

(3) Distributive:
$$A \times (B + C) = A \times B + A \times C$$
$$A + (B \times C) = (A + B) \times (A + C)$$

3. De Morgan's Theorem

$$\overline{A + B} = \overline{A} \times \overline{B}$$
$$\overline{A \times B} = \overline{A} + \overline{B}$$

This theorem is widely used in Boolean logic design. To "invert" (negate) a Boolean expression, we should replace the AND operator with the OR operator (or vice versa) and invert the individual terms.

4. Absorption Rules

$$A + A \times B = A$$
$$A \times (A + B) = A$$

In addition, the **XOR** and **NXOR** operations are usually utilized in the computer.

　　XOR　　$A \oplus B = \overline{A} \times B + A \times \overline{B}$
　　NXOR　　$A \otimes B = \overline{A} \times \overline{B} + A \times B$

The first expression implies that if A is different from B, the result will be 1, otherwise it'll be 0. And it can be inferred from the last one that if A is the same as B, the result will be 1, otherwise it'll be 0.

要点:

(1) 二进制数的逻辑运算:与、或、非。

或运算表/达式:Y = A + B,两者皆伪者则结果必伪,有一为真则结果为真。

与运算表达式:Y = A × B,两者皆真者则结果必真,有一为伪则结果为伪。

非运算表达式:$Y = \overline{A}$。

(2) 布尔代数的基本运算规律:

恒等式。

运算规律:交换律、结合律、分配律。

摩根定律:

$$\overline{A + B} = \overline{A} \times \overline{B}$$
$$\overline{A \times B} = \overline{A} + \overline{B}$$

1.3 Binary Numbers and Binary Addition/Subtraction

In order to simplify the computer hardware circuits, the binary number system is adopted, which makes a computer easy to distinguish only between 0 and 1 bit. In the computer storage system, the **bit**, a contraction of "**binary digit**", is the most fundamental building block. The values 1 or 0 represents two states **on** or **off** respectively.

A binary number could be 8-bit, 16-bit or 32-bit, even 64-bit. How many data bits the computer could handle depends on its processor architecture.

1.3.1 Negative Binary Numbers

A signed binary number is considered to be positive if its leftmost bit is a 0, whereas a signed negative binary number contains a 1-bit in its leftmost position. However, representing a binary number as negative is not as simple as setting the leftmost bit to 1, such as converting 01000001 (+65) to 11000001. Instead, a binary number value is expressed in **two's complement** notation. That is, the rule to represent a binary number as negative is: **Reverse the bit values and add 1**. For example:

【**Example 1-15**】 Please write the binary representing of the number −65.

Number +65:	01000001
Reverse the bits:	10111110
Add 1:	+ 1
Number −65:	10111111

A signed binary number is negative if its leftmost bit is 1, but adding the 1-bit value to determine the decimal value number 10111111 couldn't get 65. To determine absolute value of a negative binary number, we can simply apply the two's complement rule. That is, **reverse the bits and add 1**. As shown in **Example 1-16.**

【Example 1-16】

$$
\begin{array}{lr}
\text{Number} -65: & 10111111 \\
\text{Reverse the bits:} & 01000000 \\
\text{Add 1:} & +\underline{\qquad 1} \\
\text{Number} +65: & 01000001
\end{array}
$$

It is obvious that the sum of $+65$ and -65 should be zero. This rule can help to illustrate that the procedure above works properly. The result is shown in **Example 1-17**.

【Example 1-17】

$$
\begin{array}{rr}
+65 & 01000001 \\
\underline{-65} & +\underline{10111111} \\
0 & (1)00000000
\end{array}
$$

The sum of $+65$ and -65 is $(1)00000000$. All 8-bit value is zero, and the carry of the 1-bit on the left is lost. Note that a carry into and out of the sign bit, so the result is considered to be correct.

1.3.2 Binary Addition

The addition of two binary numbers is completed bit by bit from right to left. Suppose there are two binary numbers: $A = A_3 A_2 A_1 A_0$ and $B = B_3 B_2 B_1 B_0$. The sum of these two numbers is $S = S_3 S_2 S_1 S_0$, and each bit of the sum can be calculated by following operation. The addition process is shown below:

$S_0 = A_0 + B_0$ $\qquad$ carry C_1
$S_1 = A_1 + B_1 + C_1$ $\qquad$ carry C_2
$S2 = A_2 + B_2 + C_2$ $\qquad$ carry C_3
$S3 = A_3 + B_3 + C_3$ $\qquad$ carry C_4
$A + B = C_4 S_3 S_2 S_1 S_0$

For example, calculating the sum of 011 and 010, we should do as follow:

$$
\begin{array}{r}
011 \\
+\underline{010} \\
101
\end{array}
$$

1.3.3 Binary Subtraction

When it comes to binary subtraction, we can simply convert the subtrahend to 2's complement format and add the result to the minuend. **Example 1-18** shows the process that we subtract 42 from 65. The binary representation of 42 is 00101010, and its 2's complement is 11010110. Add $-42(11010110)$ to $65(01000001)$, the result is $23(00010111)$:

【Example 1-18】

$$
\begin{array}{rr}
65 & 01000001 \\
+\underline{(-42)} & +\underline{11010110} \\
23 & (1)00010111
\end{array}
$$

1.3.4 Adder Circuits

Adder circuits can handle both addition and subtraction if the numbers are represented in two's complement format.

1. Half Adder

In **Figure 1-9**, (**a**) is the truth table of 2 bit binary addition and (b) is the circuit. It is obviously that it needs to provide two inputs and two outputs: the Sum (S_0) and the carry (C_1).

$S_0 = A_0 + B_0$ $\qquad\qquad$ carry C_1

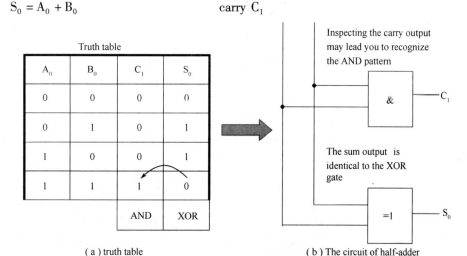

(a) truth table $\qquad\qquad$ (b) The circuit of half-adder

Figure 1-9 Truth table of half adder and the circuit

Obviously, there is a problem when we consider the addition of multi-bit binary numbers. Except the last bit, the other bit addition usually needs 3 inputs. That is why **Figure 1-9**(**b**) is referred to as **half adder**. The full adder is adopted in multi-bits addition.

2. Full Adder

Suppose $A = 011$, $B = 011$, as shown in the following, the sum of A and B is 110.

$$\begin{array}{r} 011 \\ +011 \\ \hline 110 \end{array}$$

Notice that a "carry 1" occurs when adding the two bits in the first column, which then needs to be added into the second column. It's clear that there is a need to cater for the carry-in value from the previous column. Thus we need an adder circuit with three inputs: a 3 bit full adder, not a 2 bit half adder. The truth table and the circuit are shown in **Figure 1-10**.

Note: The rule for multi-input XOR gate is that the gate outputs "1" if there are odd "1" inputs. Otherwise, the output will be "0".

The **symbols** of half adder and full adder are shown in **Figure 1-11** and **Figure 1-12** respectively:

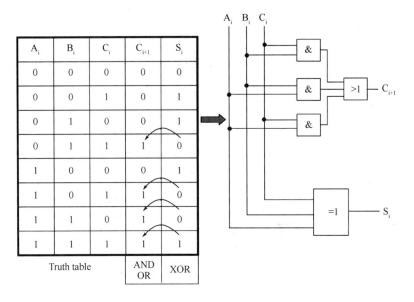

Figure 1-10 Truth table of full adder and the circuit

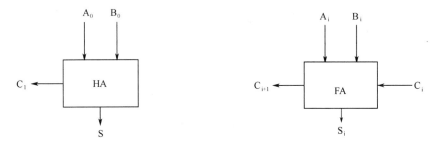

Figure 1-11 Symbol of half adder   Figure 1-12 Symbol of full adder

3. Parallel Adders

Suppose $A = A_3A_2A_1A_0$, $B = B_3B_2B_1B_0$, we can design the adder circuit as below:

In **Figure 1-13**, there are 3 full adders positioned next to each other in a row except the last one is a half-adder. Each of them is responsible for adding two bits from a four bit number.

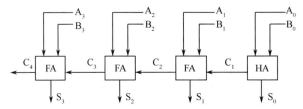

Figure 1-13 Adder circuit

1.3.5 Switchable Inverter and Binary Addition/Subtraction Circuit

Subtracting can be considered identical to addition if we switch the second number to its negative value. Thus we can use an adder to do subtraction by using XOR gate as switchable inverter as shown in **Figure 1-14**. The truth table of the inverter is shown in Table 1-2:

Table 1-2 Truth table of the inverter

SUB	B_0	Y_0	Y_0 and B_0
0	0	0	same
0	1	1	same
1	0	1	pposition
1	1	0	pposition

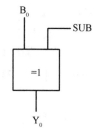

Figure 1-14 Switchable inverter

Thus the adder circuit can be designed as **Figure 1-15**:

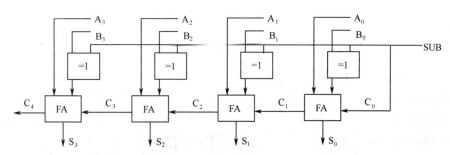

Figure 1-15 Adder circuit with switchable inverter

Notice that the actual circuits are obviously much more complicated.

要点:带符号二进制数的表示

计算机中的符号数可表示如下:

符号位 + 真值;

其中符号位"0"表示正,"1"表示负。

正数的补码同原码。负数的补码的数值部分为真值的各位按位取反加1。

1.4 Computer Data Formats

Successful programming requires a precise understanding of data formats. In this section, **ASCII** code and **BCD** format are described because they are usually appeared in computer system.

1.4.1 ACSII Code

To standardize the representation of data, ASCII (American Standard Code for Information Interchange) code is used to facilitate the transfer of data among different computer devices. The 8-bit ASCII codes that the PC uses provide 256 characters, including symbols for foreign alphabets. **For example**, the combination of bit 01000001 (41H) indicates the letter A. **Table 1-3** provides a convenient list of the 128 ASCII characters which is coded by 7-bit.

Table 1-3   ASCII characters

Dec	Hex	Char	Dec	Hex	Char	Dec	Hex	Char	Dec	Hex	Char
0	00	Null	32	20	Space	64	40	@	96	60	`
1	01	Start of heading	33	21	!	65	41	A	97	61	a
2	02	Start of text	34	22	"	66	42	B	98	62	b
3	03	End of text	35	23	#	67	43	C	99	63	c
4	04	End of transmit	36	24	$	68	44	D	100	64	d
5	05	Enquiry	37	25	%	69	45	E	101	65	e
6	06	Acknowledge	38	26	&	70	46	F	102	66	f
7	07	Audible bell	39	27	'	71	47	G	103	67	g
8	08	Backspace	40	28	(	72	48	H	104	68	h
9	09	Horizontal tab	41	29	)	73	49	I	105	69	i
10	0A	Lin feed	42	2A	*	74	4A	J	106	6A	j
11	0B	Vertical tab	43	2B	+	75	4B	K	107	6B	k
12	0C	Form feed	44	2C	,	76	4C	L	108	6C	l
13	0D	Carriage return	45	2D	-	77	4D	M	109	6D	m
14	0E	Shift out	46	2E	.	78	4E	N	110	6E	n
15	0F	Shift in	47	2F	/	79	4F	O	111	6F	o
16	10	Data link escape	48	30	0	80	50	P	112	70	p
17	11	Device control 1	49	31	1	81	51	Q	113	71	q
18	12	Device control 2	50	32	2	82	52	R	114	72	r
19	13	Device control 3	51	33	3	83	53	S	115	73	s
20	14	Device control 4	52	34	4	84	54	T	116	74	t
21	15	Neg. acknowledge	53	35	5	85	55	U	117	75	u
22	16	Synchronous idle	54	36	6	86	56	V	118	76	v
23	17	End trans. block	55	37	7	87	57	W	119	77	w
24	18	Cancel	56	38	8	88	58	X	120	78	x
25	19	End of medium	57	39	9	89	59	Y	121	79	y
26	1A	Substitution	58	3A	:	90	5A	Z	122	7A	z
27	1B	Escape	59	3B	;	91	5B	[	123	7B	{
28	1C	File separator	60	3C	<	92	5C	\	124	7C	|
29	1D	Group separator	61	3D	=	93	5D	]	125	7D	}
30	1E	Record separator	62	3E	>	94	5E	^	126	7E	~
31	1F	Unit separator	63	3F	?	95	5F	_	127	7F	□

The ASCII control characters which perform control functions in a computer system are also listed in **Table 1-3**. These functions include clear screen, backspace, line feed, and so on. To enter the control code through the computer keyboard, we should press the Control (Ctrl) key while typing a letter. To obtain the control code 01H, type a Control-A; a 02H is obtained by a Control-B, etc.

1.4.2 BCD (Binary-Coded Decimal) Format

Binary-coded decimal (BCD) information is stored in either Packed BCD data which is stored as two digits per byte or unpacked BCD data which is stored as one digit per byte. The range of a BCD digit extends from $0000(2)$ to $1001(2)$, or 0-9 decimal. Unpacked BCD data are returned from keypad or keyboard. Packed BCD data are used for some of the instructions included for BCD addition and subtraction in the instruction set of the microprocessor.

Example 1-19 shows how to convert some decimal numbers to both the packed and the unpacked BCD forms.

【Example 1-19】

Decimal	Packed		Unpacked		
12	0001 0010		0000 0001	0000 0010	
623	0000 0110	0010 0011	0000 0110	0000 0010	0000 0011
910	0000 1001	0001 0000	0000 1001	0000 0001	0000 0000

要点:

ASCII 码:美国国家标准信息交换码,用 7 位二进制编码表示 128 个字符和符号字符的编码。

BCD 码:用二进制编码表示的十进制数,计数规律与十进制相同,均为"逢十进一"。压缩 BCD 码是用 4 位二进制码表示一位十进制数。扩展 BCD 码是用 8 位二进制码表示一位十进制数。

1.5 Logic Circuit

1.5.1 Logic Gates

Logic circuit is composed by three basic gate circuits such as "AND" gate, "OR" gate and Inverter. Gate is a switching circuit (a network of interconnected switches) and it has following characteristics.

(1) It is the most basic digital device, a building block for combinational circuits.

(2) Has one or more inputs.

(3) The output is a function of the current input values.

A very large number of operations can be performed by combinations of only a very few different types of "gates". The three basic digital gates (or **Logic Gates**) have been outlined in the preceding relevant pages. On the base of these three gates, other more complicated logic gates can be made. **Figure 1-16** shows some of them.

1.5.2 Flip-Flops

A "Flip-Flop" is a circuit which is stable in one of two "states". It remains "locked" in

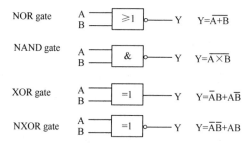

Figure 1-16 Other logic gates

each state until triggered to swap to the other. The flip-flop is made of gates. Many of the applications of Flip-Flops require that they respond only at certain times. (Such as used in computers and "A to D" converters.) To achieve this function, a "clock input" is added. Such a Flip-Flop will not respond to any changes until the clock pulse arrives.

1. RS Flip-Flop (RS Latch)

As shown in the Figure 1-17(a), an RS flip-flop (RS latch) is made of a pair of NAND gates. **Figure 1-17(b)** shows the symbol of a RS flip-flop and the truth table of a RS flip-flop is given by **Table 1-4.**

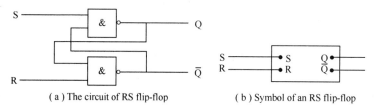

(a) The circuit of RS flip-flop   (b) Symbol of an RS flip-flop

Figure 1-17 The circuit and symobol of RS flip-flop.

Table 1-4 Truth table of RS flip-flop

S	R	output	Q
0	0	0	(old) memory
0	1	0	Reset
1	0	1	Set
1	1	?	Illegal

(1) Applying a "1" to the "S" input sets the flip-flop, i. e. it makes Q = 1 and it will remain in this state even if the "1" is removed from the "S" input.

(2) Applying a "1" to the "R" input resets the flip-flop, i. e. it makes Q = 0 and it will remain in this state even if "1" is removed from the "R" input.

(3) Applying a "1" to "R" and "S" simultaneously should be avoided, the resulting states are undefined.

When using flip flops, it is often desirable to establish the logic levels of the inputs without actually setting or resetting the flip-flops the moment the input levels change. One way is to establish the inputs first and then let the change in output occurs on the clock pulse.

Flip flops may be designed to "trigger" on either the positive-going or negative-going transition. **Figure 1-18** shows an example of a clocked RS flip-flop. Basically a pair of AND gates are employed at the inputs to trigger the flip-flop with the clock. It is clear that in this case, the input will not be allowed to "pass through" unless the clock signal goes HIGH (i. e. this flip-flop triggers on the positive going transition).

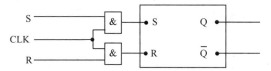

Figure 1-18 Clocked RS flip-flop

Figure 1-19 shows another example of clocked RS flip-flop. Note the triangle at the clock input. In this case, the triangle indicates that the flip-flop is activated on the transition of the signal. This means that the flip-flop is **edge triggered** as opposed to **level triggered**. Thus the flip-flop operates in the usual manner when the output changes at the clock edges. The bubble indicates that the clock signal is active low so that the flip-flop will respond to a negative going transition of the clock.

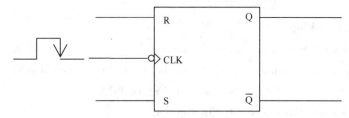

Figure 1-19 Clocked RS flip-flop

2. The "D-type" flip-flop

RS flip-flops have the problem of having an illegal state when both the R and S are activated at the same time. The **D** flip-flop provides an easy solution by having only a single input to the S input. Connection to the R input is through an inverter as shown in **Figure 1-20**.

Figure 1-21 shows a simple form of the clocked D flip-flop.

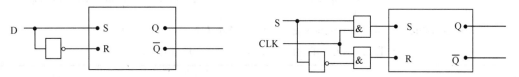

Figure 1-20 Connection to the R input through an inverter

Figure 1-21 Clocked D flip-flop

If the D input is LOW when the clock pulse is applied, the flip-flop switches (or remains in) the "reset" state. i. e. the output Q is LOW. If the D input is HIGH at the clock pulse, the flip-flop switches to the "set" state, i. e. the output Q is HIGH.

The D-type flip-flop is sometimes called a "**Latch**". Since it stores or latches the logic level presented at the D input until the arrival of the next clock pulse.

In other words, the logic state at D is transferred to Q at each clock pulse. **Table 1-5** is the truth table for the **D** flip-flop.

Table 1-5 Truth table of D flip-flop

D	CLK	Q	$\bar{Q}$
1	↑	1	0
0	↑	0	1

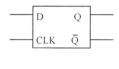

Just like the RS flip-flops, **Figure 1-22** shows the edging triggering D flip-flop.

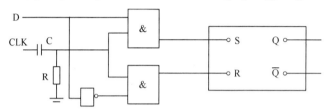

Figure 1-22 Edging triggering D flip-flop

3. The JK Flip-Flop

JK flip-flops are the most common and versatile flip-flops. They have no disallowed state in this way to eliminate the ambiguity of RS flip-flops. Adding a pair of AND gates to the RS flip flop forms the basic JK flip-flop as shown in **Figure 1-23.**

Figure 1-24 shows the symbol of a JK flip-flop.

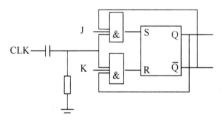 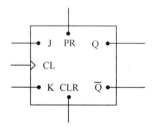

Figure 1-23 JK flip-flop   Figure 1-24 Symbol of JK flip-flop

The truth table of a JK flip-flop is given by **Table 1-6.**

Table 1-6 Truth table of clocked JK flip-flop

J	K	Q	$\bar{Q}$
0	0	No change	
1	0	1	0
0	1	0	1
1	1	Complement	

FEATURES:

(1) A clock pulse will not cause any changes in the states of the flip-flop if neither J nor

K input is activated.

(2) When the Q output is HIGH, the state of the flip-flop is the "Set" state. When the Q output is LOW, it is the "Reset" state. That is to say, the J and Clk input set the flip flop; K and Clk reset it.

(3) The Q and Q-Bar outputs are always complementary, i. e. when one of them is LOW, the other is HIGH.

(4) If both J and K inputs are 1, the flip flop will change state.

1.5.3 Registers

Any system which stores and manipulates information is called a "register". Flip-Flops are the common building-blocks of registers. According to their different function, the registers have different name, such as buffer register, shifting register, traveling wave counter, synchronous counter, ring counter, program counter, accumulator and so on. In this section, we will introduce some of them and their working principles.

1. Shifting Register

Figure 1-25 is the diagram of a shifting register which consists of 4 D flip-flops.

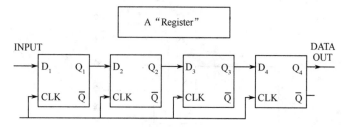

Figure 1-25　A shifting register

If we want to store a binary value of "**1001**" using this shifting register, the sequence is required as below.

When the first clock pulse arrives, D_1 is High which drives Q_1 High. (The clock pulse must end before the affection that Q_1 causes to the second latch.)

The second clock pulse coincides with D_1 going LOW, but before Q_1 goes LOW, D_2 goes HIGH, driving Q_2 HIGH.

Next clock pulse: D_1 LOW, Q_1 LOW, D_2 LOW, Q_2 LOW, D_3 HIGH, Q_3 HIGH.

Next clock pulse: D_1 HIGH, Q_1 HIGH, D_2 LOW, Q_2 LOW, D_3 LOW, Q_3 LOW, D_4 HIGH, Q_4 HIGH.

Therefore the pattern becomes:

Q_1　Q_2　Q_3　Q_4
1　0　0　1

This pattern will remain "latched" in the register for as long as power is supplied. Computer memories are based on this type of "latching" principle. (Applying four more clock pulses "shifts" the stored data to the "Data Out" pin.)

2. Buffer Register

Buffer is a device or area used to store data temporarily and deliver it at a rate different from the rate it received. **Figure 1-26** is the circuit of a 4-bits register. It consists of four clocked D flip-flops. The diagram indicates that the Clock signal is active high so that the flip-flop will respond to a positive going transition of the clock. Before the clock signal goes high, Q_0, Q_1, Q_2 and Q_3 keep their original data. The input X_0, X_1, X_2 and X_3 will not be allowed to pass through unless the clock signal goes high.

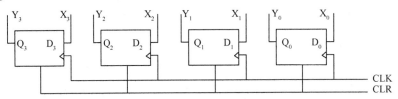

Figure 1-26 A 4-bit buffer register

Sometimes, we add a control circuit to a register. As showed in **Figure 1-27**, the **load gate** is a logic device to control registers. It's comprised by two "AND gates", one "OR gate" and a "NOT gate".

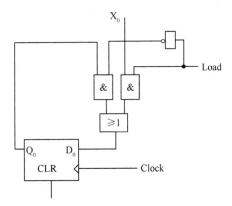

Figure 1-27 Load gate

When X_0 sends a data (0 or 1) and L ("L" is short for "Load") is **inactive low**, the right "AND gate" is blocked and X_0 cannot be sent out. For another "AND gate", it is always "ON" because one of the two inputs of this "AND gate" is always high. At this time, the load gate is in the state of "**self locking**", for that former data stored in the D-trigger will pass the "AND gate" and can be stored in D_0. On the contrary, if L is high, X_0 can be sent out through the right "AND gate" and reach D_0. X_0 will be sent to the bus at the rising edge of the clock signal, which is called "Load Data".

The **Figure 1-28** is an example of a controllable buffer register with a serial of load gates.

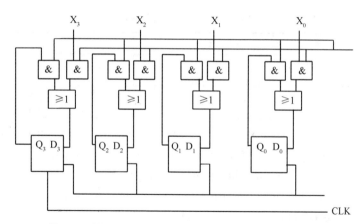

Figure 1-28 A controllable buffer register with load gate

1.5.4 Tri-state Gate and BUS Structure

A **tri-state gate** is a logic device with a special output that can take a third state. That is, the new state is neither a 0 nor a 1. Any logic gate can have a tri-state output. In this section, we only focus on the tri-state buffer.

Figure 1-29(a) gives the circuit diagram of a tri-state buffer gate and **Table 1-7** is the truth table of **Figure 1-29(a)**. The (b) is the symbol of (a). It shows a tri-state gate with an active-high control input E. As shown in **Table 1-7**, when E is asserted **high**, the output of the gate Y is **equal** to its input X. When E is **inactive low**, both G_1 and G_2 are low potential, it means that the gate's output Y is internally **disconnected** with the X, and the gate does not drive the output. That is, we cannot say what the output Y is, because it is disconnected from the gate. Output Y will usually be connected to a bus and the signal level at Y when the gate is disabled will be that of the bus. Consequently, this output state is called floating because the output floats up and down with the traffic on the bus. **Figure 1-29(c)** illustrates another tri-state gate, note that the control input is active low and the gate is enabled only when its control input is low.

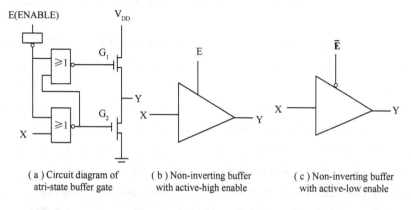

(a) Circuit diagram of a tri-state buffer gate

(b) Non-inverting buffer with active-high enable

(c) Non-inverting buffer with active-low enable

Figure 1-29 The tristate gate

Table 1-7 Input and output of the tri-state gate

E	X	Y
0	0	Floating
0	1	Floating
1	0	0
1	1	1

Figure 1-30 shows a system with four registers connected to a bus. The Q output of each register is connected to the bus by means of a buffer with a tri-state output. This buffer passes the unchanged signal at its input terminal to its output terminal when the enable input is in a 1 state. When the enable input is 0, the tri-state gate disconnects its output pin from the rest of the circuit. In **Figure 1-30** only one of the enable signals E_0, E_1, E_2, and E_3 can be set to a logical 1 at any instant.

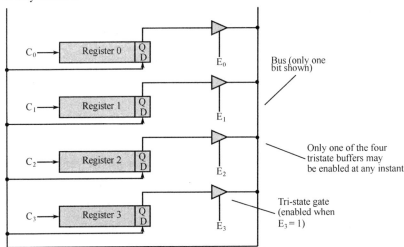

Figure 1-30 Registers and buses

When one of the registers in **Figure 1-30** is clocked, the data on the bus is copied into that register. **For example**, if we set E_2 to 1 and clock C_3, the effect is to put the contents of register R_2 on the bus and copy it into register R_3. This operation is represented in RTL by $[R_3] \leftarrow [R_2]$.

Figure 1-31 demonstrates how we can take the structure of **Figure 1-30** and create a simple functional unit that can execute the MOVE instruction. **For example**, this system can execute the machine level operation MOVE R_i, R_y that is defined as $[R_y] \leftarrow [R_i]$ in RTL.

Figure 1-32 is an example to show how to transfer the data between any two registers.

In **Figure 1-32**, there are four registers such as A, B, C and D. Each of them connects to the bus with a load gate and a tri-state gate, and therefore has L and E control signals. The data will be transferred between any two registers only when a E of one register and a L of another are effective. We can write down all the E and L of each registers in a line and name it as "control signals" which abbreviated as "CON". **For example**, in **Figure 1-32**, the "CON" is $CON = L_A E_A L_B E_B L_C E_C L_D E_D$.

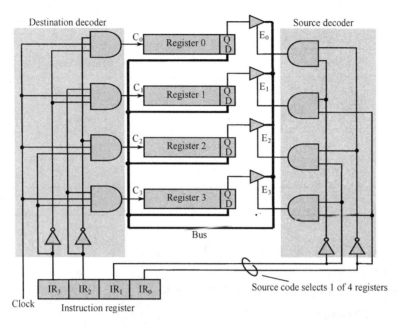

Figure 1-31 Controlling the bus

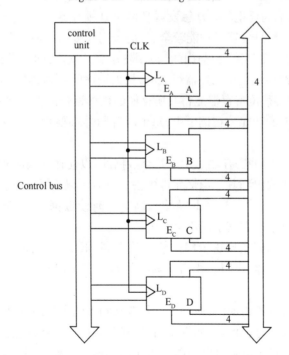

Figure 1-32 Data transform on the bus

Table 1-8 shows the data transformation between the registers on bus.

Table 1-8 Data transform between the resisters on bus

CON								Information Transform
L_A	E_A	L_B	E_B	L_C	E_C	L_D	E_D	
1	0	0	1	0	0	0	0	From B to A
0	1	1	0	0	0	0	0	From A to B
0	1	0	0	1	0	0	0	From A to C
0	1	0	0	0	0	1	0	From A to D
0	0	1	0	0	0	0	1	From D to B
1	0	0	0	0	1	0	0	From C to A

要点：

（1）三种基本门电路：非门（反向器）、或门、与门。

（2）触发器。

① RS 触发器。RS 触发器用一些与非门组成，S 端一般称为置位端，使 Q = 1；R 端一般称为复位端，使 Q = 0。

② D 触发器。D 触发器只有一个输入端口。当 D 端为高电平时，Q 端为高电平；当 D 端为低电平时，Q 端为低电平。D 触发器又称为 D 锁存器，输出 Q = D。

③ JK 触发器。JK 触发器是功能较全、目前应用较多的一种。在 RS 触发器的前面增加两个与门，并从输出端到输入（与门的输入端）作交叉反馈，即可得到 JK 触发器。

（3）寄存器。寄存器是由触发器组成的，根据用途的不同，寄存器可有不同的命名。

① 移位寄存器。移位寄存器能将其所存储的数据逐位向左或向右移动，以达到计算机在运行过程中所需的功能。在例子中，每个时钟脉冲都要把所储存的各位向右移动一个位置。

② 缓冲寄存器。缓冲寄存器是用以暂存数据的，以便在适当的时间节拍和给定的计算步骤将数据输入或输出到其他记忆元件中去。有时可以为缓冲寄存器增设一个可控的 LOAD 门（L 门）。L 门的作用是当 L 端输入高电平时使数据装入，低电平时，形成自锁，既存的数据能够可靠地存在其中而不会丢失。

（4）三态门。三态门是指逻辑门的输出除有高、低电平两种状态外，还有第三种状态——高阻状态的门电路，高阻态相当于隔断状态。三态门都有一个 EN 控制使能端来控制门电路的通断。

记忆元件是由触发器组成的，而触发器只有 0 和 1 两个状态，这样每条信息传输线也只能传送一个触发的信息。但是，如果使用三态门，就可使一条信息传输线传输随意多个触发器的信息。三态输出电路可以由两个或非门和两个 NMOS 晶体管（T_1、T_2）及一个非门组成。三态门（E 门）和装入门（L 门）一样，都可加到任何寄存器电路上去。这样的寄存器就称为三态寄存器。L 门负责对寄存器的装入数据的控制，而 E 门则负责由寄存器输出数据的控制。

Tips

absorption rules　合并规则
algorithm　运算法则
AND operation　与运算
arithmetic　算术
ASCII　美国信息交换标准码
associative　结合律
asterisk　星号"*"
audible bell　振铃
BCD　二进制编码的十进制
binary　二进制的
binary number　二进制数
Boolean Algebra　布尔代数
buffer　缓冲器
byte　字节(8位)
carriage return　回车
char　字符
commutative　交换律
data link escape　删除
decimal　十进制的
denominator　分母
distributive　分配律
dividend　被除数
double-precision　双精度数
D-trigger　D触发器
electrical disturbance　电路中断
end of transmit　传输结束
enquiry　查询
exponent　指数
exponential value　指数值
floating point numbers　浮点数(实数)
form feed　换页
fraction　分数
gate　门电路
hex number　十六进制数
hexadecimal　十六进制的
high order byte　高字节
hundreds position　百位
identical equation　恒等式
inactive low　非当前(非活动态)低电平

Kilobyte (KB)　千字节(1024字节)
line feed　换行,移行
line feed　换行
logic circuit　逻辑电路
low order byte　低字节
mantissa　尾数
Megabyte (MB)　兆字节(1024KB = 1048576 字节)
memory location　存储单元
Morgan's theorem　摩根定律
multiplication　乘法
nibble　半字节(4位)
NOT operation　非运算
numbering system　数制系统
numerator　分子
NXOR operation　异或非运算
octal　八进制的
one's complement　反码
OR operation　或运算
Packed BCD　压缩的BCD码
paragraph　节(128位)
parity　奇偶校验
position notation　位置记数法
power　幂
Quadword　四倍字长(64位)
radix　基
real Numbers　实数
remainder　余数
RTL (Resistor Transistor Logic)　电阻晶体管逻辑(电路)
shuffle　运载,传输
signed integer　有符号整数
single-precision　单精度数
space　空格
statute of "self-locking"　自锁状态
tens position　十位
the least significant byte　低字节
the most significant byte　高字节
tri-state gate　三态门
two's complement　补码

units position 个位
Un-packed BCD 非压缩的 BCD 码
unsigned integer 无符号整数
weight 权

whole number 整数
word 字(16位)
XOR operation 异或运算

Exercise

1. Fill the blanks:

(a) A binary bit stores a(n) _____ or a(n) _____ .

(b) A computer KB is equal to _____ bytes.

(c) A computer MB is equal to _____ K bytes.

(d) A computer GB is equal to _____ M bytes.

(e) A nibble is a _____ bits binary number.

(f) A number that contains 3 one bits is said to have _____ parity.

2. Convert the following binary numbers into decimal:

(a) 1001.0101   (b) 101101.1011

(c) 101010.0110   (d) 101.1011

3. Convert the following octal numbers into decimal:

(a) 637.5   (b) 46.2

(c) 1210.07   (d) 37.26

4. Convert the hexadecimal numbers into decimal:

(a) BC.D   (b) 456.7

(c) 32.7F   (d) AC12.6

5. Convert the following decimal integers into binary, and hexadecimal:

(a) 45.625   (b) 7098

(c) 4512   (d) 5693.375

6. Convert the following binary numbers to the one's complement form:

(a) 1100 1001   (b) 0111 0110

(c) 0101 0101   (d) 1011 0100

7. Define byte, word, and double word.

8. Convert the following words into ASCII-coded character strings:

(a) FROG12   (b) 7+2=9

(c) Macro Message   (d) "Hello, World!"

9. Convert the following decimal numbers into 8-bit or 16-bit signed binary numbers:

(a) +54   (b) -54

(c) +899   (d) -258

10. Show how the following 16-bit or 32-bit hexadecimal numbers are stored in the memory system (use the standard Intel format):

(a) 5230H   (b) AF78H

(c) FFCA7568H   (d) 120A7650H

11. Convert the following decimal numbers into both packed and unpacked BCD formats:
   (a) 201 (b) 66
   (c) 514 (d) 2500

12. Convert the following BCD numbers (assume that these are packed numbers) into decimal numbers:
   (a) 1000 1001 (b) 0000 1001
   (c) 0011 0010 (d) 0000 0001

Chapter 2 System Organization of Microcomputer

To be a good programmer of 80x86 assembly language requires considerable familiarity with the 80x86 family and a strong knowledge of the underlying hardware. This chapter describes the basic components that make up a computer system: the CPU, memory, I/O, and the bus that connects them. Although you can write software that is ignorant of these concepts, high performance software requires a complete understanding of this material.

2.1 The Basic System Components

The basic operational design of a computer system is called its architecture. John Von Neumann, a pioneer in computer design, is given credit for the architecture of most computers in use today. **For example**, the 80x86 family uses the Von Neumann architecture (VNA). A typical Von Neumann system has four major components: the **central processing unit** (or **CPU**), **memory**, **input/output** (or **I/O**), and **the bus** which connects them together, just as **Figure 2-1** shows:

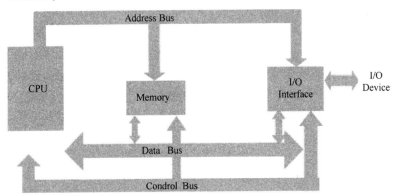

Figure 2-1 The whole structure of microcomputer

In VNA machines, like the 80x86 family, the CPU is where all the action takes place. All computations occur inside the CPU. Data and CPU instructions reside in memory until required by the CPU. To the CPU, most I/O devices look like memory because the CPU can store data to an output device and read data from an input device. The major difference between memory and I/O locations is the fact that I/O locations are generally associated with external devices in the outside world.

2.1.1 CPU (Central Processing Unit)

The heart of the microprocessor-based computer system is the microprocessor integrated circuit. The microprocessor, sometimes referred to as the CPU (central processing unit), is a controlling element in a computer system. The microprocessor controls memory and I/O through a series of connections called buses.

The CPU performs three main tasks for the computer system:

(1) Data transfer between itself and the memory or I/O subsystems;
(2) Simple arithmetic and logic operations;
(3) Program flow via simple decisions.

2.1.2 The System Bus

The system bus is a collection of wires on which electrical signals pass between components in the system. It connects the various components of a VNA machine. Just as the 80x86 family, there are always three major buses: the **address bus**, the **data bus**, and the **control bus**. These buses vary from processor to processor. However, each bus carries comparable information on all processors.

A bus may be a simple affair that connects the data paths from two registers together or it may be as complex as the PCI bus found in the PC that connects functional modules together and provides a wide range of facilities. Here we introduce some basic concepts of buses.

1. The Data Bus

Processors use the data bus to shuffle data between the various components in a computer system.

On typical 80x86 systems, the data bus contains 8, 16, 32, or 64 lines. As shown in **Table 2-1**, the 8088 and 80188 microprocessors have an eight bit data bus (eight data lines) and the 8086, 80186, 80286, and 80386SX processors have a 16 bit data bus, while the 80386DX, 80486, and Pentium Overdrive' processors have a 32 bit data bus. The Pentium' and Pentium Pro processors have a 64 bit data bus.

Having an 8-bit data bus does not limit the processor to 8-bit data types. It simply means that the processor can only access one byte of data per memory cycle. Therefore, the eight bit bus on an 8088 can only transmit half the information per unit time (memory cycle) as the 16 bit bus on the 8086. Therefore, processors with a 16 bit bus are naturally faster than processors with an eight bit bus. Likewise, processors with a 32 bit bus are faster than those with a 16 or eight bit data bus. The size of the data bus affects the performance of the system more than the size of any other bus.

Of course, the 16, 32 and 64 bit members of 80x86 family can also access smaller memory units of 8, 16, or 32 bits. They can process data up to the width of the bus. Therefore, anything you can do with a small data bus can be done with a larger data bus as well. The larger data bus, however, may access memory faster and can access larger chunks of data in one memory operation.

Table 2-1 80x86 processor data bus sizes

Processor	Data Bus Size
8088	8
80188	8
8086	16
80186	16
80286	16
80386SX	16
80386DX	32
80486	32
80586 class/ Pentium (Pro)	64

2. The Address Bus

The data bus transfers information between a particular memory location or I/O device and the CPU. To differentiate memory locations and I/O devices, the system designer assigns a unique address to each memory element and I/O device. When the software wants to access some particular memory location or I/O device, it places the corresponding address on the address bus. Circuitry associated with the memory or I/O device recognizes this address and instructs the memory or I/O device to read the data from or place data on the data bus. In either case, all other memory locations or other I/O devices ignore the request. Only the device whose address matches the value on the address bus responds.

With a single address line, a processor could create exactly two unique addresses: zero and one. With n address lines, the processor can provide 2^n unique addresses. Therefore, the number of bits on the address bus will determine the maximum number of addressable memory and I/O locations. The 8088 and 8086, **for example**, have 20 bit address buses. Therefore, they can access up to 1,048,576 (or 2^{20}) memory locations. Larger address buses can access more memory. As shown in **Table 2-2**, different processors have different address buses and therefore have different addressable memory locations.

Table 2-2 80x86 Family address bus sizes

Processor	Address Bus Size	Max Addressable Memory	In English
8088	20	1,048,576	One Megabyte
8086	20	1,048,576	One Megabyte
80188	20	1,048,576	One Megabyte
80186	20	1,048,576	One Megabyte
80286	24	16,777,216	Sixteen Megabytes
80386SX	24	16,777,216	Sixteen Megabytes
80386DX	32	4,294,976,296	Four Gigabytes
80486	32	4,294,976,296	Four Gigabytes
80586/Pentium (Pro)	32	4,294,976,296	Four Gigabytes

The 80x86 family, unlike many other processors, provides two distinct address spaces: one for memory and one for I/O. While the memory address buses on various 80x86 processors vary in size, the I/O address bus on all 80x86 CPUs is 16 bits wide. This allows the processor to address up to $65,536(2^{16})$ different I/O locations. As it turns out, most devices (like the keyboard, printer, disk drives, etc.) require more than one I/O location. Nonetheless, 65,536 I/O locations are more than sufficient for most applications. The original IBM PC design only allowed the use of 1,024 of these.

3. The Control Bus

The control bus is an eclectic collection of signals that control how the processor communicates with the rest of the system. For example, for a data bus, the CPU sends data to memory and receives data from memory on the data bus. This prompts the question, "Is it sending or receiving?" There are two lines on the control bus, read and write, which specify the direction of data flow. Other signals include system clocks, interrupt lines, status lines, and so on. The exact make up of the control bus varies among processors.

As we know, the read and write control lines control the direction of data on the data bus. When both contain a logic one, the CPU and memory-I/O are not communicating with one another. If the read line is low (logic zero), the CPU is reading data from memory (that is, the system is transferring data from memory to the CPU). If the write line is low, the system transfers data from the CPU to memory.

The byte enable lines are another set of important control lines. These control lines allow 16, 32 and 64 bit processors to deal with smaller chunks of data.

Although the 80x86 family supports two address spaces: one for memory and one for I/O, it does not have two address buses (for I/O and memory). Instead, the system shares the address bus for both I/O and memory addresses. Additional control lines decide whether the address is intended for memory or I/O. When such signals are active, the I/O devices use the address on the low 16 bits of the address bus. When inactive, the I/O devices ignore the signals on the address bus (the memory subsystem takes over at that point).

2.1.3 The Memory Subsystem

A processor addresses a maximum of 2^n different memory locations, where n is the number of bits on the address bus. As you've seen already, 80x86 processors have 20, 24, and 32 bit address buses. **Figure 2-2** is an example of a 16×8 memory. Because n is equal to four, the max addressable memory is sixteen.

The 80x86 supports byte addressable memory. Therefore, the basic memory unit is a byte. So with 20, 24, and 32 address lines, the 80x86 processors can address one megabyte, 16 megabytes, and four gigabytes of memory, respectively. Think of memory as a linear array of

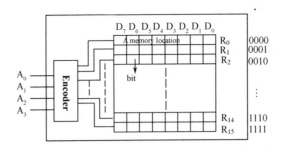

Figure 2-2   16 × 8 Memory

bytes. The address of the first byte is zero and the address of the last byte is $2^n - 1$.

"What exactly is a memory location?"

For an 8088 with a 20 bit address bus, the following pseudo-C array declaration is a good approximation of memory:

char Memory[2^{20}];

To execute the equivalent of the C statement "Memory[125] = 0;", as shown in Figure 2-3, the CPU places the value zero on the data bus, the address 125 on the address bus, and asserts the write line (since the CPU is writing data to memory).

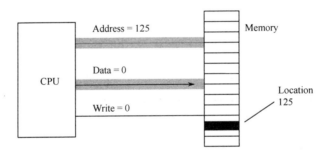

Figure 2-3   Example of memory write

To execute the equivalent of "Data = Memory[125];" as shown in Figure 2-4, the CPU first places the address 125 on the address bus, asserts the read line (since the CPU is reading data from memory), and then reads the resulting data from the data bus.

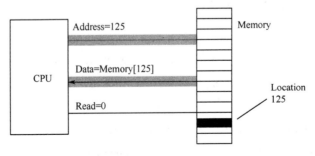

Figure 2-4   Example of memory read

1. ROM (Read Only Memory)

An acronym for Read Only Memory, ROM is computer memory on which data has been pre-recorded. The programming code and/or data on a ROM chip are/is written to the chip at the factory. It can be read, but it cannot be erased or removed. It's permanent. ROM retains its data or content even when the computer is turned off, unlike a computer's main memory (RAM), which needs a constant charge of electricity to keep its information. For this reason, ROM is considered to be "non-volatile" and RAM is "volatile".

ROM chips are used in all kinds of electronic devices from calculators to video games. Most personal computers have several applications of ROM memory. These chips often store permanent and critical information and programs that don't need to be changed, or don't need to be written to. Most personal computers have a small amount of ROM that stores the code that starts up or boots the computer. Early computers also used ROM to store the BIOS (Basic Input Output System) which acts as a translator between the PC's hardware and the operating system.

The symbol of the ROM is shown as **Figure 2-5** and it usually works with a MAR (Memory Address Register). **Figure 2-6** shows the connection between the ROM and MAR.

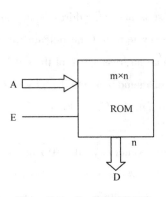

Figure 2-5 Symbol of ROM

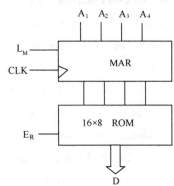

Figure 2-6 Mar and ROM

2. RAM (Random Access Memory)

Random Access Memory is the workspace of your computer. It's the area that stores the information and data for processing by the computer's CPU. Retrieving data from the hard driver is much slower, but once the data is loaded into RAM, it can be accessed instantly and randomly. RAM is volatile, which means that once you turn off the power, all data in memory disappears. RAM is stored in chips on small circuit boards called SIMMs (single in-line memory module) or DIMMs (dual in-line memory modules), that attach to the motherboard.

Figure 2-7 is the symbol of RAM. Different from the ROM, the RAM works together with MAR and MDR. **Figure 2-8** shows a memory subsystem consisting of RAM. The M_E is enable signal for the RAM while it is high. The control line W_E is responsible for the data to be sent or to be received. In this diagram, the machine's RAM memory is accessed by first placing the 8-bit address in the Memory Address Register (MAR), while the data to the RAM appears in the

Memory Data Register(MDR).

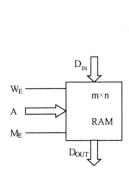

Figure 2-7 Symbol of RAM

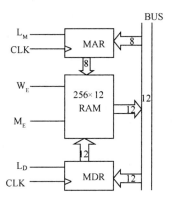

Figure 2-8 Memory sub system

2.1.4 The I/O Subsystem

There are three basic forms of input and output that a typical computer system will use: **I/O-mapped I/O**, **memory-mapped input/output**, and **direct memory access** (**DMA**). As shown in **Figure 2-9**, I/O-mapped input/output has separate I/O address space and uses special instructions to transfer data between the computer system and the outside world; memory-mapped I/O uses special memory locations in the normal address space of the CPU to communicate with real-world devices; DMA is a special form of memory-mapped I/O where the peripheral device reads and writes memory without going through the CPU. Each I/O mechanism has its own set of advantages and disadvantages.

Besides the 20, 24, or 32 address lines which access memory, the 80 × 86 family provides a 16 bit I/O address bus. This gives the 80 × 86 CPUs two separate address spaces: one for memory and one for I/O operations. Lines on the control bus differentiate between memory and I/O addresses. Other than separate control lines and a smaller bus, I/O addressing behaves exactly like memory addressing. Memory and I/O devices both share the same data bus and the low 16 lines on the address bus.

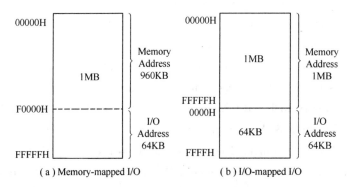

Figure 2-9 Memory-mapped I/O and I/O-mapped I/O

An I/O port is a device that looks like a memory cell to the computer but contains connections to the outside world. An I/O port typically uses a latch rather than a flip-flop to implement the memory cell. When the CPU writes to the address associated with the latch, the latch device captures the data and makes it available on a set of wires external to the CPU and memory system.

A perfect example of an output port is a parallel printer port. The CPU typically writes an ASCII character to a byte-wide output port that connects to the DB-25F connect on the back of the computer's case. A cable transmits this data to a printer where an input port (to the printer) receives the data. A processor inside the printer typically converts this ASCII character to a sequence of dots it prints on the paper.

Generally, a given peripheral device will use more than a single I/O port. A typical PC parallel printer interface, **for example**, uses three ports: a read/write port, an input port, and an output port. The read/write port is the data port (it is read/write to allow the CPU to read the last ASCII character it wrote to the printer port). The input port returns control signals from the printer; these signals indicate whether the printer is ready to accept another character, is off-line, is out of paper, etc. The output port transmits control information to the printer such as whether data is available to print.

To the programmer, the difference between I/O-mapped and memory-mapped input/output operations is the instruction to use. For memory-mapped I/O, any instruction that accesses memory can access a memory-mapped I/O port. On the x86, the MOV, ADD, SUB, CMP, AND, OR, and NOT instructions can read memory; the MOV and NOT instructions can write data to memory. I/O-mapped input/output uses special instructions to access I/O ports. **For example**, the x86 CPUs use the GET and PUT instructions, the Intel 80x86 family uses the IN and OUT instructions. The 80x86 IN and OUT instructions work just like the MOV instruction except they place their address on the I/O address bus rather than the memory address bus.

Memory-mapped I/O subsystems and I/O-mapped subsystems both require the CPU to move data between the peripheral device and main memory. **For example**, to input a sequence of ten bytes from an input port and store these bytes into memory, the CPU must read each value and store it into memory. For very high-speed I/O devices the CPU may be too slow when processing this data a byte at a time. Such devices generally contain an interface to the CPU's bus so it directly read and write memory. This is known as direct memory access (DMA) since the peripheral device accesses memory directly, without using the CPU as an intermediary. This often allows the I/O operation to proceed in parallel with other CPU operations, thereby increasing the overall speed of the system. Note, however, that the CPU and DMA device cannot both use the address and data buses at the same time. Therefore, concurrent processing only occurs if the CPU has a cache and is executing code and accessing data found in the cache (so the bus is free). Nevertheless, even if the CPU must halt and wait for the DMA operation to complete, the I/O is still much faster since many of the bus operations during I/O or memory-mapped input/output consist of instruction fetches or I/O port accesses which are not present during DMA operations.

要点:

(1) 微处理器(Microprocessor)。微处理器是微型机算计的运算及控制部件,也称为中央处理单元(CPU)。它本身不构成独立的工作系统,因而,它也不能独立地执行程序。微处理器由算术逻辑单元(ALU)、寄存器阵列、控制单元和片内总线组成,是计算机系统的核心,一般由大规模集成电路构成。

(2) 总线。总线可分为数据总线、地址总线及控制总线三类。

数据总线:用于传输数据。从结构上看,数据总线是双向的,即数据即可从 CPU 送到其他部件,也可从其他部件送到 CPU。

地址总线:专门用来传送地址信息。由于地址总是从 CPU 送出的,所以和数据线不同,地址总线是单向的。地址总线的位数决定了 CPU 可以直接寻址的内存范围。

控制总线:用来传输控制信号。其中包括 CPU 送往存储器和 I/O 接口电路的控制信号,如读信号、写信号和中断响应信号等;还包括其他部件送到 CPU 的信号,如时钟信号、中断请求信号和准备就绪信号。

(3) 存储器。存储器又叫内存或主存,是微型计算机的存储和记忆部件,用于存放数据和当前执行的程序。每个单元都对应一个地址,以实现对单元内容的寻址。内存容量的大小依 CPU 的寻址范围而定(即 CPU 地址信号线的位数)。存储器由寄存器组成,可以看作是一个寄存器堆,每个存储单元实际上相当于一个缓冲寄存器。

存储器可分为 ROM 和 RAM 两类。

ROM 是用以存放固定程序的存储器,一旦程序放进去之后,即不可改变,不能再"写"入新的字节,而只能从中"读"出其所有存储的内容,因此称为只读存储器。

随机存储器 RAM 又叫读/写存储器,它不但能读取已存放在其各个存储单元中的数据,而且还能够随时写进新的数据,或者改写原来的数据。因此,RAM 的每一个存储单元相当于一个可控缓冲寄存器。

存储地址寄存器(MAR)是存储器的一个附件,它将所要寻找的存储单元的地址暂存下来,以备下一条指令之用。

存储器数据寄存器(MDR)是一个可控缓冲寄存器,用于暂存要写入 RAM 的数据。

(4) I/O 接口。I/O 接口是 CPU 与外部设备间的桥梁,I/O 接口的功能是:数据缓冲寄存,信号电平或类型的转换,实现主机与外设间的运行匹配。

I/O 的编址方式有两种:存储器映射方式(Memory-Mapped)和 I/O 映射(I/O Mapped)方式。

存储器映射方式的特点是:把一个 I/O 端口看成存储器统一编址,I/O 端口空间是存储空间的一部分,因此所有访问存储器单元的指令都可以用来访问端口而无须设置专门的 I/O 类指令,如 M6800、R6502 系列微机采用这种方式。

I/O 映射方式的特点是:I/O 端口构成单独的 I/O 空间,不占用存储器地址,CPU 设置专门的 I/O 指令来访问端口。存储器地址和 I/O 端口地址可重叠,由于二者访问指令不同,因此不会产生混淆。Z80、8086/8088 等系列微机都采用这种方式。

2.2 A Simple Computer

2.2.1 Introduction

For too many students, the level of understanding of how digital computers really execute programs never quite makes it beyond that of "black magic". Many of them know what the basic parts of the computer are and what kinds of actions occur in each. But, in too many cases, they cannot put it all together and formulate a clear "big picture" of what is really going on inside the machine. In this section we present simple computer architecture and describe in detail the way in which its control section may be organized.

2.2.2 Architecture of the Simple Computer

Let's first take a look at the structure diagram of our simple computer.

As shown in **Figure 2-10**, it also has four major parts: the **CPU**, **memory**, **input/output** (or **I/O**), and **the bus**. A single 8-bit wide bus provides for exchange of information between the each part of the computer within the data path section. Note that some components are connected to the bus with Load Gate or tri-state Gate.

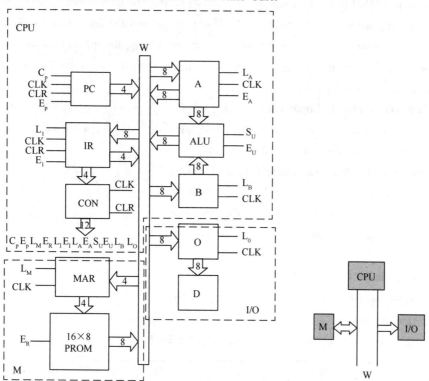

Figure 2-10 Simple computer system structure

1. CPU

The constituent parts of CPU in this simple computer are mentioned in brief as following:

1) **CPU Registers**—(**A and B register**)

CPU registers are very special memory locations constructed from flip-flops (triggers). They are not part of main memory, the CPU implements them on-chip. Various members of the 80x86 family have different register sizes. The simple CPU has exactly two registers-accumulator A and register B, both 8 bits wide.

Because registers are on-chip and handled specially by the CPU, they are much faster than memory. Accessing a memory location requires one or more clock cycles. Accessing data in a register usually takes zero clock cycles. Therefore, you should try to keep variables in the registers. Register sets are very small and most registers have special purposes which limit their use as variables, but they are still an excellent place to store temporary data.

The registers and the 16 ×8 bit PROM memory are controlled by 12 bits long control signals. The register A has Load (L) and Enabled (E) signals. An active L signal of the register A causes the data on the bus to be loaded on the next rising pulse from the system clock. An active E signal of the register B enables the data out. Therefore, the data transfer is accomplished from register B to register A.

2) **Program Counter (PC)**

Program Counter (PC) always contains the address of the next instruction to execute. Its value ranges from 0000 to 1111 in binary (0-F in hex). It (The CLK signal) initializes its value to 0000 when program starts running and adds by one after CPU fetches an instruction from the memory. The instructions on this machine are one byte long, this provides a simple mechanism for sequential instruction execution. The C_P signal is used to control the PC self-increasement by 1 when it is ineffective.

3) **The Arithmetic & Logic Unit (ALU)**

Processing of data is done by the Arithmetic-Logic-Unit (ALU), a circuit that is capable of adding or subtracting the 8-bit numbers contained in its two input registers: the accumulator (A) and register B. The operation performed by the ALU is selected by the Add (A) or Subtract (S) control signals. The processing procedure of ALU can be explained as the **Figure 2-11**.

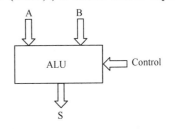

Figure 2-11 Symbol of ALU

4) **Instruction Register (IR)**

The Instruction Register (IR) is used to hold the instruction that is about to be executed and provides its op-code to the controller/sequencer. The format of 8-bit wide instruction is given as **Figure 2-12.**

X X X X	X X X X
Most Significant Bits (MSB)	Least Significant Bits (LSB)

Figure 2-12 Instruction format

As shown in **Figure 2-12**, the high half of the instruction word called MSB connects to the instruction decoder, and the low half called LSB denotes the operand's address.

5) The Control Unit (CU)

Basically there are two types of control units: hard-wired controllers and micro-programmed controllers. In order to appreciate the difference and see how computers really work, we present a very simple hardwired controller as shown in **Figure 2-13**, whose main function includes:

(1) Initialize the PC and IR at the beginning of running a program, and make:
$$PC = 0000$$
$$IR = 0000\ 0000$$

(2) Generate a 12-bit wide control signals at every clock cycle.

(3) Synchronize all parts of computer by using clock signal.

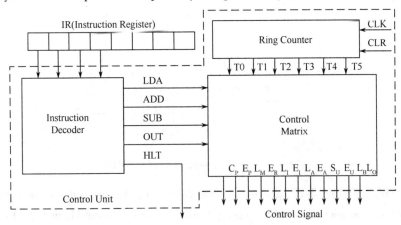

Figure 2-13 Organization of the control unit

Figure 2-13 is a block diagram showing the internal organization of a hard-wired control unit for our simple computer. Input to the controller consists of the 4-bit op-code of the instruction currently contained in the Instruction Register. The controller's output is a set of 12 control signals that go out to the various registers and to the memory of the computer, in addition to a HLT signal that is activated whenever the leading bit of the op-code is one. It is often to use "CON" as an abbreviation for "Control signals", as shown in **Figure 2-13**.

$$CON = C_P E_P L_M E_R \quad L_I E_I L_A E_A \quad S_U E_U L_B L_O$$

According to each bit of the control signals, computer can operate the instructions automatically and orderly.

2. Memory

Programs in early (pre-Von Neumann) computer systems were often "hard-wired" into the circuitry. That is, the computer's wiring determined what problem the computer would solve. One had to rewire the circuitry in order to change the program. It is a very difficult task. The next advance in computer design was the programmable computer system, one that allowed a computer programmer to easily "rewire" the computer system using a sequence of sockets and plug wires. A computer program consisted of a set of rows of holes (sockets), each row representing one operation during the execution of the program. The programmer could select one of several instructions by plugging a wire into the particular socket for the desired instruction. **Figure 2-14** is a diagram of 4 ×4 PROM which is simpler and easier to understand.

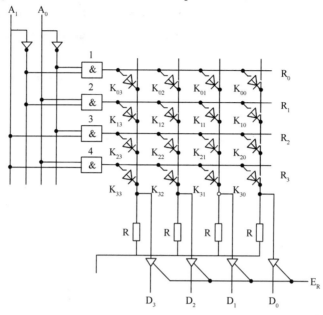

Figure 2-14   Diagram of 4 ×4 PROM

As shown in **Figure 2-10**, for the simple computer, the machine's memory is 16 ×8 PROM which can be accessed by first placing the 4-bit address in the Memory Address Register (MAR). Moreover, these 16 units of the memory are divided into two sections. One is the code segment, another is the data segment. The allocation of the memory is given in Table 2-3.

Table 2-3   Allocation of the memory

Area Function	Code Segment	Data Segment
Storing Unit	$R_0 \sim R_7$	$R_8 \sim R_{15}$
Binary Address	0000 ~ 0111	1000 ~ 1111
Hexadecimal Address	0H ~ 7H	8H ~ FH

3. I/O

There are no input ports in this simple computer but an output register (O) connected to the display (D).

4. Bus

In this simple computer, like that shown in **Figure 2-10**, the data bus contains 8 lines to shuffle data between the different components, while the address bus contains 4 lines.

要点：

（1）存储地址寄存器 MAR 接收来自 PC 的二进制程序号,作为地址码送至 PROM。

可编程只读存储器 PROM 实际上同时具有 RAM 和 ROM 的功能。

（2）简化的微型计算机的组成部件：

① 累加器 A 和寄存器 B。

累加器 A 用以存储计算机运行的中间结果，可接收总线送来的数据（$L_A = 1$），也能将数据送到总线上去（$E_A = 1$）。它还有一个双态输出端（数据），可将数据送至 ALU 进行算术运算。此输出端不受 E 门的控制。

寄存器 B 到 ALU 的输出是双态的，即无 E 门控制。主要用于存储将要与 A 相加减的数据。

② 程序计数器 PC。

程序计数器 PC 的计数范围是 0000～1111。每次运行之前，先复位为 0000，取出一条指令后 PC 加 1。

③ 算术逻辑部件 ALU。

ALU 为一个二进制补码加法器/减法器。

当 $S_u = 0$ ALU → A + B。

当 $S_u = 1$ ALU → A − B。

④ 指令寄存器 IR。

IR 从 PROM 接收到指令字（当 $L_I = 1$，$E_P = 1$），同时将指令字分送到控制部件 CON 和 W 总线上去。指令是 8 位的，分为高四位和低四位两部分，高四位为指令段，低四位为地址字段。

X	X	X	X	X	X	X	X
高四位（指令段）				低四位（地址字段）			

⑤ 控制部件 CON。

每次运行之前，CON 先发出 CLR = 1，使有关的部件清零，即

$$PC = 0000; IR = 0000\ 0000$$

CON 有一个同步时钟，能发出脉冲 CLK 到各个部件去，使他们同步运行。

在 CON 中有一个控制矩阵 CM，能根据 IR 送来的指令发出 12 位的控制字：

$$CON = C_P E_P L_M E_R L_I E_I L_A E_A S_U E_U L_B I_O$$

根据控制字中各位的 置 1 或置 0 情况，计算机就能自动的按指令有秩序地运行。

2.2.3 Instruction Set of the Simple Computer

A fair question to ask at this point is "How exactly does a CPU perform assigned chores?" This is accomplished by giving the CPU a fixed set of commands, or instructions, to work on. Keep in mind that CPU designers construct these processors using logic gates to execute these instructions. To keep the number of logic gates to a reasonably small set (tens or hundreds of thousands), CPU designers must necessarily restrict the number and complexity of the commands the CPU recognizes. This small set of commands is the CPU's instruction set.

Our simple CPU provides five basic instruction classes, such as LDA, ADD, SUB, OUT and HLT. Three of these instructions have a single operand, and two instructions have no operands at all. The following paragraphs describe the way that each of these works.

The LDA and OUT are instructions for data movement, they take the following form:

LDA memory
OUT

where memory is an operand specifying a memory location.

The arithmetic instructions take the following forms:

ADD memory
SUB memory

The ADD instruction adds the value of the operand specified by memory location address to the data in A register. The SUB instruction subtracts the value of the operand from A. These two instructions both leave the result in register A.

The remaining instruction HLT does not require any operands. HLT means to halt or terminate the program execution.

2.2.4 Encoding Instructions

Although we could arbitrarily assign operation codes (op-codes) to each of the instructions, keep in mind that a real CPU uses logic circuitry to decode the op-codes and acts appropriately on them. A typical CPU op-code uses a certain number of bits in the op-code to denote the instruction class (e.g., MOV, ADD, SUB), and a certain number of bits to encode each of the operands.

As shown in **Figure 2-15**, in this simple computer, the leading four bits form the operation code (op-code) which specifies the action to be taken, and the remaining 4 bits, when used, indicate the memory address of the instruction's operand.

I	I	I	I	M	M	M	M
Operation Code				Operand Address			

Figure 2-15 Instruction format of the simple computer

For some instructions that must have two operands, the other operand is always contained within the accumulator. **Table 2-4** shows the unique op-code for every different instruction classes.

Table 2-4 Instruction set of simple computer

Instruction	Op-code
LDA	0000
OUT	1110
ADD	0001
SUB	0010
HLT	1111

Regardless of the programming language used, the CPU can only execute instructions encoded in binary. Needless to say, each instruction in the program you write will be translated into corresponding destination codes and stored into the assigned part of the memory. The implementation procedure is indicated in **Example 2-1**.

【**Example 2-1**】 Develop a program which can accomplish the equation: D = (R_9) + (R_A) + (R_B) − (R_C).

To accomplish this equation, you can write the instruction sequences as following:

LDA　　R_9; *load the data in R_9 to A register*
ADD　　R_A; *add the content in R_A to A register*, (R_9) + (R_A) →A
ADD　　R_B; (R_9) + (R_A) + (R_B) →A
SUB　　R_C; (R_9) + (R_A) + (R_B) − (R_C) →A
OUT
HLT

When you got the program, in the simple computer, you should allocate the memory to store the instructions and data respectively. For example, the locations R_0-R_7 are designated as code segment to store the instructions, while the R_8-R_F are allocated as data segment. Then encode the instructions one by one and store them into the memory locations. **Table 2-5** gives us the program encoding process and their storage.

Table 2-5　Instruction encoding and memory allocation

	Source Program		Destination Code	Storage Cell	Cell Address
Encoding instruction	Code Segment	LDA　R_9	0000 1001	R_0	0000
		ADD　R_A	0001 1010	R_1	0001
		ADD　R_B	0001 1011	R_2	0010
		SUB　R_C	0010 1100	R_3	0011
		OUT	1110 XXXX	R_4	0100
		HLT	1111 XXXX	R_5	0101
	Date Segment	R_9 = 16(10)	0001 0000	R_9	1001
		R_A = 20(10)	0001 0100	R_A	1010
		R_B = 24(10)	0001 1000	R_B	1011
		R_C = 28(10)	0001 1100	R_C	1100

In the simple computer, you can load the destination program by setting each bit of the 16×8 PROM to 1 or 0.

要点：

(1) 每一个指令助记符都有与其对应的二进制码,如表2-4所列。

(2) 在简化计算机中,将16个存储单元分配成程序存放区(指令区)和数据存放区(数据区)两个区。

(3) 指令存放在存储器的指令区。每条指令存放时,先根据助记符与二进制的对照表将每一条指令的助记符译成二进制码,并将存储单元符号写成地址码,再将翻译好的机码存放到对应的地址单元中去,如表2-5所列。

2.2.5　Organization of the Control Unit

The control unit, as shown in **Figure 2-13**, also called as the controller is composed of the following functional parts: A ring counter, an instruction decoder, and a control matrix.

1. The Ring Counter

The ring counter provides a sequence of six consecutive active signals that cycle continuously. As shown in **Figure 2-16**, synchronized by the system clock, the ring counter first activates its T_0 line, then its T_1 line, and so forth. After T_5 is active, the sequence begins again with T_0.

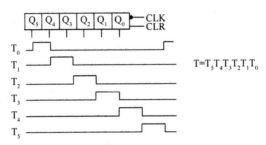

Figure 2-16 The ring counter

2. Instruction Decoder (ID)

The instruction decoder takes its four-bit input from the op-code field of the instruction register (IR) and activates one and only one of its 8 output lines. Each line corresponds to one of the instructions in the computer's instruction set. **Figure 2-17** shows the organization of this decoder.

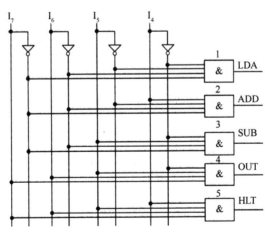

Figure 2-17 Internal organization of ID

3. Control Matrix (CM)

The most important part of the hard-wired controller is the control matrix. It receives input from the ring counter and the instruction decoder and provides the proper sequence of control signals. **Figure 2-18** is a diagram of how the control matrix for our simple machine might be wired. To understand how this diagram was obtained, we must look carefully at the machine's instruction set.

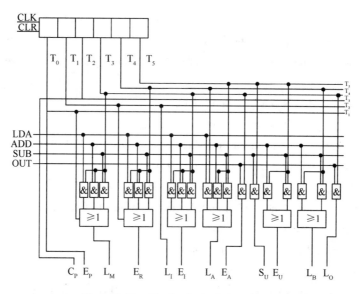

Figure 2-18 The hard-wired control unit

要点：

控制部件主要包括环形计数器、指令译码器、控制矩阵等部分。

（1）环形计数器。环形计数器不是用来计数用的，而是用来发出环形字，产生机器节拍，形成顺序控制信号的。

（2）指令译码器。指令译码器主要用于指令翻译。进入指令寄存器（IR）的数据的高 4 位被送入控制部件的指令译码器，这高 4 位编码经过译码后，使得某一个动作的控制线为高电平，从而实现该动作。

（3）控制矩阵。控制矩阵是控制部件的核心，产生 12 位的控制字。

2.2.6 Step-by-Step Instruction Execution

The simple CPU does not complete execution of an instruction in a single clock cycle. The CPU executes several steps for each instruction. **For example**, the CPU issues the following commands to execute an instruction:

(1) Fetch the instruction byte from memory.

(2) Update the PC register to point at the next byte.

(3) Decode the instruction to see what it does.

(4) If required, fetch an 8-bit instruction operand from memory.

(5) Store the fetched value into the destination register.

Table 2-6 shows which control signals must be active at each ring counter pulse for each of the instructions in the computer's instruction set (and for the instruction fetch operation). The table was prepared by simply writing down the instructions in the left-hand column. (In the circuit these will be the output lines from the decoder.) Entries into the table consist of the moments (ring counter pulses T_0, T_1, T_2, T_3, T_4, or T_5) at which each control signal must be active in order to have the instruction executed. In order to help you understand the computer working process clearly, now we will describe how the instruction LDA R_9 is executed in detail.

(1) At ring count 1, $T_0 = 1$. The E_P and L_M control signals should be active. The instruction fetch operation will be done. The content of PC (address code) will be sent to MAR and then sent to PROM.

$E_P = 1$, PC is ready to send data.

$L_M = 1$, MAR is ready to receive data.

The control signals of the control unit should be: CON = $C_P E_P L_M E_R$, $L_I E_I L_A E_A S_U E_U L_B L_O$

= 0 1 1 0   0 0 0 0   0 0 0 0

(2) At ring count 2, $T_1 = 1$. The content of address defined by the address code should be sent to IR and the MSB (instruction code) should be sent to CU through IR. The E_R and L_I control signals will be active.

$E_R = 1$, PROM is ready to send data.

$L_I = 1$, IR is ready to receive data.

So, the CON should be   0 0 0 1   1 0 0 0   0 0 0 0

(3) At ring count 3, $T_2 = 1$. The PC should accomplish self-increasement by 1 to make sure it is ready for the next instruction fetching.

$C_P = 1$, the C_P control signal should be active. PC is ready to fetch next instruction.

CON = 1 0 0 0   0 0 0 0   0 0 0 0

(4) At ring count 4, $T_3 = 1$. The op-code (high four bits) of the instruction code has been analyzed and it is confirmed as LDA instruction, then the E_I and L_M control signals will be active and a command will be sent as follows.

$E_I = 1$, The LSB of IR will be sent to bus.

$L_M = 1$, MAR obtains LSB as address and sends it to PROM.

CON = 0 0 1 0   0 1 0 0   0 0 0 0

(5) At ring count 5, $T_4 = 1$. The content of address denoted by the LSB will be sent to the accumulator A. The E_R and L_A control signals will be active.

$E_R = 1$, PROM is ready to send data.

$L_A = 1$, A is ready to receive data. CON = 0 0 0 1   0 0 1 0   0 0 0 0

(6) At ring count 6, $T_5 = 1$. As the data has been sent to A, the LDA instruction has finished, and there is nothing to do at T_5. No control signal will be active.

CON = 0 0 0 0   0 0 0 0   0 0 0 0

Then, what is T_5 used for? Although LDA doesn't use T_5, there are other instructions that use T_5, such as ADD, SUB, etc.. To make sure the machine cycle of each instruction is the same, here (6), a void step will be inserted to. Such kind of machine is called fix cycle computer.

Table 2-6 may help us clarify what the CPU is doing. When CPU fetches the instruction from memory, it copies the value of the PC register to the MAR and reads the byte at that address. This will take two clock cycles.

After fetching the instruction byte, the CPU updates PC so that it points at the next byte in the instruction stream. This takes one clock cycle.

The next step is to decode the instruction to see what it does. This will tell the CPU, among other things. This takes one clock cycle.

During decoding, the CPU determines if it needs to fetch additional operand bytes from memory. If the instruction requires an 8 bit operand (i.e., ADD R_9) then the CPU fetches that operand from memory. This step may require two clock cycles.

In the end, the CPU instructs the ALU to perform the action (add or subtract) and store the result in the destination register (A). This takes one clock cycle.

Table 2-6 Execution proceeding of Example 2-1

Instruction \ Action Address			Machine Cycle (Instruction Cycle)					
			Instruction Fetch Cycle			Execution Cycle		
			T_0	T_1	T_2	T_3	T_4	T_5
0000	LDA 0000	R_9 1001	MAR←PC(ROM) L_M, E_P	IR←ROM L_I, E_R	(0001) PC←PC+1 C_P	MAR←IR(ROM) L_M, E_I	A←ROM L_A, E_R	—
0001	ADD 0001	R_A 1010	MAR←PC(ROM) L_M, E_P	IR←ROM L_I, E_R	(0010) PC←PC+1 C_P	MAR←IR(ROM) L_M, E_I	B←ROM L_B, E_R	A←A+B L_A, E_U
0010	ADD 0001	R_B 1011	MAR←PC(ROM) L_M, E_P	IR←ROM L_I, E_R	(0011) PC←PC+1 C_P	MAR←IR(ROM) L_M, E_I	B←ROM L_B, E_R	A←A+B L_A, E_U
0011	SUB 0010	R_C 1101	MAR←PC(ROM) L_M, E_P	IR←ROM L_I, E_R	(0100) PC←PC+1 C_P	MAR←IR(ROM) L_M, E_I	B←ROM L_B, E_R	A←A−B L_A, E_U, S_U
0100	OUT 1110	×××	MAR←PC(ROM) L_M, E_P	IR←ROM L_I, E_R	(0101) PC←PC+1 C_P	O←A L_O, E_A	—	—
0101	HLT 1111	×××	MAR←PC(ROM) L_M, E_P	IR←ROM L_I, E_R	(0110) PC←PC+1 C_P	HLT	—	—

Tips

abbreviation　缩写
address bus　地址总线
aforementioned　上述的，前述的
anemic address space　有限的地址空间
arbitrary　任意的
arithmetic & logic unit (ALU)　算术逻辑单元
buffer　缓冲器
bus　总线
cable　电缆

cache　高速缓冲存储器
central processing unit　中央处理器(CPU)
chunk of data　数据块
circuitry　电路，线路
clock frequency　时钟频率
clock period　时钟周期
consecutive　连续的
control bus　控制总线
control matrix (CM)　控制矩阵

control unit (CU) 控制部件	op-code 操作码
data bus 数据总线	operand 操作数
data locality 数据局部性	Pascal statement Pascal 语句
denote 指示,表示	PCI 周边元件扩展接口
direct memory access(DMA) 直接存储器存取方式	periodic rate 周期率
encoding instruction 编译指令	peripheral device 外围设备
erase 擦除	plug 塞子,插头,插销
erroneous 错误的,不正确的	program counter (PC) 程序计数器
fabricated 制作好的,已制作完成的	program flow via simple decision. 程序流程控制
falling edge 下降沿	PROM 可编程的只读存储器
flip-flop 触发器	pseudo-C array declaration C 语言数组声明(语句)
I/O-mapped I/O 映射方式	RAM(Random Access Memory) 随机存储器
inactive low 非当前(非活动态)低电平	reciprocal 倒数
initialize 初始化	reside 驻留
instruction 指令	ring counter (RC) 环形计数器
instruction decoder (ID) 指令译码器	rising edge 上升沿
instruction register (IR) 指令寄存器	ROM(Read Only Memory) 只读存储器
integrated circuit 集成电路(IC)	RTL (Resistor Transistor Logic) 电阻晶体管逻辑(电路)
interrupt line 中断线	
latch 锁存器	shuffle 运载,传输
memory address register (MAR) 存储地址寄存器	socket 插孔,插座,
memory cell 存储单元	status line 状态线
memory data register (MDR) 存储器数据寄存器	statute of "self-locking" 自锁状态
memory-mapped 内存映射方式	synchronization 同步
microprocessor 微处理器	system clock 系统时钟
microsecond 微秒 (10^{-6}s)	tri-state gate 三态门
mild controversy 轻微的争论	volatile 易失性
module 模块	Von Neumann architecture (VNA) 冯·诺依曼体系结构
nanosecond 纳秒 (10^{-9}s)	
non-volatile 非易失性	wait state 等待状态

Exercise

Figure 2-19 is the part of a computer. The control signals is CON = C_P E_P L_M W_E M_E L_D L_I E_I. Suppose that the sequence of the control signals is that:

CON_1 = 0110 0000

CON_2 = 0000 0010

CON_3 = 0000 0101

CON_4 = 0001 1000

CON_5 = 1000 0000

what does the computer operate? Suppose the computer starts work after the CLR signal is

1 and the $I_0 = 1100\ 0001\ 1001$.

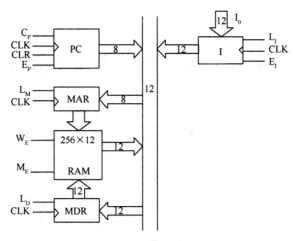

Figure 2-19 Data store process

Chapter 3 · Intel 8086 Microprocessor

3.1 The History of Intel Microprocessor Family

The idea of a computing system is not new—it has been around long before modern electrical and electronic devices were developed. The idea of calculating with a machine dates to 500 B. C. when the Babylonians invented the abacus, the first mechanical calculator.

A microprocessor is one of the most central parts of a modern personal computer or, in fact, any advanced computer device. The world's first microprocessor, the Intel 4004, was a 4-bit microprocessor—a programmable controller on a chip. It addressed a mere 4096 4-bit wide memory locations. The 4004 instruction set contained only 45 instructions. It was fabricated with the then-current state-of-the-art p-channel MOSSFET technology that only allowed it to execute instructions at the slow rate of 50 KIPS (kilo-instructions per second). This was slow when compared to the 100,000 instructions executed per second by the 30-ton ENIAC computer in 1946. The main difference was that the 4004 weighed much less than an ounce.

Late in 1971, realizing that the microprocessor was a commercially viable product, Intel Corporation released the 8008—an extended 8-bit version of the 4004 microprocessor. The 8008 addressed an expanded memory size (16K bytes) and contained additional instructions (a total of 48) that provided an opportunity for its application in more advanced systems.

In 1973, Intel introduced the 8080 microprocessor, the first of the modern 8-bit microprocessors, for they discovered the 8008 microprocessor's somewhat small memory size, slow speed, and instruction set limited its usefulness. About six months after, Motorola Corporation released its MC6800 microprocessor. Both of them ushered in the age of the microprocessor.

What was special about the 8080? The reasons why the 8080 was called as the first of the modern 8-bit microprocessor are that the 8080 not only could address more memory and execute additional instructions, but also it executed them 10 times faster than the 8008. Moreover, the 8080 was compatible with TTL (transistor-transistor logic), whereas the 8008 was not directly compatible. This made interfacing much easier and less expensive.

Incidentally, the first personal computer, the MITS Altair 8800, was released in 1974. (Note that the number 8800 was probably chosen to avoid copyright violations with Intel.) The BASIC language interpreter, written for the Altair 8800, was developed by Bill Gates who was the founder of Microsoft Corporation, which once produced DR-DOS for the personal computer.

In 1977, Intel Corporation introduced an updated version of the 8080-the 8085. This was to be the last 8-bit, general-purpose microprocessor developed by Intel. In 1978, Intel released the 8086 microprocessor. A year or so later, it released the 8088. Both devices weve 16-bit microprocessors, which executed instructions in as little as 400 ns (2.5 MIPs, or 2.5 millions of instructions per second). This represented a major improvement over the execution speed of the 8085. In addition, the 8086 and 8088 addressed 1M bytes of memory, which was 16 times more memory than the 8085. This higher execution speed and larger memory size allowed the 8086 and 8088 to replace smaller minicomputers in many applications. One other feature found in the 8086/8088 was a small 4-or 6-byte instruction cache or queue that prefetched a few instructions before they were executed. The queue sped the operation of many sequences of instructions and proved to be the basis for the much larger instruction caches found in modern microprocessors.

The 80286 microprocessor (also a 16-bit architecture microprocessor) was almost identical to the 9096 and 8088, except it addressed a 16M byte memory system instead of a 1M byte system. The instruction set of the 80286 was almost identical to the 8086 and 8088, except for a few additional instructions that managed the extra 15M bytes of memory. The clock speed of the 80286 was increased, so it executed some instructions in as little as 250 ns (4.0 MIPs) with the original release 8.0 MHz version. Some changes also occurred in the internal execution of the instructions, which led to an eight-fold increase in speed for many instructions when compared to 8086/8088 instructions.

As we know, the more powerful the processor is, the more instructions it can carry out at one time, leading to faster processing and more effectiveness at complex tasks. It was the application demands for faster microprocessor speeds, more memory, and wider data paths that led to the arrival of the 80386 in 1986, by Intel Corporation. The 80386 represented a major overhaul of the 16-bit 8086-80286 architecture. The 80386 was Intel's first practical 32-bit microprocessor that contained a 32-bit data bus and a 32-bit memory address. Through these 32-bit buses, the 80386 addressed up to 4G bytes of memory. A 4G byte memory can store an astounding 1,000,000 typewritten, double-spaced pages of ASCII text data.

In 1989, Intel released the 80486 microprocessor, which incorporated an 80386-like microprocessor, an 80387-like numeric coprocessor, and an 8K byte cache memory system into one integrated package. The internal structure of the 80486 was modified from the 80386 so that about half of its instructions executed in one clock instead of two clocks. The average speed improvement for a typical mix of instructions was about 50 percent over the 80386 that operated at the same clock speed.

The Pentium, introduced in 1993, was similar to the 80386 and 80486 microprocessors. Probably the most ingenious feature of the Pentium is its dual integer processors. The Pentium executes two instructions, which are not dependent on each other, simultaneously because it contains two independent internal integer processors called superscalar technology. This allows the Pentium to often execute two instructions per clocking period. Another feature that enhances performance is a jump prediction technology that speeds the execution of programs that include loops. As with the 80486, the Pentium also employs an internal floating-point coprocessor to handle floating-point data, albeit at five times the speed improvement. These features portend continued success for the Intel family of microprocessors.

The Pentium II was introduced on May 7, 1997. It refers to Intel's sixth-generation Microarchitecture ("P6") and x86-compatible microprocessors. Containing 7.5 million transistors (27.4 million in the case of the mobile Dixon with 256 KB L2 cache), the Pentium II featured an improved version of the first P6-generation core of the Pentium Pro, which contained 5.5 million transistors. However, its L2 cache subsystem was a downgrade when compared to Pentium Pros. In early 1999, the Pentium II was superseded by the almost identical Pentium III, which basically only added SSE instructions to the CPU.

The Pentium III (marketed as Intel Pentium III Processor, informally PIII, also stylized as Pentium) brand, introduced on February 26, 1999, refers to Intel's 32-bit x86 desktop and mobile microprocessors based on the sixth-generation P6 microarchitecture. The brand's initial processors were very similar to the earlier Pentium II -branded microprocessors. The most notable differences were the addition of the SSE instruction set (to accelerate floating point and parallel calculations), and the introduction of a controversial serial number embedded in the chip during the manufacturing process.

Pentium 4 was introduced by Intel on November 20, 2000 and shipped through August 8, 2008. It is a line of single-core desktop, laptop and entry level server central processing units (CPUs). They had a seventh-generation x86 microarchitecture, called NetBurst, which was the company's first all new design since the introduction of the P6 microarchitecture of the Pentium Pro CPUs in 1995. NetBurst differed from P6 (Pentium III, II, etc.) by featuring a very deep instruction pipeline to achieve very high clock speeds. Intel claimed that NetBurst would allow clock speeds of up to 10 GHz in future chips. However, severe problems with heat dissipation (especially with the Prescott Pentium 4) limited CPU clock speeds to a much lower 3.8 GHz.

The Pentium M is a family of mobile 32-bit single-core x86 microprocessors (with the modified Intel P6 microarchitecture) introduced in March 2003 and forming a part of the Intel Carmel notebook platform under the then new Centrino brand. The Pentium M processors had a maximum thermal design power (TDP) of 5-27 W depending on the model, and were intended for use in laptops (thus the "M" suffix standing for mobile). They evolved from the core of the last Pentium III-branded CPU by adding the front-side bus (FSB) interface of Pentium 4, an improved instruction decoding and issuing front end, improved branch prediction, SSE2 support, and a much larger cache. The first Pentium M-branded CPU, code-named Banias, was followed by Dothan. The Pentium M-branded processors were succeeded by the Core-branded dual-core mobile Yonah CPU with a modified microarchitecture.

The Pentium D/Extreme Edition brand introduced in 2005. It refers to two series of desktop dual-core 64-bit x86-64 microprocessors with the NetBurst microarchitecture, which is the dual-core variant of Pentium 4 "Prescott" manufactured by Intel. Each CPU comprised two dies, each containing a single core, residing next to each other on a multi-chip module package. The final shipment date of this brand microprocessor was 2008, which marked the end of the Pentium D brand and also the NetBurst microarchitecture.

Core 2 is a brand, introduced on 27 July 2006, encompassing a range of Intel's consumer 64-bit x86-64 single-, dual-, quad-core and the Extreme (dual- or quad-core CPUs for enthusiasts) microprocessors based on the Core microarchitecture. The single- and dual-core models are single-die, whereas the quad-core models comprise two dies, each containing two cores, packaged in a multi-chip module. The introduction of Core 2 relegated the Pentium brand to the mid-range market, and reunified laptop and desktop CPU lines, which previously had been divided into the Pentium 4, Pentium D, and Pentium M brands. The majority of the desktop and mobile Core 2 processor variants are Core 2 Duo with two processor cores on a single Merom, Conroe, Allendale, Penryn, or Wolfdale chip.

The successors to the Core 2 brand are a set of Nehalem microarchitecture based processors called Core i3, i5, and i7. Core i7 was officially launched on 17 November 2008 as a family of three quad-core processor desktop models, further models started appearing throughout 2009. The last Core 2 processors to be released were the Core 2 Quad Q9500 in January 2010, and the Core 2 processor line was removed from the official price lists in July 2011.

Arrandale, released on 7 January 2010, is the code name for a family of mobile Intel processors, sold as mobile Intel Core i3, i5 and i7 as well as Celeron and Pentium. It is closely related to the desktop Clarkdale processor; both use dual-core dies based on the Westmere 32 nm die shrink of the Nehalem microarchitecture, and have integrated Graphics as well as PCI Express and DMI links. Arrandale is the successor of the 45 nm Core microarchitecture based Penryn processor that is used in the many mobile Intel Core 2, Celeron and Pentium Dual-Core processors. While Penryn typically used both a north bridge and a south bridge, Arrandale already contains the major north bridge components, which are the memory controller, PCI Express for external graphics, integrated graphics and the DMI connector, making it possible to build more compact systems without a separate northbridge or discrete graphics as Lynnfield.

Sandy Bridge is the codename for a microarchitecture developed by Intel beginning in 2005 for central processing units in computers to replace the Nehalem microarchitecture. Intel demonstrated a Sandy Bridge processor in 2009, and released first products based on the architecture in January 2011 under the Core brand. Developed primarily by the Israeli branch of Intel, the codename was originally "Gesher" (meaning "bridge" in Hebrew).

Intel demonstrated the Ivy Bridge processors in 2011, which is the codename for a line of processors based on the 22 nm manufacturing process developed by Intel. The name is also applied more broadly to the 22 nm die shrink of the Sandy Bridge microarchitecture based on FinFET ("3D") tri-gate transistors, which is also used in the Xeon and Core i7 Ivy Bridge-EX (Ivytown), Ivy Bridge-EP and Ivy Bridge-E microprocessors released in 2013.

Ivy Bridge processors are backwards compatible with the Sandy Bridge platform, but such systems might require a firmware update (vendor specific). In 2011, Intel released the 7-series Panther Point chipsets with integrated USB 3.0 to complement Ivy Bridge.

Volume production of Ivy Bridge chips began in the third quarter of 2011. Quad-core and dual-core-mobile models are launched on 29 April 2012 and 31 May 2012 respectively. Core i3 desktop processors, as well as the first 22 nm Pentium, were announced and available the first week of September, 2012.

As the successor to the Ivy Bridge microarchitecture, Intel officially announced CPUs based on Haswell microarchitecture on June 4, 2013 at Computex Taipei 2013, while a working Haswell chip was demonstrated at the 2011 Intel Developer Forum. With Haswell, which uses a 22 nm process, Intel also introduced low-power processors designed for convertible or "hybrid" ultrabooks, designated by the "Y" suffix.

Broadwell is Intel's codename for the 14 nanometer die shrink of its Haswell microarchitecture. It is a "tick" in Intel's tick-tock principle as the next step in semiconductor fabrication. Unlike the previous tick-tock iterations, Broadwell will not completely replace the full range of CPUs from the previous microarchitecture (Haswell), as there will be no low-end desktop CPUs based on Broadwell. Broadwell's H and C variants will be used in conjunction with Intel 9 Series chipsets (Z97, H97 and HM97), in addition to retaining backward compatibility with some of the Intel 8 Series chipsets.

On September 5, 2014, Intel launched the first three Broadwell-based processors that belong to the low-TDP Core M family, Core M 5Y10, Core M 5Y10a and Core M 5Y70. On October 9, 2014, the first laptop with Broadwell Intel Core M 5Y70 CPU, Lenovo Yoga 3 Pro, was launched. On October 31, 2014, four more Broadwell based CPUs were launched belonging to Core M family, increasing the number of launched Broadwell CPUs to seven.

Figure 3-1 shows the development of microprocessor of Intel family, starting with the Intel 4004. The information in this section was taken from many sources, including other overviews of history of the microprocessor and some internet websites.

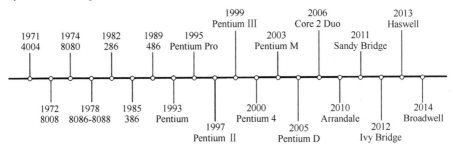

Figure 3-1 The microprocessor timeline of Intel family

Today's microprocessors are immensely powerful, capable of executing complex instructions at a faster rate than ever before. As computers grow ever more sophisticated, engineering science begins to run up against puzzling challenges. As increasingly globalized societies demand better computing technology, more great advances are sure to be made within the near future of microprocessor technology.

3.2　8086 CPU Architecture

Figure 3-2 shows the internal architecture of the 8086. As illustrated in **Figure 3-2**, the processor is partitioned into two logic units: an execution unit (EU) and a bus interface unit (BIU).

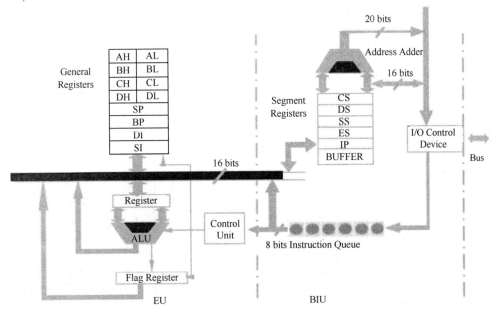

Figure 3-2　8086 CPU architecture

3.2.1　Execution Unit and Bus Interface Unit

The role of the EU is to execute instructions, whereas the BIU delivers instructions and data to the EU. The EU contains an arithmetic and logic unit (**ALU**), a control unit (**CU**), and a number of registers. These features provide for execution of instructions and arithmetic and logic operations.

The most important function of the BIU is to manage the bus control unit, segment registers, and instruction queue. The BIU controls the buses that transfer data to the EU, to memory, and to external input/output devices, whereas the segment registers control memory addressing.

Another function of the BIU is to provide access to instructions. Because the instructions for a program that is executing are in memory, the BIU must access instructions from memory and place them in an instruction queue, which varies in size depending on the processor. This feature enables the BIU to look ahead and pre-fetch instructions so that there is always a queue of instructions ready to execute.

The EU and BIU work in parallel, with the BIU keeping one step ahead. The EU notifies the BIU when it needs access to data in memory or an I/O device. Also the EU requests machine instructions from the BIU instruction queue. The top instruction is the currently executable one and, while the EU is occupied executing an instruction, the BIU fetches another instruction from memory. This fetching overlaps with execution and speeds up processing.

As shown in **Figure 3-3**, except for the instruction queue, which is actually a 6-byte queue, the working registers are divided into three groups according to their functions. There are the data group (AX, BX, CX, DX), which is essentially the set of arithmetic registers; the pointer group, which includes base and index registers, but also contains the program counter (For the 8086 the program counter is called the Instruction Pointer (IP)) and stack pointers; and the segment group (CS, DS, SS, ES), which is a set of special purpose base registers. All of the registers are 16 bits wide.

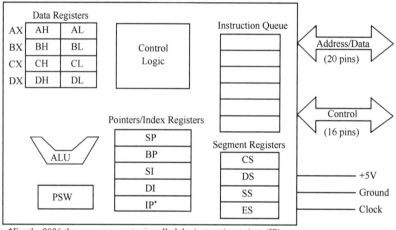

*For the 8086 the program counter is called the instruction pointer(IP).

Figure 3-3　8086's internal registers

要点:

(1) 8086 从功能上分为两部分,即总线接口部件(BIU)和执行部件(EU)。

BIU 主要负责与存储器、I/O 端口传送数据。即 8086/8088 CPU 与存储器之间的信息传送,都是由 BIU 进行的。BIU 负责从内存的指定部分取出指令,送到指令流队列中排队;在执行指令时所需的操作数,也由 BIU 从内存的指定区域取出,传送给 EU 部分去执行。

EU 部分负责指令的执行。完成指令译码并利用内部的寄存器和 ALU 对数据进行处理。

(2) 总线接口部件和执行部件的动作管理原则。

① EU 和 BIU 的工作不同步,二者以并行方式工作。

② 8086 的指令队列是 6 字节的,8088 则是 4 字节。每当 8086 的指令队列中有 2 个空字节,或 8088 的指令队列中有 1 个空字节时,BIU 就会自动把指令取到指令队列中。

③ 每当 EU 准备执行一条指令时,它会从 BIU 的指令队列前部取出指令的代码,然后用几个时钟周期去执行指令。

④ 如需取数据送结果,EU 向 BIU 申请,送出偏移、段地址。当 BIU 空闲时(队列

满),该操作立即执行,若 BIU 正在取指,则等 BIU 取完当前字节指令后再执行。

⑤ 当指令队列已满,而且 EU 对 BIU 又没有总线访问请求时,BIU 便进入空闲状态。

⑥ 在执行转移、调用、返回指令时,原指令队列中已装入的指令被清空,指令队列中将重新装入新的指令。

3.2.2 Organization of Execution Unit

1. Data Register-4 General-Purpose Registers: AX, BX, CX, &DX

The data group consisted of the AX, BX, CX, and DX registers are also called General-Purpose Registers. All of them can be used for computations to store operands and results.

They are the workhorses of the system, and are unique in that you can address them as one word or as 1-byte portion. The leftmost byte is the "high" portion and rightmost byte is the "low" portion. That is to say, each of them can be accessed as a whole, or the upper and lower bytes can be accessed separately. **For example**, the AX register consists of an AH (high) and an AL (low) portion, and you can reference any portion by its name. The 80386 and later processors support all the general-purpose registers, plus 32-bit extended versions of them: the EAX, EBX, ECX, and EDX.

AX register: The AX register, the primary accumulator, is used for operations involving input/output and most arithmetic. **For example**, the multiply, divide, and translate instructions assume the use of the AX. Also, some instructions generate more efficient code if they reference the AX rather than another register.

BX register: The BX is known as the base register since it is the only general-purpose register that can be used as an indirect addressing mode register.

CX register: The CX register is known as the counter register. It may contain a value to control the number of times a loop is repeated or a value to shift bits left or right. Sometimes it is used as an implied counter by certain instructions.

DX register: The DX register is known as the data register. It is used to hold the I/O address during I/O operations. And multiply and divide operations that involve large values assume the use of the DX and AX together as a pair.

2. Pointer and Index Registers

The point and index groups consist of the SI, DI, BP, SP, and IP.

SI (Source Index) register: The 16-bit source index register is required for some string (character) operations. In this context, the SI is associated with the DS register. The 80386 and later processor support a 32-bit extended register, the ESI.

DI (Destination Index) register: The 16-bit destination index register is also for some string operations. In this context, the DI is associated with ES register. The 80386 and later processor support a 32-bit extended register, the EDI.

SP (Stack Pointer) register: The 16-bit SP register provide an offset value, which, when associated with the SS register, refers to the current work being processed in the stack. The 80386 and later processors have an extended 32-bit stack pointer, the ESP register. The system automatically handles these registers.

BP (Base Pointer) register: The 16-bit BP facilitates referencing parameters, which are data and addresses that a program passes via the stack. The processor combines the address in the SS with the offset in the BP. The 80386 and later processor have an extended 32-bit BP called EBP.

The default segment registers are listed in **Table 3-1**. It indicated that the SP and BP registers are associated with the SS register and permit the system to access data in the stack segment. The SI and DI registers are available for indexed addressing and for use in addition and subtraction.

Table 3-1 Default segment of registers

Register	Default Segment
BP or SP	SS
SI or DI	DS
DI strings	ES
SI strings	DS

IP (Instruction Pointer) register: The 16-bit IP is the essential program counter, which is used to form Instruction address with the contents of CS (Code Segment) register.

The 16-bit IP register contains the offset address of the next instruction that is to execute. The IP is associated with the CS register in that the normally reference the IP register in a program, but you can change its value when using the DEBUG program to test a program. The 80386 and later processor have an extended 32-bit IP called EIP.

A data address may be formed by adding together a combination of the contents of the registers. The result of such an address computation is called an effective(EA) or indicate a quantity that is added to the contents of a register to form an EA. **For example**, the EA of the instruction MOV AX, [BX][DI] is a combination of the contents of BX and DI.

3. Flags Register

The flags register contains 16 bits, with various instructions set to indicate the status of an operation. In all cases, a flag remains set until another instruction change it.

Although the flags register is 16 bits long, as shown in **Table 3-2**, 7 of them are not used, the rest nine bits are common to all 8086 family processors to indicate the current status of the computer and the results of processing. Many instructions involving comparisons and arithmetic change the status of the flags, which some instructions may test to determine subsequent action. The 8086 flags are divided into the conditional flags, which reflect the result of the previous operation involving the ALU, and the control flags, which control the execution of special functions. The flags are in the flags register in the following locations (which you need not memorize).

Table 3-2 Flags' location in the flag register

Bit No.	15	14	13	12	11	10	9	8	7	6	5	4	3	2	1	0
Flag Register					OF	DF	IF	TF	SF	ZF		AF		PF		CF

1) The Conditional Flags

OF (Overflow Flag) : Indicates over of a high-order (leftmost) bit following arithmetic. Is set if an overflow occurs. i. e. , a result is out of rang. More specially, for addition this flag is set when there is a carry into the MSB and no carry out of the MSB or vice versa. For subtraction, it is set when the MSB needs a borrow and there is no borrow from the MSB, or vice versa.

SF (Sign Flag) : Contains the resulting sign of an arithmetic operation. It is equal to the MSB of the result. Since in 2's complement negative numbers have a 1 in the MSB and for nonnegative numbers this bit is 0, this flag indicates whether the previous result was negative or nonnegative.

ZF (Zero Flag) : Indicates the result of an arithmetic or comparison operation, which is set to 1 if the result is zero and 0 if the result is nonzero.

AF (Auxiliary Carry Flag) : Contains a carry out of bit 3 on 8-bit data, for specialized arithmetic. Is set if there is a carry out of the bit 3 during an addition or a borrow by bit 3 during a subtraction. This flag is used exclusively for BCD arithmetic.

PF (Parity Flag) : Indicates even or odd parity of a low-order (rightmost) 8-bit data operation. It is set to 1 if the low order 8 bits of the result contain an even number of 1s; otherwise it is cleared.

CF (Carry Flag) : An addition causes this flag to be set if there is a carry out of the MSB, and a subtraction causes it to be set if a borrow is needed. Other instructions also affect this flag and its value will be discussed when this instructions are defined.

2) The Control Flags

DF (Direction Flag) : Used by string manipulation instructions. If clear, the string is processed from its beginning with the first element having the lowest address. Otherwise, the string is processed from the high address.

IF (Interrupt Flag) : Indicates that all external interrupts, such as keyboard entry, are to be processed or ignored. If set, a certain type of interrupt (a mask-able interrupt) can be recognized by the CPU; otherwise, these interrupts are ignored.

TF (Trap Flag) :   Permits operation of the processor in single-step mode. Debugger programs such as DEBUG set the trap flag so that you can step through execution a single instruction at a time to examine the effect on registers and memory. If it is set, a trap would be executed after each instruction.

The flags most relevant to assembly programming are OF, SF, ZF and CF for comparisons and arithmetic operations, and DF for the direction of string operation.

4. ALU

The Arithmetic and Logic Unit is the "**core**" of any processor. It's the unit that performs the calculations. As shown in Figure 3-4, a typical ALU will have two input ports (A and B) and a result port (Y). It will also have a control input telling it which operation (add, subtract, and, or, etc) to perform and additional outputs for **conditional** flags **codes** (carry, overflow, negative, zero result).

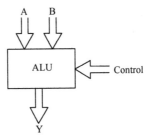

Figure 3-4　Block diagram of ALU

要点：
(1) 8086 的标志寄存器是 16 位的,但其中只有 9 位被使用。
(2) 8086 的标志可分为两类：
① 状态标志。它是操作在执行后,决定算术逻辑处在执行何种状态,这种状态会影响后面的操作。
② 控制标志。它是人为设置的,指令系统中有专门的指令用于控制标志的设置和清除,每个控制标志都对每一种特定的功能起控制作用。

3.2.3　Organization of Bus Interface Unit (BIU)

1. 4 Segment Registers

A segment register is 16 bits long and provides for addressing an area of memory known as the current segment. Because a segment aligns on a paragraph boundary, its address in a segment register assumes 4 0-bit to its right. The segment registers consists of CS, DS, SS, and ES.

CS (Code Segment) **register**: Contains the starting address of a program's code segment. This segment address, plus an offset value in the instruction pointer (IP) register, indicates the address of an instruction to be fetched for execution. For normal programming purposes, you need not reference the CS register.

【**Example 3-1**】 If the CS register contains 39B4[0]H and the IP contains 514H. To find the next instruction to be executed, the processor combines the address in the CS with the offset in the IP:

$$
\begin{array}{lr}
\text{Segment address in CS} & 39\text{ B40H} \\
\underline{\text{Plus offset address in IP}} & \underline{+\ 514\text{H}} \\
\text{Address of next instruction} & 3\text{A054H}
\end{array}
$$

DS (Data Segment) **register**: Contains the starting address of a program's data segment.

Instructions use this address to locate data. This address, plus an offset value in an instruction, causes a reference to a specific byte location in the data segment.

SS (Stack Segment) **register**: Permit the implementation of a stack in memory, which a program uses for temporary storage of addresses and data. The system stores the starting address of a program's stack segment in the SS register. This segment address, plus an offset value in the stack pointer (SP) register, indicates the current word in the stack being addressed. For normal programming purposes, you need not directly reference the SS register.

ES (Extra Segment) **register**: Used by some string (character data) operations to handle memory addressing. In this context, the ES (extra segment) register is associated with the DI (index) register. A program that requires the use of the ES may initialize it with an appropriate segment address.

(**FS and GS registers**: Additional extra segment registers on the 80386 and later processors.)

2. 20-bit Address Adder

20-bit Address Adder is used to offer 20 bits physical address. As indicated above, the registers that can be used for addressing are only 16 bits wide and, therefore, an effective address has only 16 bits. On the other hand, for 8086 CPU the address put on the address bus, called physical address, must contain 20 bits. If a segment aligns on a paragraph boundary, it's very simple by appending four 0 bits to the right of the number in the segment register. Then the extra 4 bits are obtained by adding the effective address to the contents of one of the segment registers as shown in **Figure 3-4**. Thus a 20-bit result is produced.

【Example 3-2】 If (CS) = 1250H and (IP) = 25A0H, then the next instruction will be fetched from:

$$
\begin{array}{ll}
25A0 & \text{Effective address} \\
+\ 12500 & \text{Beginning segment address} \\
\hline
14AA0 & \text{Physical address of instruction}
\end{array}
$$

An illustration of **Example 3-2** is given in **Figure 3-5**.

Note that it is standard notation for parentheses around an entity to mean "contents of," e. g., (IP) means the contents of IP. Also, all addresses are given in hexadecimal.

The reason why we append four 0 bits to the register of the content of the segment register is that a segment address is on a paragraph boundary (evenly divisible by 16, or hex 10), the rightmost 4 bits of its address are zero. A segment address is stored in a segment register, and the processor assumes 4 rightmost 0 bits, as hex nnnn[0]H allows addressing up to 1,048,560 bytes. Therefore the extra 4 bits are obtained by adding the effective address to the contents of one of the segment registers. The address accumulator is used to generate the 20-bit physic address of data. **Figure 3-6** illustrates how a physical 20-bit address is generated:

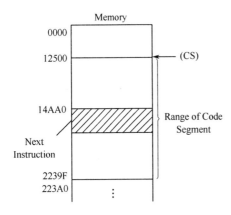
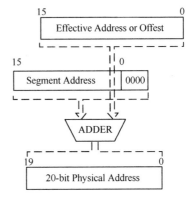

Figure 3-5 Address with in segment   Figure 3-6 Memory address generation

3. 6-byte Instruction Queue

In 8086 CPU, the BIU accesses instructions from memory and place them in the 6-bytes instruction queue. This feature enables the BIU to look ahead and pre-fetch instructions so that there is always a queue of instructions ready to execute.

3.3 Internal Memory

The 8086/8088 CPU has the memory capacity to be 1 Megabyte. The two types of internal memory on the PC are random access memory (RAM) and read-only memory (ROM). Bytes in the memory are numbered consecutively, beginning with 00, so that each location has a uniquely numbered address. **Figure 3-7** shows a physical memory map of an 8086-type PC. The first 640K is base RAM, most of which is available for your own use.

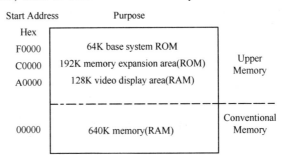

Figure 3-7 Map of base memory

3.3.1 Addressing Data in Memory

The different processor can access one or more bytes of memory at a time. Consider the decimal number 21238. The hex representation of this value, 08BEH, requires 2 bytes, or 1 word, of memory. It consists of a high-order (most significant) byte, 08H, and a low-order (least significant) byte, 0BEH. As the description that introduced in **Figure 1-4**, Chapter 1, the system stores the data in memory in reverse-byte sequence: the low-order byte in the low memory address and the high-order byte in the high memory address.

【Example 3-3】 Please describe how does a processor transfer 5060H from AX register into memory addresses 6500H and 6501H.

The data transfer like this:

Register(AX)	50H(AH)	60H(AL)
Memory	60H	50H
Address	6500H	6501H
	Least Significant Byte	Most Significant Byte

The processor expects numeric data in memory to be in reverse-byte sequence and processes the data accordingly. When the processor retrieves the word from memory, it again reverse the bytes, restoring them correctly in the register as hex 50 60. Although this feature is entire automatic, you have to alert to it when programming and debugging assembly language programs.

3.3.2 Memory Segment

As we know, all registers in 8086/8088 are 16 bits wide. It's obviously that there are not enough bits to store the 20-bit address. Memory Segment is needed to deal with the problem. Now, the question is that "What is a segment?" A segment is a special area defined in a program that begins on a paragraph boundary, that is, at a location evenly divisible by 16, or hex 10. Although a segment may be located almost anywhere in memory and in real mode may be up to 64K bytes, it requires only as much space as the program requires for its execution.

There may be any number of segments, to address a particular segment, it is necessary only to change the address in an appropriate segment register. The four main segments are the code, data, extra and stack segments.

1. Code Segment

The code segment contains the machine instructions that are to execute. Typically, the first executable instruction is at the start of this segment, and the operating system links to that location to begin program execution. As the name implied, the code segment (CS) register addresses the code segment. If your code area requires more than 64K, your program may need to define more than one code segment.

2. Data Segment

The data segment contains a program's defined data, constants, and work areas. The data segment (DS) register addresses the data segment. Data are accessed in the data segment by an offset address or the contents of other registers that hold the offset address. If your data requires more than 64K, your program may need to define more than one data segment.

3. Stack Segment

The stack segment defines the area of memory used for stack. In simple terms, the stack contains any data and address that you need to save temporarily or for use by your own "called" subroutines. The stack segment (SS) register addresses the stack segment.

4. Extra Segment

The extra segment is an additional data segment that is used by some of the string instructions to hold destination data.

3.3.3 Segment Boundary

A segment register contains the starting address of a segment. **Figure 3-8** presents a graphic view of ES, SS, DS, and CS registers and their relationships to the extra, stack, data, and code segments. (The registers and segments are not necessarily in the order shown.) Other segment registers are on the 80286 and later processors, the FS and GS registers, which have specialized uses.

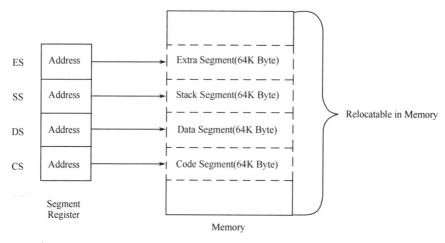

Figure 3-8 Segment and register

As mentioned earlier, a segment begins on a paragraph boundary, which is an address evenly divisible by decimal 16, or hex 10. Consider a data segment that begins at memory location 038E0H. Because in this and all other cases the rightmost hex digit is zero, the computer designers decided that it would be unnecessary to store the zero digit in the segment register. Thus 038E0H is stored as 038E, with the rightmost zero understood. Where appropriate, this text uses square brackets to refer to the rightmost zero, such as 038E [0].

3.3.4 Segment Offset

Within a program, all memory locations are relative to a segment's starting address. The distance in byte from the segment address to another location within the segment is expressed as an offset (or displacement). A 2-byte (16-bit) offset can range from 0000H to FFFFH, or zero through 65,535. Thus the first byte of the code segment is at offset 00, the second byte is at offset 01, and so forth, through zero to offset 65,535. To reference any memory location in a segment, the processor combines the segment address in a segment register with an offset value.

[Example 3-4] Consider the data segment that begins at location 038E0H. The DS register contains the segment address of the data segment, 038E [0] H, and an instruction references a location with an offset of 0032H bytes within the data segment. Please calculate the physical address of the actual memory location of the byte referenced by the instruction.

The Figure 3-9 is a displacement diagram of a memory location within a data segment. So the physical address can be calculated like this:

$$\begin{aligned}\text{DS segment address:} \quad & 038\text{E0H} \\ \text{Offset:} \quad & +0032\text{H} \\ \hline \text{Actual address:} \quad & 03912\text{H}\end{aligned}$$

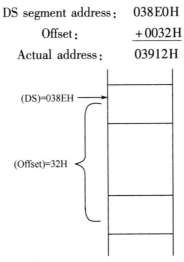

Figure 3-9 Memory location for 038E[0]

Note that a program contains one or more segments which may begin almost anywhere in memory, may vary in size, and may be in any sequence.

The utilization of segment registers essentially divides the memory space into overlapping segments, with each segment being 64K bytes long and beginning at a 16-byte, a paragraph boundary. We will hereafter refer to the contents of a segment register as the segment address, and the segment address multiplied by 16_{10} as the beginning physical segment address, or simply, the beginning segment address. The overall segmentation of memory is shown in **Figure 3-10(a)**. The advantages of using segment registers are that they:

(1) Allow the memory capacity to be 1 megabyte even though the addresses associated with the individual instructions are 16 bits wide.

(2) Allow the instruction, data, or stack portion of a program to be more than 64K bytes long by using more than one code, data, or stack segment.

(3) Facilitate the use of separate memory areas for a program, its data, and the stack.

(4) Permit a program and/or its data to be put into different areas of memory each time the program is executed.

Figure 3-8 shows how a program's code and its associated data and stack can be separated in memory. The simpler and conventional approach is to let both the code and the data reside in one contiguous area in memory and put the stack in some fixed area which always begins at, say, address 08000H. This is satisfactory if there is only one program in memory at a time, but in a multiprogramming environment there may be several programs in memory simultaneously. For multiprogramming it is better to keep the contiguous chunks of memory as small as pos-

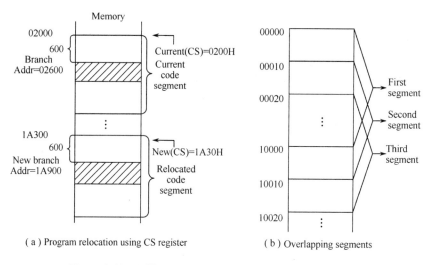

Figure 3-10　Address computations and memory segmentation

sible and, therefore, it is better to separate the code from the data. Analogously, the space in a pail is more fully utilized by filling it with small stones than with large stones. The option of making the code and data contiguous is always present, because the contents of CS, DS, and ES can always be made equal.

　　Advantage (4) is also related to multiprogramming. For example, consider a time-sharing system in which several users are executing different programs. If there is not enough memory space, these programs are shuttled back and forth between the memory and mass storage. As they take turns using memory they are dynamically put into different places in memory. A given program may begin at physical address 02000H while it is currently executing, but the next time it is brought into memory it may begin at 1A300H. The question is: if the program contains a branch instruction that would currently cause a branch to 02600H, would the branch be correctly taken to 1A900H when the program is relocated to 1A300H? The answer is yes provided that the contents of CS are changed to 1A300 when the program is relocated. This situation is depicted in **Figure 3-10(a)**. Similar statements can be made regarding the relocation of a block of data or stack.

　　Although segments are 64K bytes long, they can be overlapped to better utilize the memory, if, as shown in **Figure 3-10(b)**, a program fills only part of a segment, another code, data, or stack segment could overlap the program's segment and the beginning address of the second segment could be within 16 bytes of the end of the program. The only wasted space would be the few bytes between the end of the program and the next 16-byte boundary.

要点：

（1）8086/8088 系统中存储器按字节编址，可寻址的存储器空间为 1M 字节，因此每个字节所对应的地址是 20 位。

（2）存储器分段。在 8086/8088 中，CPU 内部寄存器均为 16 位的，无法存放 20 位地址，存在指针位数不够用的矛盾。为此，在 8086/8088 系统中，把 1M 存储空间分成若干个逻辑段,并允许它们在整个存储空间浮动,即段与段之间可以部分重叠、连续排列或断续排列。

存储器分段时要求各个逻辑段从节(16字节)的整数边界开始,即段首地址低四位为"0",把段首地址的高16位称为"段基址"存放在段寄存器DS或CS或SS或ES中。

(3) 在采用分段结构的存储器中,每一个存储单元可以看成具有两种类型的地址:物理地址和逻辑地址。

① 逻辑地址由段基址和偏移地址两个部分构成,它们都是无符号的16位二进制数,如5500H:0100H。

② 物理地址也称为绝对地址,它是20位的,是唯一标识1M字节空间的某一字节的地址。它是由逻辑地址变换得来的。二者的换算关系如下:

$$单元物理地址 = 段首地址 \times 10H + 段内偏移量$$

(4) 对于任何一个物理地址,可以唯一地被包含在一个逻辑段中,也可包含在多个相互重叠的逻辑段中,只要有段地址和段内偏移地址就可以访问到这个物理地址所对应的存储空间。

3.3.5 About Stack Segment

A stack segment is a special area which contains any data and address that you need to save temporarily. It's a LIFO (last-in, first-out) memory, which describes the way that data are stored and removed the stack.

The processor uses the stack when interrupts strike, or procedure calls for return addresses. It is convenient to have one around for temporary storage. The operation principle of the Stack Segment is as follows:

The stack is maintained by two registers: SP and SS. It grows backwards through memory towards the start of the stack segment. Data are placed onto the stack with a PUSH instruction and removed with a POP instruction. The process can be described with the **Figure 3-11** below.

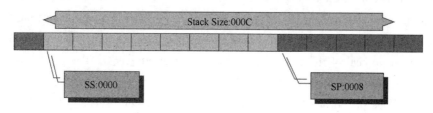

Figure 3-11 The execution process of the stack segment

要点:

堆栈是以"后进先出"方式工作的一个存储区,它必须存在于堆栈段中,其段地址应存放于SS寄存器中。它只有一个出口,所以只有一个堆栈指针寄存器SP,SP的内容在任何时候都指向当前的栈顶。

3.4 System Timing

Although modern computers are quite fast and getting faster all the time, they still require a finite amount of time to accomplish even the smallest tasks. On Von Neumann machines, like the 80x86, most operations are serialized. This means that the computer executes commands in a prescribed order.

Clearly we need some way to control which statement executes first and which executes second.

The CPU is a very complex piece of circuitry. To ensure that all operations occur at just the right moment, the 80x86 CPUs use an alternating signal called the **system clock.**

3.4.1 The System Clock

As shown in **Figure 3-12**, the system clock is an electrical signal on the control bus which alternates between zero and one at a periodic rate. It's used to handle all synchronization within computer system.

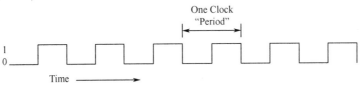

Figure 3-12 System clock signal

For example, it works to ensure many of the logic gates that make up the CPU to operate in a synchronized fashion.

The frequency with which the system clock alternates between zero and one is the system **clock frequency**. The time it takes for the system clock to switch from zero to one and back to zero is the **clock period**. One full period is also called a **clock cycle**. The clock frequency is simply the number of clock cycles which occur each second. "Hertz" (Hz) is the technical term meaning one cycle per second. Typical frequencies for 80x86 parts range from 5 MHz up to 200 MHz and beyond. Note that one clock period (the amount of time for one complete clock cycle) is the reciprocal of the clock frequency. **For example**, a 1 MHz clock would have a clock period of one microsecond (1/1,000,000th of a second). Likewise, a 10 MHz clock would have a clock period of 100 nanoseconds (100 billionths of a second). A CPU running at 50 MHz would have a clock period of 20 nanoseconds. Note that we usually express clock periods in millionths or billionths of a second.

To ensure synchronization, most CPUs start an operation on either the **falling edge** (when the clock goes from one to zero) or the **rising edge** (when the clock goes from zero to one). The system clock spends most of its time at either zero or one and very little time switching between the two. Therefore clock edge is the perfect synchronization point.

Since all CPU operations are synchronized around the clock, the CPU cannot perform tasks any faster than the clock. However, just because a CPU is running at some clock frequency doesn't mean that it is executing that many operations each second. Many operations take multiple clock cycles to complete so the CPU often performs operations at a significantly lower rate.

3.4.2 Memory Access Time

Memory devices have various ratings, but the two major ones are capacity and speed (access time). Memory access is probably the most common CPU activity. Memory access is definitely an operation synchronized around the system clock. That is, reading a value from memory or writing a value to memory occurs no more often than once every clock cycle. Indeed, on many 80x86 processors, it takes several clock cycles to access a memory location. The memory access time is the number of clock cycles the system requires to access a memory location; this is an important value since longer memory access time results in lower performance.

Different 80x86 processors have different memory access times ranging from one to four clock cycles. **For example**, the 8088 and 8086 CPUs require four clock cycles to access memory; the 80486 requires only one. Therefore, the 80486 will execute programs which access memory faster than an 8086, even when running at the same clock frequency.

Memory access time is the amount of time between a memory operation request (read or write) and the time the memory operation completes. On a 5 MHz 8088/8086 CPU, the memory access time is roughly 800 ns (nanoseconds). On a 50 MHz 80486, the memory access time is slightly less than 20 ns. Note that the memory access time for the 80486 is 40 times faster than the 8088/8086. This is because the 80486's clock frequency is ten times faster and it uses one-fourth the clock cycles to access memory.

When reading from memory, the memory access time is the amount of time from the point that the CPU places an address on the address bus and the CPU takes the data off the data bus.

Note that the CPU doesn't wait for memory. The access time is specified by the clock frequency. If the memory subsystem doesn't work fast enough, how can a system designer get away with using 70 ns memory? The answer is wait states.

3.4.3 Wait States

A wait state is nothing more than an extra clock cycle to give some device time to complete an operation. **For example**, a 50 MHz 80486 system has a 20 ns clock period. This implies that you need 20 ns memory. In fact, in most computer systems there is additional circuitry between the CPU and memory, such as decoding and buffering logic which could introduce additional delays into the system. For example, in **Figure 3-13**, if the CPU needs the data back in 20 ns, the memory must respond in less than 10 ns, because the system loses 10ns to buffering and decoding.

You can actually buy 10ns memory. However, it is very expensive, bulky, consumes a lot of power, and generates a lot of heat.

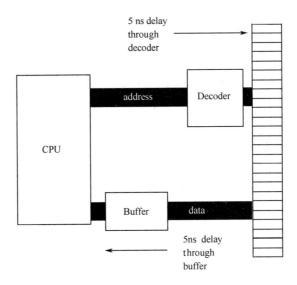

Figure 3-13   The connection diagram of CPU and memory

If cost-effective memory won't work with a fast processor, the wait state is insected. **For example, if you have a** 20 MHz processor with a memory cycle time of 50 ns and you lose 10 ns to buffering and decoding, you'll need 40 ns memory. What do you do if you can only afford 80 ns memory in a 20 MHz system? Adding a wait state to extend the memory cycle to 100 ns (two clock cycles) will solve this problem. Subtracting 10ns for the decoding and buffering leaves 90 ns. Therefore, 80 ns memory will respond well before the CPU requires the data.

Figure 3-14 is an example of wait state insertion.

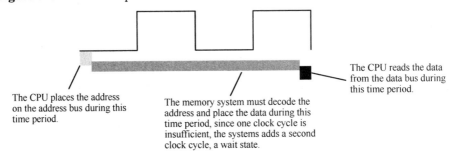

Figure 3-14   The example of wait state insertion

Almost every general purpose CPU in existence provides a signal on the control bus to allow the insertion of wait states. Generally, the decoding circuitry asserts this line to delay one additional clock period, if necessary. This gives the memory sufficient access time, and the system works properly. Sometimes a single wait state is not sufficient.

Needless to say, from the system performance point of view, wait states are not a good thing. While the CPU is waiting for data from memory it cannot operate on that data. Adding a single wait state to a memory cycle on an 80486 CPU doubles the amount of time required to access the data. This, in turn, halves the speed of the memory access. Running with a wait state on every memory access is almost like cutting the processor clock frequency in half. You're going to get a lot less work done in the same amount of time.

3.4.4 Bus Cycle

The bus cycle is the number of clock cycles the system requires to access a memory location or an I/O port or to place instructions in the instruction queue. The length of a bus cycle in an 8086/8088 system is four clock cycles, denoted T_1 through T_4, plus an indeterminate number of wait state clock cycles, denoted T_w. If the bus is to be inactive after the completion of a bus cycle, then the gap between successive cycles is filled with idle state clock cycles represented by T_i. Wait states are inserted between T_3 and T_4 when a memory or I/O interface is not able to respond quickly enough during a transfer. A typical succession cycles is given in **Figure 3-15.**

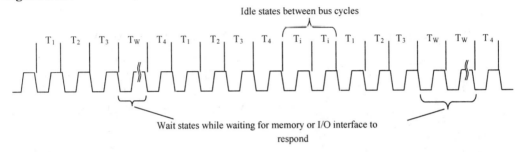

Figure 3-15 Typical sequence of bus cycles

【Example 3-5】 For following two instructions, whose execution will lead CPU enter a bus cycle?

    MOV   AX,BX
    MOV   AX,[1000H]

The answer is that the first instruction will not lead CPU enter a bus cycle, because AX and BX are internal registers of CPU. While the second one will do, because CPU fetches the data from memory location at DS:1000H.

要点：

(1) 存储器存取时间又称存储器访问周期，是指从启动一次存储器操作到完成该操作所经历的时间。

(2) 总线周期。为了取得指令和传输数据的协调工作，就需要 CPU 的总线接口部件执行一个总线周期，一个总线周期由 4 个基本的时钟周期及数量不确定的等待周期组成。只有在 CPU 和内存或 I/O 接口之间传输数据，以及填充指令队列时，CPU 才执行总线周期。

3.5 8086 Pin Assignments and Working Modes

The Intel 8086, a 16-bit microprocessor, contains approximately 29,000 transistors and is fabricated using the HMOS technology. Its throughput is a considerable improvement over that of the Intel 8080, its 8-bit predecessor. Although some attempt at compatibility with the 8080 CPU architecture was made, the designers decided not to sacrifice sophistication in order to attain compatibility. By increasing the number of address pins from 16 to 20, the memory addressing capacity was increased from 64K bytes to $2^{20} = 1$ megabyte. The expanded memory capacity made multiprogramming feasible and several multiprogramming features have been incorporated into the 8086's design. The 8086 also includes a number of features which enhance its multiprocessing capabilities, thus allowing it to be used with other processing elements such as the 8087 numeric data processor.

3.5.1 Pins and Their Function Descriptions

A pin assignment diagrams for the 8086 and 8088 are given in **Figure 3-16**. Pin 33 ($\overline{MN/MX}$) determines the configuration option.

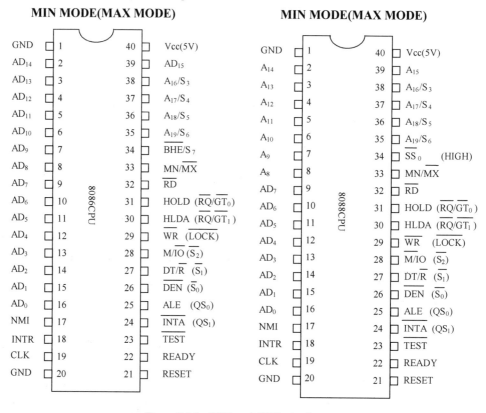

Figure 3-16 8086 and 8088 pin diagram

Pins and their Functions are described as following:

1. AD_{15}-AD_0 (Address Data Bus) (Input/Output /3 States)—pins 2 ~ 16, 39

Outputs address during T_1 and inputs or outputs data during the remaining part of the bus cycle. (On 8088, AD_{15}-AD_8 and A_{15}-A_8 are only for outputting address bits.) The 8086 has 20 address pins, 16 of which are also used as data pins. The use of pins for both address and data means that both an address and datum cannot be sent to the system bus at the same time.

This multiplexing of addresses and data reduces the number of pins needed, but does slow down the transfer of data. However, because of the timing on the bus, the transfer rate is not decreased as much as one might expect. AD_{15}-AD_0 enter a high-impedance state when the CPU responds to an interrupt or when HLDA = 1.

2. A_{19}/S_6-A_{16}/S_3 (Address/Status) (Output /3 States) —pins 35-38

They are multiplexed to provide address signals A_{19}-A_{16} and also status bits S_6-S_3. During T_1 the upper 4 bits of the address are output and during the remainder of the bus cycle status is output. S_3 and S_4 indicate the segment register being used during the current bus cycle as follows:

S_4	S_3	Register
0	0	ES
0	1	SS
1	0	CS or none
1	1	DS

Note that S_5 gives the current setting of IF. S_6 is always 0.

3. $\overline{BHE}/S_7$ (Bus High Enable/Status) (Output/3 States) —pin 34

The bus high enable pin is used in 8086 to enable the most-significant data lines (AD_{15}-AD_0) during a read or a write operation. That is to say $\overline{BHE}/S_7$ designates whether or not at least 1 byte of a transfer is to be made on AD_{15} through AD_8. A zero on this pin indicates that the AD_{15}-AD_8 are to be used; otherwise, only AD_7 through AD_0 are used. Status S_7 is output during the latter part of bus cycle, but, presently, S_7 has not been assigned a meaning. (On 8088, this pin is denoted $\overline{SS_0}$ and is used in minimum mode to denote status, which is logically equivalent to $\overline{S_0}$. It is always 1 in maximum mode.) The $\overline{BHE}$ and AD_0 are used together to indicate the way that data transfers for the memory or the interface device connected to the bus. Four possible combinations are defined in **Table 3-3**.

Table 3-3 $\overline{BHE}$, AD_0 signal and data pin used for write/read operation

Operation	$\overline{BHE}$	AD_0	Data Pins Used
Write/read a word at an even address	0	0	AD_{15}-AD_0
Write/read a byte at an even address	1	0	AD_7-AD_0
Write/read a byte at an odd address	0	1	AD_{15}-AD_8
Write/read a word at an odd address	0	1	AD_{15}-AD_8 (First bus cycle: puts the least significant data byte on AD_{15}-AD_8)
	1	0	AD_7-AD_0 (Second bus cycle: puts the most significant data byte on AD_7-AD_0)

Where 0 is low and 1 is high.

4. NMI (Non-maskable Interrupt) (Input) —pin 17

Non-maskable interrupt request positive-going edge triggered. If NMI is actived, CPU will respond no matter the IF flag bit is logical or not. The interrupt vector of NMI is 2.

5. INTR (Interrupt Request) (Input) —pin 18

The pin is used to request a hardware maskable Interrupt. If INTR is held high, CPU will respond and enters an interrupt acknowledge cycle when IF is set.

6. $\overline{RD}$ (Read) (Output) —pin 32

The pin indicates a memory or I/O read is to be performed. In minimum mode, it is used along with pin 28 ($M/\overline{IO}$), which distinguishes a memory transfer from an I/O transfer.

7. CLK (Clock) (Input) —pin 19

The clock pin provides the basic timing signal to the microprocessor.

Clock—33% duty cycle, maximum rate depends on CPU model: 5MHz for 8086, 10MHz for 8086-1, and 8MHz for 8086-2; 10MHz for 8088-1, 25MHz for 8088, and 8MHz for the 8088-2.

8. RESET (Reset) (Input) —pin 21

The reset input causes the microprocessor to reset itself if this pin is held high at lest 4 clock cycles.

Terminates activity, clears PSW, IP, DS, SS, ES and the instruction queue and sets CS to FFFFH. Processing begins at FFFF0H when signal is dropped. Signal must be 1 for at least 4 clock cycles.

9. READY (Ready) (Input) —pin 22

This input is controlled to insert wait states into the timing of the microprocessor . If READY signal is placed at a logic 0 level the CPU enters into wait states and remains idle . If it set 1, it has no effect on the operation of the CPU.

The pin inputs acknowledgement from memory or I/O interface that input data will be out on the data bus or output data will be accepted from the data bus within the next clock cycle. In either case, the CPU and its bus control logic can complete the current bus cycle after the next clock cycle.

10. $\overline{TEST}$(Test) (Input) —pin 23

The pin is used in conjunction with the WAIT instruction in multiprocessing environments. A WAIT instruction will cause the CPU to idle, except for processing interrupts, until a 0 is applied to this pin.

11. MN/$\overline{MX}$(Minimum/Maximum Mode Control) (Input) —pin 33

The pin selects either minimum mode or maximum mode operation for the microprocessor. CPU is in minimum mode when strapped to +5V and in maximum mode when grounded.

12. GND (Ground—pins 1&20) & VCC (Voltage—pin 40)

The ground pin is the return for the power supply. Note that the 8086/8088 have two pins labled GND. Ground =0, Vcc Supply voltage +5V (±10%).

3.5.2 Working Modes

In order to adapt to as many situations as possible both the 8086 and 8088 have been given two modes of operation, the minimum mode and the maximum mode. The minimum mode is used for a small system with a single processor, a system in which the 8086/8088 generates all the necessary bus control signals directly (thereby minimizing the required bus control logic).

The maximum mode is for medium-size to large systems, which often include two or more processors. In the maximum mode, the 8086/8088 encodes the basic bus control signals into 3 status bits, and the remaining control pins to provide the additional information that is needed to support a multiprocessor configuration.

1. Minimum Mode

Pin 33 ($MN/\overline{MX}$) determines the configuration option. When it is strapped to +5V the processor is to be operated in its minimum mode.

The definitions for pins 24 through 31 for the minimum mode are given in **Table 3-4** and a typical minimum mode configuration is shown in **Figure 3-17**.

Table 3-4 Pins definitions for the minimum mode

Pin(s)	Symbol	In/Out(3-State)	Description
24	$\overline{INTA}$	O-3	$\overline{INTA}$(Interrupt Acknowledge): Indicates recognition of an interrupt request. Consists of two negative going pulses in two consecutive bus cycles. It's used to gate the interrupt vector number onto the data bus in response to an interrupt request
25	ALE	O	ALE (Address Latch Enable): Outputs a pulse at the beginning of the bus cycle and is to indicate an address is available on the address/data pins. Note that the ALE signal does not float during a hold acknowledge
26	$\overline{DEN}$	O-3	$\overline{DEN}$(Data Enable): The pin activates external data bus buffers. Output during the latter portion of the bus cycle and is to inform the transceivers that the CPU is ready to send or receive data
27	$DT/\overline{R}$	O-3	$DT/\overline{R}$ (Data Transmit/Receive): Indicates to the set of transceivers whether they art to transmit (1) or receive (0) data
28	$M/\overline{IO}$	O-3	$M/\overline{IO}$(Memory/Input and Output): Distinguishes a memory transfer from an I/O transfer. For a memory transfer it is 1. (For the 8088, the symbol is $IO/\overline{M}$ and a 1 indicates an I/O transfer. It's at its high-impedance state during a hold acknowledge
29	$\overline{WR}$	O-3	$\overline{WR}$(Write): The write line is a stroke to indicate that the 8086/8088 CPU is outputting data to a memory or I/O device. When 0, it indicates a write operation is being performed. It is used in conjunction with pins 28 ($M/\overline{IO}$) and 32 ($\overline{RD}$) to specify the type of transfer
30	HLDA	O	HLDA(Hold Acknowledge): Outputs a bus grant to a requesting master and indicate that the 8086/8088 has entered the hold state. Pins with tristate gates are put in high impedance state while HLDA = 1
31	HOLD	I	HOLD(Hold Request): Receive bus requests from bus masters. If HOLD signal is logic 1, the microprocessor steps executing software and places it's address, data and control bus at the high impedance state. The 8086/8088 will not gain control of the bus until this signal is dropped

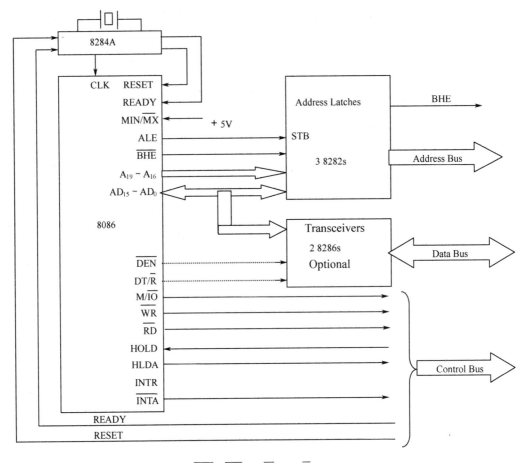

Note: in an 8088 system $\overline{BHE}$ is $\overline{SSO}$, $M/\overline{IO}$ is $IO/\overline{M}$, and only one 8286 is needed.

Figure 3-17 Minimum mode system

  The address must be latched since it is available only during the first part of the bus cycle. To signal that the address is ready to be latched a 1 is put on pin 25, the address latch enable (ALE) pin. Typically, the latching is accomplished using Intel 8282s, as shown in **Figure 3-18**. Because 8282 is an 8-bit latch, in an 8086 system, three are needed for a full 20-bit address is used. Note that $\overline{BHE}$ would also have to be latched. For a small 8088 system that has only 64K bytes of memory, only two 8282s would be required. A signal on the STB pin latches the bits applied to the input data lines DI_7-DI_0. Therefore, STB is connected to the 8086's ALE pin and DI_7-DI_0 are attached to eight of the address lines. An active low signal on the $\overline{OE}$ enables the latch's output DO_7-DO_0, and a 1 at this pin force the output into the high-impedance state. In an 8086/8088 single-processor system that does not include a controller this pin is grounded.

  If a system includes several interfaces, then drivers and receivers, which may not be needed on small, signal-board systems, will be required for the data lines. The Intel IC device for implementing the transceiver (driver/receiver) block shown in **Figure 3-17** is the 8286 transceiver device. The 8286 contains 16 tri-state elements, eight receivers and eight drivers.

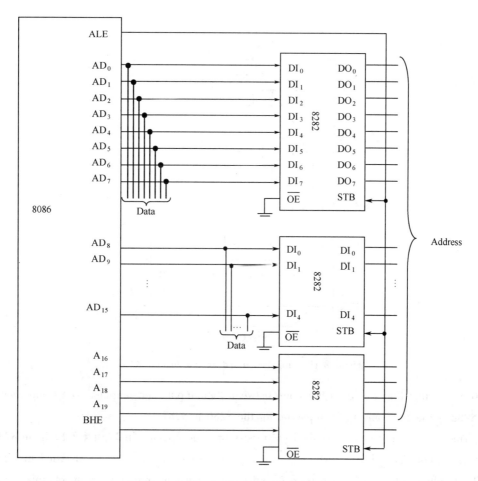

Figure 3-18 Application of 8282 latches

Therefore, only one 8286 is needed to service all of the data lines for an 8088, but two are required in an 8086 system. **Figure 3-19** shows how 8286s are connected into a system.

The 8286 is symmetric with respect to its two sets of data pins, either the pins A_7-A_0 can be the inputs and B_7-B_0 the outputs, or vice versa. The output enable ($\overline{OE}$) pin determines whether or not data are allowed to pass through the 8286 and the transmit (T) pin controls the direction of the data flow. When $\overline{OE} = 1$, data are not transmitted through the 8286 in either direction. If it is 0, then T = 1 causes A_7-A_0 to be the inputs and T = 0 results in B_7-B_0 being the inputs. In an 8086/8088-based system the $\overline{OE}$ pin would be connected to the $\overline{DEN}$ pin, which is active low whenever the processor is performing an I/O operation.

The A_7-A_0 pins are connected to the appropriate address/data lines and the T pin is tied to the processor's DT/$\overline{R}$ pin. Thus, when the processor is outputting, the data flow is from A_7-A_0 to B_7-B_0, and when the processor is inputting, the flow is in the other direction. The processor floats $\overline{DEN}$ and DT/$\overline{R}$ pins in response to a bus request on the HOLD pin.

Sometimes a system bus is designed so that the address and/or data signals are inverted. Therefore, the 8282 and 8286 both have companion chips that are the same as the 8282 and

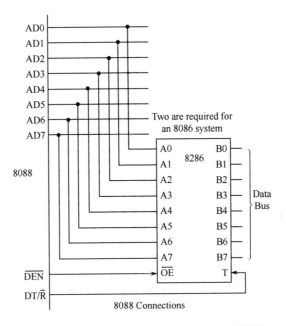

Figure 3-19 Application and internal logic of 8286

8286 except that they cause an inversion between their inputs and outputs. The companion for the 8282 is the 8283 and the companion for the 8286 is 8287.

The third component, other than the processor, that appears in **Figure 3-17** is an 8284A clock generator. This device, which is actually more than just a clock, is detailed in **Figure 3-20.** In addition to supplying a train of pulses at a constant frequency is synchronizes ready ($\overline{RDY}$) signals, which indicate an interface is ready to complete a transfer, and reset ($\overline{RES}$) signals, which initialize the system, with the clock pulses. Although these two signals may be sent at any time, the 8284A will not reflect them in its READY and RESET outputs until the trailing edge of the clock pulse in which they are received.

The frequency source applied to the 8284A may be from a pulse generator that is connected across X1 and X2. If the input to $F/\overline{C}$ is 1, then the EFI input determines the frequency; otherwise, it is the oscillator input. In either case the clock output CLK is one-third of the input frequency.

All three of the devices considered above, the 8282, 8286, and 8284A, require only +5V supply voltages. Their inputs and outputs are TTL compatible and, therefore, the devices are compatible with each other and with the 8286 and 8088. Many of the details of these devices have not been included here and, for more information, one should refer to the Intel manuals. This comment applies particularly to the 8284A clock.

In a minimum system the control lines do not need to be passed through transceivers, but can be used directly. The M/$\overline{IO}$, $\overline{RD}$, and $\overline{WR}$ lines specify the type of transfer according to **Table 3-5** and **Table 3-6**, where 0 is low and 1 is high.

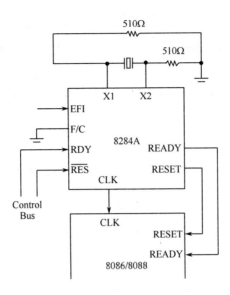

Figure 3-20 Typical 8284A clock connection

Table 3-5 Type of transfer and their signal lines

M/$\overline{\text{IO}}$	$\overline{\text{RD}}$	$\overline{\text{WR}}$	Operation
0	0	1	I/O read
0	1	0	I/O write
1	0	1	Memory read
1	1	0	Memory write

Table 3-6 8088 $\overline{\text{M/IO}}$, DT/$\overline{\text{R}}$, $\overline{\text{SS}}_0$ and its operations

$\overline{\text{M/IO}}$	DT/$\overline{\text{R}}$	$\overline{\text{SS}}_0$	Operation
1	0	0	Send interrupt response
1	0	1	Read I/O
1	1	0	Write I/O
1	1	1	Pause
0	0	0	Fetch instruction
0	0	1	Read memory
0	1	0	Write memory
0	1	1	—

2. Maximum Mode

A processor is in maximum mode when its MN/$\overline{\text{MX}}$ pin is grounded. The maximum mode definitions of pins 24 through 31 are given in **Table 3-7** and a typical maximum mode configuration is shown in **Figure 3-21**.

Table 3-7 Maximum mode pin definitions

Pin(s)	Symbol	In/Out(3-State)	Description
24, 25	QS_1, QS_0	O	Reflects the status of the instruction queue. This status indicates the activity in the queue during the previous clock cycle
26, 27, 28	$\overline{S}_0 \overline{S}_1 \overline{S}_2$	O-3	Indicates the type of transfer to take place during the current bus cycle: $\overline{S}_0$ $\overline{S}_1$ $\overline{S}_2$ 0　0　0　Interrupt acknowledge 0　0　1　Read I/O port 0　1　0　Write I/O port 0　1　1　Halt 1　0　0　Instruction fetch 1　0　1　Read memory 1　1　0　Write memory 1　1　1　Inactive-passive (1 represents high and 0 represents low.) The status becomes active prior to the beginning of a bus cycle and returns to inactive during the later part of the cycle
29	$\overline{LOCK}$	O-3	Indicates the bus is not to be relinquished to other potential bus masters. It is initiated by a LOCK instruction prefix and is maintained until the end of the next instruction. It is also active during and between the two $\overline{INTA}$ pulses
30	$\overline{RQ}/\overline{GT}_1$	I/O	For inputting bus requests and outputting bus grants
31	$\overline{RQ}/\overline{GT}_0$	I/O	Same as $\overline{RQ}/\overline{GT}_1$ except that a request on $\overline{RQ}/\overline{GT}_0$ has higher priority

It is clear from **Figure 3-21** that the main difference between minimum and maximum mode configurations is the need for additional circuitry to translate the control signals. The circuitry is for converting the status bits S_0, S_1 and S_2 into the I/O and memory transfer signals needed to direct data transfers and for controlling the 8282 latches and 8286 transceivers. It is normally implemented with an Intel 8288 bus controller. Also included in the system is an interrupt priority management device; however, its presence is optional.

Note: In maximum mode the 8086 and 8088 pins have the same definitions except for pin 34, which on the 8088 is always 1.

The $\overline{S}_0$, $\overline{S}_1$, and $\overline{S}_2$ status bits specify the type of transfer that is to be carried out and when used with an 8288 bus controller they obviate the need for the $M/\overline{IO}$ (or $IO/\overline{M}$), $\overline{WR}$, $\overline{INTA}$, ALE, $DR/\overline{R}$, and $\overline{DEN}$ signals that are output over pins 24 through 29 when the processor is operating in minimum mode. Except for the case $\overline{S}_1 = \overline{S}_0 = 1$, $\overline{S}_2 = 0$ indicates a transfer between an I/O interface and the CPU and $\overline{S}_2 = 1$ implies a memory transfer. The $\overline{S}_1$ bit specifies whether an input or output is to be performed. From the status the 8288 is able to originate the address latch enable signal to the 8282s, the enable and direction signals to the 8286 transceivers and the interrupt acknowledge signal to the interrupt controller.

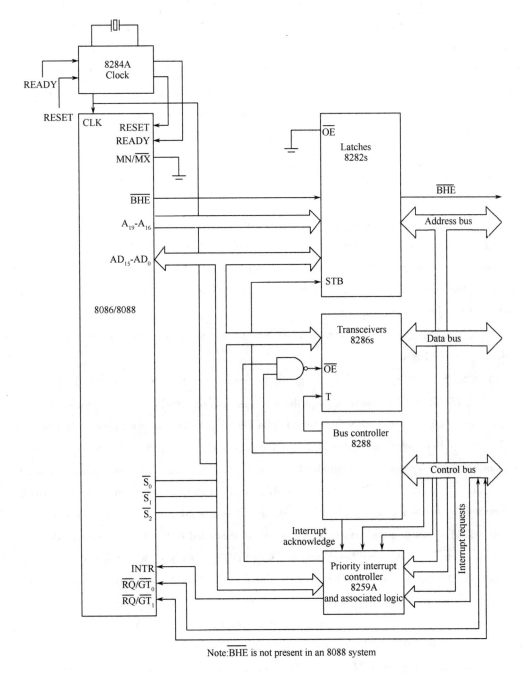

Note: $\overline{BHE}$ is not present in an 8088 system

Figure 3-21 Typical maximum mode configuration

The QS_0 and QS_1 pins are to allow the system external to the processor to interrogate the status of the processor instruction queue so that it can determine which instruction it is currently executing, and the $\overline{LOCK}$ pin indicates that an instruction with a LOCK prefix is being executed and the bus is not to be used by another potential master. These pins are needed only in multi-processor systems and, along with the LOCK prefix.

The HOLD and HLDA pins become the $\overline{RQ}/\overline{GT_0}$ and $\overline{RQ}/\overline{GT_1}$ pins. Both bus requests and bus grants can be given through each of these pins. They are exactly the same except that if requests are seen on both pins at the same time, then one on $\overline{RQ}/\overline{GT_0}$ is given higher priority. A request consists of a negative pulse arriving before the start of the current bus cycle. The grant is negative pulse that is issued at the beginning of the current bus cycle provided that:

(1) The previous bus transfer was not the low byte of a word to or from an odd address if the CPU is an 8086. For 8088, regardless of the address alignment, the grant signal will not be sent until second byte of a word reference is accessed.

(2) The first pulse of an interrupt acknowledgement did not occur during the previous bus cycle.

(3) An instruction with a LOCK prefix is not being executed.

If condition 1 or 2 is not met, then the grant will not be given until the next bus cycle and if condition 3 is not met, the grant will wait until the locked instruction is completed. In response to the grant the three-state pins are put in their high-impedance state and the next bus cycle will be given to the requesting master.

An expanded view of a maximum mode system which shows only the connections to an 8288 is given in **Figure 3-22**.

The $\overline{S_0}, \overline{S_1}$, and $\overline{S_1}$ pins are for receiving the corresponding status bits from the processor. The ALE, DT/$\overline{R}$ and DEN pins provide the same outputs that are sent by the processor when it is minimum mode (except DEN is inverted from $\overline{DEN}$). The CLK input permits the bus controller activity to be synchronized with that of the processor. The $\overline{AEN}$, IOB, and CEN pins are for multiprocessor systems. In a single-processor system $\overline{AEN}$ and IOB are normally grounded and a 1 is applied to CEN. The meaning of MCE/$\overline{PDEN}$ output depends on the mode, which is determined by the signal applied to IOB. When IOB is grounded it assumes its master cascade enable (MCE) meaning and can be used to control cascaded 8259As. In the event that +5V is connected to IOB, the peripheral data enable ($\overline{PDEN}$) meaning, which is used in multiple-bus configurations, is assumed. The reaming pins given in **Figure 3-22** have the following definitions as shown in **Table 3-8**:

These signals are active low and are output during the middle portion of a bus cycle. Clearly, only one of them will be issued during any given bus cycle.

Not shown in **Figure 3-22** are the $\overline{AIOWC}$ (advanced I/O write command) and $\overline{AMWC}$ (advanced memory write command) pins. They serve the same purposes as $\overline{IOWC}$ and $\overline{MWTC}$ pins except that they are activated one clock pulse sooner. This gives slow interfaces an extra clock cycle to prepare to input the data.

As with the other 8086 supporting devices, the 8288 requires a +5V supply voltage and has TTL-compatible inputs and outputs. For more detailed information one should refer to the Intel manuals.

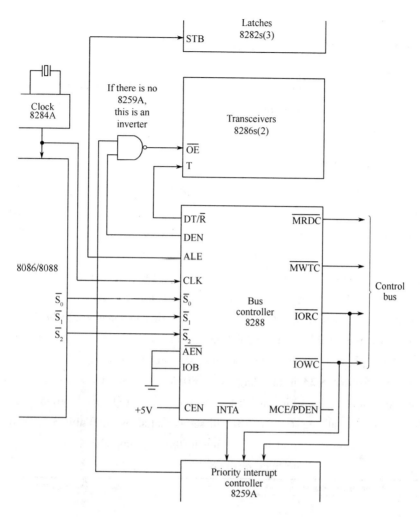

Figure 3-22 Connection to an 8288 bus controller

Table 3-8 Definition of the remaining pins given in Figure 3-22

Pin	Description
$\overline{INTA}$	Issues the two interrupt acknowledgment pulses to a priority interrupt controller or an interrupting device when $\overline{S0} = \overline{S1} = \overline{S2} = 0$
$\overline{IORC}$	(**I/O Read Command**)—Instructs an I/O interface to put the data contained in the address port on the data bus
$\overline{IOWC}$	(**I/O Write Command**)—Instructs an I/O interface to accept the data on the data bus and put the data into the addressed port
$\overline{MRDC}$	(**Memory Read Command**)—Instructs the memory to put the contents of the addressed location on the data bus
$\overline{MWTC}$	(**Memory Write Command**)—Instructs the memory to accept the data on the data bus and put the data into the addressed memory location
\multicolumn{2}{} Not shown in **Figure 3-22** are the $\overline{AIOWC}$ (advanced I/O write command) and $\overline{AMWC}$ (advanced memory write command) pins. They serve the same purposes as $\overline{IOWC}$ and $\overline{MWTC}$ pins except that they are activated one clock pulse sooner. This gives slow interfaces an extra clock cycle to prepare to input the data

3.6 Basic Operations of 8086/8088

In this section we mainly focus on some operations in a 8086 system such as reset, input and output, bus request and bus grant timing in minimum mode/maximum mode systems, and Interrupt operation.

3.6.1 Reset Operation

Pin 21 (RESET) is used for inputting a system reset signal. Most systems include a line that goes to all system components and a pulse is automatically sent over this line when the system is turned on, or the reset pulse is manually generated by a switch that allows the operator to reinitialize the system. As shown in **Table 3-9**, a high level signal on the reset line causes the components to go to their "turn on" state. For the processor this operation has the PSW, IP, DS, SS, ES, and instruction queue cleared and CS set to 0FFFFH. For (IP) = 0000H, the processor will begin executing at FFFF0H. Normally, this location would be in a read—only section of memory and would contain a JMP instruction to a program for initializing the system and loading the application software or operating system. Such a program is referred to as a bootstrap loader. **Figure 3-23** is the diagram of 8086/8088 Reset operation. In 8086/8088, the reset signal should be kept high at least 4 clock cycles. At ① point, 8086/8088 will be in internal Reset state and all registers in CPU will set as that shown in **Table 3-9**. Notice that one clock cycle later all tri-state output lines are put in the high impedance states.

Table 3-9 The State of the 8086/8088 registers when reset

Name of the Registers	State of the Registers
Flag register (FR)	clear
Instruction point register (IP)	0000H
CS	FFFFH
DS	0000H
SS	0000H
ES	0000H
Instruction queue	empty
Others	0000H

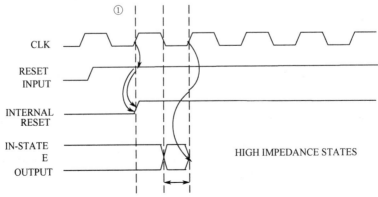

Figure 3-23 8086/8088 reset signal

要点：

复位操作：在复位的时候，代码段寄存器 CS 和指令指针寄存器 IP 分别初始化为 FFFFH 和 0000H。所以，8086/8088 在复位之后再重新启动时，便从内存的 FFFF0H 处开始执行指令，使系统再启动时，能自动进入系统程序。

3.6.2 Input and Output for 8086 Minimum Mode

The timing diagrams for 8086 minimum mode input and output transfer are shown in **Figure 3-24**. Both input and output operation times are consisted of 4 clock cycles such as T_1, T_2, T_3 and T_4.

1. Input Operation Bus Timing

As shown in **Figure 3-24(a)**, the bus timing of an input operation can be described as following according to different clock cycle.

T_1: When the state of the processor is such that is ready to initiate a bus cycle it applies a pulse to ALE pin during T_1. Before the trailing edge of the ALE signal the address, $\overline{BHE}$, M/$\overline{IO}$, $\overline{DEN}$, and DT/$\overline{R}$ signals should be stable, with the $\overline{DEN} = 0$ and DT/$\overline{R} = 0$ for an input. $\overline{BHE}$ is zero to indicate the most significant data lines are to be used. M/$\overline{IO}$ is used to designate memory transfer or I/O transfer. DT/$\overline{R} = 0$ informs the transceivers to receive data. $AD_0 \sim AD_{15}$ and $AD_{16} \sim AD_{19}$ send the reference address of a memory location or an I/O port. At the trailing edge of the ALE signal the 8282s latch the address.

T_2: During T_2 the address is dropped and S_3 through S_7 are output on $A_{16}/S_3 \sim A_{19}/S_6$ and $\overline{BHE}/S_7$, and $\overline{DEN}$ is lowered to enable the 8286 transceivers. $\overline{RD}$ is activated low during T_2 and $AD_{15} \sim AD_0$ enter a high-impedance state in preparation for input.

T_3: If the memory or I/O interface can perform the transfer immediately, there are no wait states and the data are put on the bus $AD_0 \sim AD_{15}$ during T_3.

The bus timing has been designed so that the memory or I/O interface involved in a transfer can control when data are to be placed on or taken from the bus by the interface. This is done by having the interface send a READY signal to the processor (perhaps via an 8284A) when it has made data available or accepted data. If a READY signal has not been received by the processor at the beginning of the T_3, then one or more T_W state will be inserted between T_3 and T_4 until a READY signal has been received. The bus activity during Tw is the same as during T_3. A signal applied to an RDY input of an 8254A will cause a READY output to the processor at the trailing edge of the current clock cycle. Therefore, if a wait state is to be avoided, an RDY input must be received before the beginning of the T_3 clock cycle.

T_4: After the input data are accepted by the processor, $\overline{RD}$ is raised to 1 at the beginning of T_4 and, upon detecting this transition, the memory or I/O interface will drop its data signals.

2. Output Operation Bus Timing

As shown in **Figure 3-24(b)**, when an output is being conducted, the bus timing can be described as following:

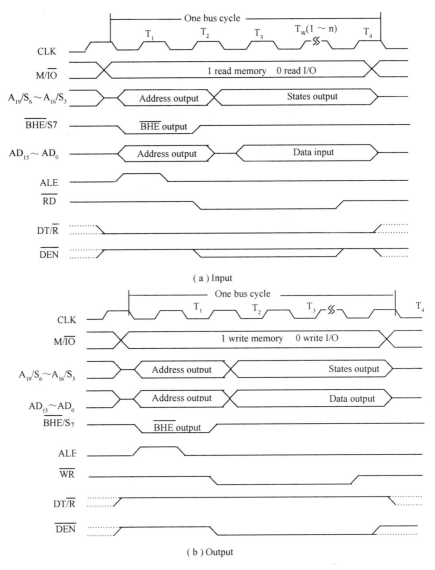

Note: For an 8088, $M/\overline{IO}$ is $IO/\overline{M}$ and $\overline{BHE}/S7$ becomes $\overline{SS_0}$ which is present throughout the bus cycle (i.e., it changes at the same time as $IO/\overline{M}$). Also, only AD_7-AD_0 carry data.

Figure 3-24　8086 minimum mode bus timing diagrams

T_1: 20 address lines ($AD_0 \sim AD_{15}$ and $A_{19}/S_6 \sim A_{16}/S_3$) offer the address of a memory location or an I/O port. $\overline{BHE}$, $M/\overline{IO}$, and ALE are the same with that in input operation except $DT/\overline{R} = 1$, which informs the transceivers to transmit data.

T_2: During T_2, the address is dropped and the data are put on $AD_0 \sim AD_{15}$ immediately. The processor applies the $\overline{WR} = 0$ signal and then the data during T_2.

T_3: The data is on $AD_0 \sim AD_{15}$ and $S_3 \sim S_7$ are output on $A_{16}/S_3 \sim A_{19} \sim S_{16}$ continuously. $\overline{WR}$, $M/\overline{IO}$ and $\overline{DEN}$ keep effectively in T_3.

T_4: In T_4, $\overline{WR}$ is raised and the data signals are dropped.

For either an input or output, $\overline{DEN}$ is raised during T_4 to disable the transceivers and the $M/\overline{IO}$ signal is set according to the next transfer at this time or during a subsequent T_i state. As shown in **Figure 3-23**, the main differences between read and write timing are minimal. The $\overline{RD}$ stroke is replaced by the $\overline{WR}$ stroke, the data bus contains information for the memory rather than information from the memory rather than information from the memory, and **DT/$\overline{R}$** remains a logic 1 instead of a logic 0 throughout the bus cycles.

[Example 3-6] Please analyze the executing process of the instruction of
'MOV AL,[1000H]'

When the instruction executes, the following steps will take plack in CPU:

(1) BIU fetches the instruction bytes from memory and store it in the instruction queue.

(2) EU takes the instruction of the queue, and then sends it to the control unit to decode it to see what it does.

(3) BIU chooses DS automatically, and generates 20-bits physics address by Address Adder.

(4) BUS cycle initiates, which is denoted T_1, T_2, T_3 &T_4.

T_1: Outputting the 20-bit wide address of the data.

T_2: A_{16}/S_3-A_{19}/S_6 send the status signals, and $AD_{15} \sim AD_0$ enter a high-impedance state in preparation for input.

T_3: The data stored in location DS: 1000H are put on the bus.

T_4: The BUS cycle is over.

(5) EU copies the data which are brought from DS: 1000H by BIU to AL register, the instruction executing ends.

(Suppose the content of DS register is 2000H)

① EU gets instruction coding and executes. BIU calculates the physics address on proper time.

② physics address = segment address: offset address = DS: 1000H = 2000H × 10H + 1000H = 21000H

It's to say, CPU should send the byte of 21000H Unit to AL.

3.6.3 Bus Request and Bus Grant Timing in Minimum Mode

Figure 3-25 shows the timing of a bus request and bus grant in a minimum mode system. The HOLD pin is tested at the leading edge of each clock pulse. If a HOLD signal is received by the processor before T_4 or during a T_1 state, then the CPU activates HLDA and the succeeding bus cycle will be given to the requesting master until that master drops its request. The lowered request is detested at the rising edge of the next clock cycle and the HLDA signal is dropped at the trailing edge of that clock cycle. While HLDA is 1, all of the processor's three-state outputs are put in their high-impedance state.

Instructions already in the instruction queue will continue to be executed until one of them

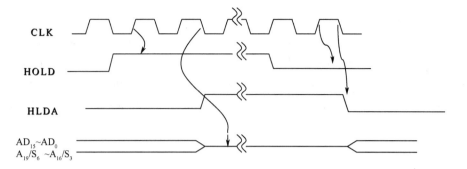

Figure 3-25 Bus request and bus grant timing on a minimum mode system

requires the use of the bus. The instruction

MOV AX, BX

could execute completely, but

MOV AX, [BX]

would only execute until it is necessary to bring in data from the memory location whose address is given by BX register.

3.6.4 Interrupt Operation

We are all confronted with interruptions from time to time. Some are pleasant, some are unpleasant and some are neutral. Some interruptions can be ignored, **for example** a telephone or doorbell ringing, but some must not be ignored, you must deal with them as soon as possible.

Whatever their cause, interrupts are essentially requests for attention. In the same way, peripherals in a computer system can request the attention of the processor. The event that makes the processor stop executing its program to perform some requested activity is called an **INTERRUPT**.

Comparing with a polling system, interrupts increase the overall efficiency of a computer system, because the external devices request the attention of the processor as needed. But for system using Polling, the processor would have to POLL (look at) every device in the system periodically, to see if any of them required attention.

1. Types of Interrupts

The processor can process two kinds of interrupts. Interrupts generated by external DEVICES, such as disk drives, parallel or serial interfaces are called External Interrupt. For those caused by internal interrupt-generating INSTRUCTIONS are called Internal Interrupt.

External Interrupt and **Internal Interrupt** are also called as the **Hardware Interrupt** and **Software Interrupt** respectively.

As shown in **Figure 3-26**, external interrupts are raised from outside the CPU. Many I/O devices raise interrupts (e.g., keyboard, timer, disk drives, CD-ROM and sound cards). These external devices can be chips on the system board or they can be on cards plugged into

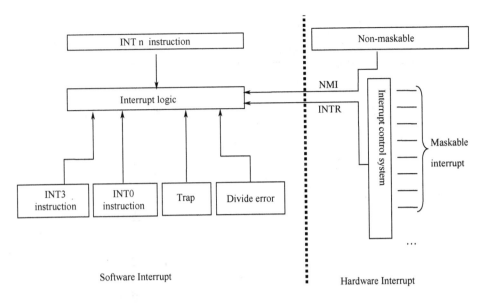

Figure 3-26 Types of interrupt

the computers external I/O bus slots. The 8086 family of processors have two type hardware interrupts, one can be ignored and another must be served as soon as they occurred.

INTR: The **Interrupt Request**, an interrupt that can be delayed if the processor is busy with some more important processes.

NMI: The **Non Maskable Interrupt**, an interrupt that must be processed when it occurs.

Internal interrupts are raised from within the CPU, either from an error or the interrupt instruction. Error interrupts are also called traps. They are generated by machine code instructions within programs. DOS uses these types of interrupts to implement its API (Application Programming Interface). More modern operating systems (such as Windows and UNIX) use a C based interface. Many interrupt handlers return control back to the interrupted program when they finish. They restore all the registers to the same values they had before the interrupt occurred. Thus, the interrupted program runs as if nothing happened (except that it lost some CPU cycles).

2. The Interrupt Response of CPU

The follow chart of an interrupt response of 8086/8088 CPU is showed as **Figure 3-27.**

According to **Figure 3-27**, when an interrupt occurs, the processor will differentiate whether it is internal interrupt or not at first. If not, NMI will be priority identified, then the INTR. Notice that when the interrupt is responded, we usually disable the maskable interrupt by using CLI. This prevents the other non-maskable interrupt from itself being interrupted. Programmer may override this behavior by executing STI within the ISR. Then save the IP program counter, CS code segment register, and Flags register on the stack. Reloading CS and IP with the address found in memory locations 4 * N, where N is the number of the interrupt, an ISR (Interrupt Service Routine) found at that address is therefore be executed. The **IRET** instruction will pull **IP** at the end of the ISR, **CS** and the **Flags** register off the stack, restoring the

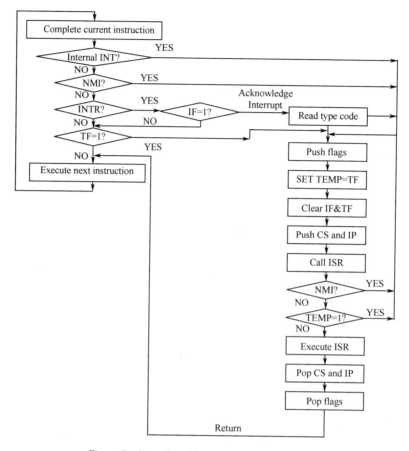

Figure 3-27 Flow Chart of Interrupt Response

CPU to the status it had before the interrupt occurred.

3. INTR and $\overline{\text{INTA}}$

The interrupt request input (INTR) is level-sensitive, which means that it must be held at logic 1 level until it is recognized. The INTR pin is set by an external event and cleared inside the interrupt service procedure.

If an interrupt request has been recognized during the previous bus cycle and an instruction has just been completed, then a negative pulse will be applied to $\overline{\text{INTA}}$ during the current bus cycle and the next bus cycle. As shown in **Figure 3-28**, each of these pulses will extend from T_2 to T_4. Upon receiving the second pulse, the interface accepting the acknowledgment will put the interrupt type on AD_7-AD_0, which are floated the rest of the time during the two bus cycles. The type will be available from T_2 to T_4.

$\overline{\text{INTA}}$ signal consists of two negative pulses output during two consecutive bus cycles. The first pulse informs the interface that its request has been recognized, and upon receipt of the second pulse, the interface is to send the interrupt type to the processor over the data bus.

4. Interrupt Vector & Interrupt Vector Table

The 8086 family of microprocessors can recognize 256 different interrupts, each with a unique **Interrupt Vector** (number) that identifies it to the microprocessor. The processor uses

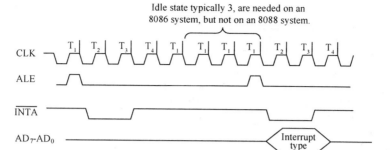

Figure 3-28 The timing of the $\overline{INTA}$ output Interrupt acknowledgements

this type code (a number between 0 and 255) to point to a location in an **INTERRUPT VECTOR TABLE**. This table is stored in the first 1024 bytes of the memory space, starting at address 00000 hex, 0000:0000 hex in Segment and Offset format. If contains 256 different 4-byte interrupt vectors stored at addresses 00000H-0003FFH. The first five interrupt is denoted. For example, interrupt 0 is for divide error. The 1 is for trap interrupt. The 2 is for Non-maskable interrupt. The 4 is for overflow interrupt.

AS indicated in **Figure 3-29**. The **INTERRUPT VECTOR TABLE** points to the locations of the INTERRUPT ROUTINES that carry out the function associated with that interrupt. As the old 8086/8088 processors had a total memory space of one Megabyte, these vectors must point to 20 bit address—this requires the use of SEGMENT and OFFSET address format and hence, each vector is FOUR BYTES long and contains the starting address of the interrupt service procedure.

The first two bytes of the vector of the offset address, and the last two bytes contain the segment address.

The INTO instruction checks the overflow flag(OF). If OF = 1, the INTO instruction calls the procedure whose address is stored in interrupt vector type number 4. If OF = 0, then the INTO instruction performs no operation and the next sequential instruction in the program executes.

The INT n instruction calls the interrupt service procedure that begins at the address represented in vector number n. To determine the vector address, just multiply the vector type number(n) by 4, which gives the beginning address of the 4-byte long interrupt vector.

The relation between the interrupt type and offset is shown in the **Figure 3-29**:

[Example 3-7] Suppose the ISR of the vector number 17H is stored in the memory location 2345:7890H, please describe the way how to store this ISR address in the interrupt vector table.

(1) Calculating where the ISR address is stored in memory by given the vector number
   17H = 00010111B   17H × 4 = 01011100 = 005CH
(2) Store the address in the memory like this:
   90H = >0000:005CH

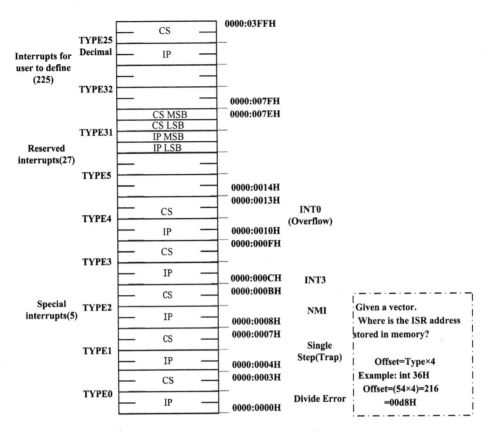

Figure 3-29 Interrupt vector table

78H = >0000:005DH
45H = >0000:005EH
23H = >0000:005FH

要点：

（1）中断是一种使 CPU 中止正在执行的程序而转去处理特殊事件的操作。它可用于解决快速的 CPU 与慢速的外设之间交换数据的矛盾。

（2）每一个微机系统中都有一个中断系统，8086/8088 的中断系统可以处理 256 种不同类型的中断。

（3）8086/8088 系统的中断可分为硬件中断和软件中断两类。

① 硬件中断。外部硬件产生的中断叫做硬件中断，也叫做外部中断。硬件中断又可以分为两类：一类叫做非屏蔽中断；另一类叫做可屏蔽中断。非屏蔽中断 NMI 是通过 CPU 的 NMI 引脚接入，不受 IF（中断允许标志位）的影响。但是系统内只有一个非屏蔽中断。可屏蔽中断 INTR 是通过 CPU 的 INTR 引脚接入，只有 IF 为 1 时中断请求才能被响应。在中断控制器（如 8259A）的配合下，一个系统可以有几十个可屏蔽中断。

② 软件中断。软件中断是 CPU 根据软件中的某条指令或软件对标记寄存器中某个标志的设置而产生的中断。进入中断时，不需要执行中断响应周期，也不从数据总线读取中断类型码，不受中断允许标志位 IF 的影响，但是它和 TF 的状态有关（只有 TF 为 1 才执行单步中断）。

(4) 中断向量。中断向量是中断子程序的入口地址,每个中断类型对应一个中断向量,它和断点不同。所谓断点,就是指响应中断时,主程序中当前指令下面的一个指令地址。

(5) 中断向量表是指根据中断类型和中断服务子程序对应关系建立的列表。8086/8088 系统是把内存中 0 段的 0000-03FFH 区域设置为一个中断向量表。每一个中断向量占 4 个存储单元。前两个单元存放中断子程序入口地址的偏移量(IP),后两个单元存放中断子程序入口地址的段地址(CS)。

(6) 可屏蔽中断。8086/8088 的可屏蔽中断响应需要两个总线周期。如果前一个总线周期中,CPU 接收到外界的中断请求信号,而中断的允许标志 IF 为 1,且上一条指令执行完毕,那么,CPU 会在当前总线周期和下一总线周期,从 $\overline{INTA}$ 引脚上往外设接口各发一个负脉冲。外设接口收到第二个负脉冲之后,立即把中断类型码送到数据线的低 8 位 $D_7 \sim D_0$ 上,通过 CPU 的地址/数据引脚 $AD_7 \sim AD_0$ 传输给 CPU。在这两个总线周期的其余时间,$AD_7 \sim AD_0$ 是浮空的。

中断响应的第一个总线周期用来通知发中断请求的设备,CPU 准备响应中断,现在应准备好中断类型码;在第二个总线响应周期中,CPU 接收外设接口发来的中断类型码,从而得到中断处理子程序的入口地址。外设的中断类型码必须通过 16 位数据总线的低 8 位传送给 8086。

3.6.5 Interrupt Operations in Maximum Mode

The timing diagram for input and output on a maximum mode system are given in **Figure 3-30**. The $\overline{S}_0$, $\overline{S}_1$ and $\overline{S}_2$ bits are set just prior to the beginning of the bus cycle. Upon detecting a change from the passive $\overline{S}_0 = \overline{S}_1 = \overline{S}_2 = 1$ state, the 8288 bus controller will output a pulse on its ALE pin and apply the appropriate signal to its $DT/\overline{R}$ pin during T_1. In T_2, the 8288 will set DNE = 1, thus enabling the transceiver, and for an input, will activate either $\overline{MDRC}$ or $\overline{IORC}$. These signals will be maintained until T_4. For an output, the AMWC or AIOWC is activated from T_2 to T_4 and the $\overline{MWTC}$ or $\overline{IOWC}$ is activated from T_3 to T. The status bit S_0, S_1, and S_2 will remain active until T_3 and will become passive (all 1s) during T_3 and T_4. As while the minimum mode, if the READY input is not activated before the beginning of T_3, wait states will be inserted between T_3 and T_4.

Interrupt acknowledgment signals are the same as in the minimum mode case except that a 0 is applied to the $\overline{LOCK}$ pin from T_2 of the first bus cycle to T_2 of the second bus cycle. Bus requests and grants are handled differently, however, and the timing on an $\overline{RQ/GT}$ pin is shown in **Figure 3-31**. A request/grant/release is accomplished by a sequence of three pulses. The $\overline{RQ/GT}$ pins are examined at the rising edge of each clock pulse and if a request is detected (and the necessary conditions discussed previously are met) the processor will apply a grant pulse to the $\overline{RQ/GT}$ immediately following the next T_4 or T_i state. When the requesting master receivers this pulse it seizes control of the bus. This master may control the bus it will send the processor the release pulse over the same line that it made its request. As noted before, $\overline{RQ/}$

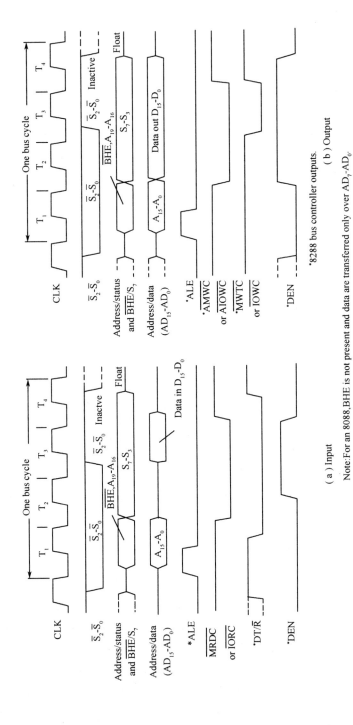

Figure 3-30 Timing diagrams for a maximum mode system

$\overline{GT_0}$ and $\overline{RQ/GT_1}$ are the same except the $\overline{RQ/GT_0}$ has higher priority.

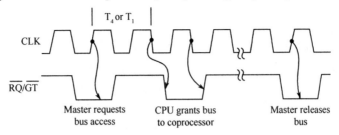

Figure 3-31　Timing for maximum mode bus requests and grants

Tips

20-bit address adder　20 位地址加法器
6-byte instruction queue　6 字节指令队列
abacus　算盘
acknowledgement　应答
albeit　虽然，即使
align　排列
alignment　调整
Altair 8800　1957 年 4 月，MITS 发布第一个通用型 Altair 8800，是世界上第一台微型计算机。
analogously　类似地
append　添加
assembly language　汇编语言
auxiliary carry flag（AF）　辅助进位标志
Babylonians　古巴比伦人
base pointer register　基数指针寄存器
BIOS（basic input/output system）　基本输入输出系统
bus cycle　总线周期
bus grant　总线授予
bus interface unit（BIU）　总线接口部件
bus request　总线请求
cache　高速缓冲存储器
carry flag（CF）　进位标志
CLI（clear interrupt）　清除中断标志
compatibility　兼容性
compatible　谐调的，一致的，兼容的
confront　使面临
contiguous　相邻的
copyright violation　侵犯版权
CS（code segment）register　指令段寄存器

debugger　调试器
detest　憎恶
direction flag（DF）　方向标志
DOS（disc operate system）　磁盘操作系统
DS（data segment）register　数据段寄存器
duty cycle　占空比
effective address　有效地址
ENIAC（Electronic Numerical Integrator And Computer）　电子数字积分计算机，世界上第一台电子计算机，于 1946 年 2 月 15 日在美国诞生。
entity　实体
ES（extra segment）register　附加段寄存器
execution unit（EU）　执行部件
external bus　外部总线
external interrupt　外部中断
facilitate　使容易，促进
feasible　可行的
flags register　标志寄存器
general-purpose register　通用寄存器
handshaking signal　握手信号（同步交换信号）
hardware interrupt　硬件中断
high-impedance state　高阻状态
HLDA　对 HOLD 信号的应答信号
HMOS　高密度金属氧化物半导体工艺（一种单片机封装工艺）
idle state　空闲状态
index register　变址寄存器
ingenious　有独创性的，精制的，具有创造才能
instruction queue　指令队列

integrated 综合的，完整的	predecessor 先驱
internal bus 内部总线	PSW 程序状态标志寄存器
internal interrupt 内部中断	receipt 接收
interrupt flag (IF) 中断标志	relinquish 放弃
interrupt type 中断类型	sacrifice 牺牲
interrupt vector 中断向量	saga 传奇
interrupt vector table 中断向量表	segment 段
interruption 中断	segment address 段地址
INTR (maskable interrupt request) 可屏蔽中断请求	segment register 段寄存器
	sign flag (SF) 符号标志
inverted 反相	single-chip 单片的
IP (instruction pointer) register 指令指针寄存器	software interrupt 软件中断
	sophistication 复杂性
IRET (interrupt return) 中断返回	square brackets 方括号
master cascade enable(MCE) 主控级联允许信号	SS (stack segment) register 堆栈段寄存器
	stack pointer 堆栈指针
maximum mode 最大工作模式	stack pointer register 堆栈指针寄存器
minimum mode 最小工作模式	state-of-the-art 艺术级的
most significant bit (MSB) 最高有效位	STI (set interrupt) 设置中断标志
multiplexing 多路传输	strap 跨接
negative pulse 负脉冲	subroutines 子程序
NMI (non-maskable interrupt) 非屏蔽中断	superscalar 超标量体系结构
numeric coprocessor 浮点运算处理器(数值协同处理器)	symmetric 对称的
	synchronize 同步
obviate 排除	throughput 产量
odd address 奇地址	trailing edge of clock pulse 时钟脉冲下降沿
offset address 偏移地址	transceiver 收发器(总线驱动器)
oscillator 振荡器	trap flag (TF) 跟踪标志
overflow flag (OF) 溢出标志	tristate element 三态门
overhaul 改进,革新	TTL (transistor-transistor logic) 晶体管-晶体管逻辑
overlap 重叠	
parentheses 圆括号	usher 引领
parity flag (PF) 奇偶标志	vaguely 含糊地
partition 分割,划分	vice versa 反之亦然
peripheral data enable (PDEN) 外设数据允许信号	wait state 等待状态
	workhorse 重负荷机器
physical address 物理地址	zero flag (ZF) 零标志
pin 管脚	
portend 预示	

Exercise

1. In 8086 CPU, what are the main functions of EU and BIU?
2. Explain each of the following terms：
   (a) segment

(b) offset

(c) address boundary

3. In 8086 CPU, what are:

(a) the four kinds of segments

(b) their maximum size

(c) the address boundary on which they begin?

4. What is the purpose of each of the four segment registers?

5. What is the purpose of IP register?

6. Explain which registers are used for the following purposes:

(a) addressing segments

(b) offset address of an instruction that is to execute

(c) addition and subtraction

(d) multiplication and division

(e) counting for looping

(f) indication of a zero result

7. Please calculate the physical address designated by following logical address of each segment located by the following segment register values:

(a) 1000H: 0050H (b) 1234H: 0732H

(c) 2300H: 01A5H (d) E000H: 00A8H

(e) AB00H: 025AH

8. Find the memory address of the next instruction executed by the microprocessor when operated in the real mode, for the following CS:IP combinations:

(a) CS = 1000H and IP = 2000H

(b) CS = 2000H and IP = 1000H

(c) CS = 2300H and IP = 1A00H

(d) CS = 1A00H and IP = B000H

(e) CS = 3456H and IP = ABCDH

9. Which register or registers are used as an offset address for string instruction destination in the microprocessor?

10. The stack memory is addressed by a combination of the _____ segment plus _____ offset.

11. If the base pointer (BP) addresses memory, the _____ segment contains the data.

12. Determine the memory location addressed by the following register combinations:

(a) DS = 1000H and DI = 2000H (b) DS = 2000H and SI = 1002H

(c) SS = 2300H and BP = 3200H (d) DS = A000H and BX = 1000H

13. Assume that the instruction:

XCHG BL, DATA

is already in the queue and is ready to execute, and give a time diagram of bus activity during its execution for each of the following cases:

(a) Minimum mode 8086 with no wait states.

(b) Minimum mode 8088 with no wait states.

(c) Maximum mode 8086 with no wait states.

(d) Maximum mode 8086 if a ready signal two clock periods long is received in the middle of T_3. (Also include the timing diagram of the ready signal.)

(e) Maximum mode 8086 with no wait states, but assumes that a bus request is detected during the input transfer and is dropped after only one bus cycle.

14. Consider a maximum mode 8086 system that is executing the instructions

    MOV AX, DATA; (DATA has an even address)

    CMP AX, 1

Suppose that an interrupt request arrives while first instruction is executing and draw a timing diagram of the bus activity from the beginning of the first instruction until the interrupt type is received by the processor.

15. Assuming that the instruction is already in the queue, give the number of the bus cycles needed to execute each of the following instructions:

(a) PUSH AX

(b) CALL NEAR PTR PROC_A

(c) CALL FAR PTR PROC_B

(d) MOV DATA, AX; (DATA has an even address)

(e) MOV DATA, AX; (DATA has an odd address)

16. What is the minimum number of bus cycles that can occur between the time an interrupt request is recognized and the first instruction in the interrupt routine is fetched?

17. What is the bus cycle? When does the 8086 CPU initiate a bus cycle?

18. Identify which of the following instructions will initiate a bus cycle?

(a) MOV AX, BP

(b) MOV AX, 100

(c) MOV AX, [BX]

(d) MOV AX, [100]

19. Give the reason why the T_w is needed to insert between T_3 and T_4 in a bus cycle.

20. Explain each of the following terms:

(a) Interrupt

(b) Interrupt vector

(c) ISR

(d) Interrupt vector table

(e) Interrupt vector number

21. Please describe the difference between the INTR and NMI.

22. Please describe the way how to obtain the interrupt vector number in 8086/8088 system.

Chapter 4 8086 Address Mode and Assembly Instructions

An assembly language program is stored as text (just as a higher level language program). Each assembly instruction represents exactly one machine instruction.

An assembler is a program that reads a text file with assembly instructions and converts the assembly into machine code. Compilers are programs that do similar conversions for high-level programming languages. An assembler is much simpler than a compiler. Every assembly language statement directly represents a single machine instruction. High-level language statements are much more complex and may require many machine instructions. Another important difference between assembly and high-level languages is that since every different type of CPU has its own machine language, it also has its own assembly language.

4.1 8086 Assembly Instruction Format

The general format of an assembler instruction is

    Label: **Mnemonic Operand1**, **Operand2** ; **Comments**

Where the inclusion of spaces is arbitrary, except that at least one space must be inserted if no space would lead to an ambiguity. Also, there can be no spaces within a mnemonic or identifier and spaces within string constants or comments will be included as space characters.

Label: A label is an identifier that is assigned the address of the first byte of the instruction in which it appears. The presence of a label in an instruction is optional, but, if present, the label provides a symbolic name that can be used in branch instructions to branch to the instruction. If there is no label, then the colon must be deleted.

When you define a Label, you should pay more attentions to following items:

(1) It consists of a string of up to 31 characters, which must be **letters** (a-z, A-Z), **digits** (0-9), **question marks** (?), **underscores** (__), or "@" **signs** except that the first character cannot be a digit.

(2) It cannot be keywords used by the assembler, such as instruction mnemonics, register names, and special operators.

(3) We can't use a single question mark as identifier.

Mnemonic: All instructions must contain a mnemonic.

Operand(s): The presence of the operands depends on the instruction. Some instructions have no operands, some have one, and some have two. If there are two operands, they are separated by a comma, and the **destination operand** appears first and **the source operand** second. When an operand is a word in memory, the low-order byte of the word will have the lower address and the high-order byte the higher address, e. g. , if the identifier COST is used as a word operand, then COST is associated with the low-order byte and COST + 1 with the high-order byte.

Comments: The comments field is for commenting the program and may contain any combination of characters. It is optional and if it is deleted the semicolon may also be deleted. A comment may appear on a line by itself provided that the first character on the line is a semicolon. An instruction may be continued to other lines by placing an ampersand (&) at the beginning of each of the continuation lines.

For example, a data transfer Instruction can be written as following:

$$\text{MOV} \quad \text{AX}, \quad \text{BX}$$

The word MOV is a mnemonic for data transfer. It is an opcode, or operation code, tells the microprocessor which operation to perform. AX and BX are operands. The right operand AX, is the destination operand appeared first and the left BX, is the source operand.

This instruction shows us the direction of data how it transfers the word contents of BX into the destination register AX.

It should be emphasized that a segment register designation can appear in an operand of only certain instructions. A segment register can be explicitly specified in only the MOV, PUSH, and POP instructions. However, the branch instructions may implicitly reference the segment registers. Except for string operations, for a double-operand instruction, one of the operands must be a register unless the source operand is immediate. In no case can IP be specified as an operand because only a branch or other control transfer instruction can change the contents of this register.

4.2 8086 Addressing Modes

An assembler language instruction must have an operand for each machine language instruction operand and the notation for each operand must be sufficient to indicate the operand's addressing mode.

The way in which an operand is specified is called its addressing mode. About **Instruction operands**, different machine code instructions have different number operands; however, in general, each instruction itself will have a fixed number of operands (0 to 2). Operands can be found in following items:

Register: These operands refer directly to the contents of the CPU's registers.

Memory: These refer to data in memory. The address of the data may be a constant hard-coded into the instruction or may be computed using values of registers. Address are always offsets from the beginning of a segment.

Immediate: These are fixed values that are listed in the instruction itself. They are stored in the instruction itself (in the code segment), not in the data segment.

Implied: These operands are not explicitly shown. For example, the increment instruction adds one to a register or memory. The one is implied.

The 80x86 processors let you access memory in many different ways. They provide flexible access to memory, allowing you to easily access variables, arrays, records, pointers, and other complex data types. Learning the 80x86 addressing modes is the first step to study 80x86 assembly language. There are a total of 7 different addressing modes on the 8086, for example, immediate addressing, direct addressing, register addressing, register indirect addressing, register relative addressing, base-plus-index addressing and base relative-plus-index addressing, just as shown in the following figures.

4.2.1 Immediate Addressing

The most common addressing mode, and the one that's easiest to understand, is the immediate addressing mode. As shown in **Figure 4-1**, the operand is either 8 bits or 16 bits long and is part of the instruction.

Figure 4-1   Immediate addressing

[Example 4-1] Examples of immediate addressing mode.

MOV AL, 5   ; *Transfers the source immediate byte 5 into the destination register AL*

MOV AX, 30AFH   ; *Move the hexadecimal immediate number 30AF to AX (accumulator register)*

4.2.2 Direct Addressing

The direct addressing mode (or displacement-only addressing mode) consists of a 16-bit constant that specifies the address of the target location. As indicated in **Figure 4-2**, the 16-bit effective address of the operand is part of the instruction.

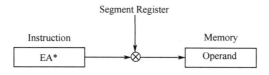

Figure 4-2   Direct addressing

[Example 4-2] Given that (DS) = 3000H, (AX) = 30AFH, for the instruction MOV AX, [2000H], please determine the effective address and it's addressing mode.

In this instruction, the effective address 2000H of the operand is given directly, so it is a direct addressing mode. The source operand of the instruction MOV AX, [2000H] is stored in 2 consecutive memory locations such as 32000H and 32001H. The instruction will copy a word from these 2 memory locations to register AX. As shown in **Figure 4-3**, the content of AX register will be set as 3050H after executing.

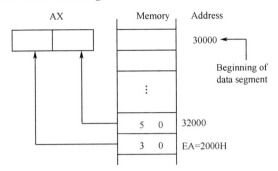

Figure 4-3 Example of direct addressing

Comparing with the instruction MOV AL, [2000H], the assembler will assume that MOV AX, [2000H] is a word transfer because of the destination operand AX is 16 bits long.

By default, all direct values provide offsets into the data segment. If you want to provide an offset into a different segment, you must use a segment override prefix before your address. **For example**, to access location 1234H in the extra segment (ES) you would use an instruction of the form MOV AX, ES: [1234H]. Likewise, to access this location in the code segment you would use the instruction MOV AX, CS: [1234H].

4.2.3 Register Addressing

As shown in **Figure 4-4**, register addressing mode is the mode that specifies the name of the register as an operand to the instruction. You may access the contents of that register. For a 16-bit operand, a register may be AX, BX, CX, DX, SI, DI, SP, or BP, and for an 8-bit operand a register may be AL, AH, BL, BH, CL, CH, DL, or DH.

[Example 4-3] Given that (AX) = 30AFH, (BX) = 123DH, for the instruction MOV AX, BX, please determine the effective address and it's addressing mode.

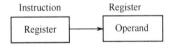

Figure 4-4 Register addressing

For the instruction MOV AX, BX, the operand is given by BX register. So it is register addressing mode. There is no effective address for the datum in specified register BX. It transfers a copy of a word from the source register BX to the destination register AX. That is after execution, we get (AX) = (BX) = 123DH.

4.2.4 Register Indirect Addressing

Figure 4-5 is a diagram of register indirect addressing mode. The effective address of the operand is in a base register or an index register that is specified by the instruction. That is to say, the operand is in a memory location addressed by an index or base register. There are totally four forms of the addressing modes on the 8086 CPU. These four addressing modes reference the byte at the offset found in the BX, BP, SI or DI register respectively. The BX, SI and DI use the data segment (DS) address as reference by default, but BP refers as the stack segment (SS) reference.

The base addressing mode and indexed addressing mode are functionally equivalent.

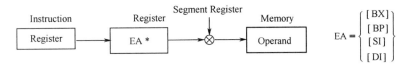

Figure 4-5 Register indirect addressing

【Example 4-4】 Given that (AX) = 30AFH, (BX) = 1000H, (DS) = 2000H, for the instruction MOV AX, [BX], please determine the effective address and it's addressing mode.

For the instruction MOV AX, [BX], the destination operand AX is 16 bit long, so the source operand is stored in 2 consecutive memory locations such as 21000H and 21001H. It is an indirect addressing mode. The instruction will copy a word from these 2 memory locations to register AX. As shown in **Figure 4-6**, the effective address of the source operand is 1000H, which will combine with DS to form the 20-bit physical address. **Example 4-4** transfers a word 50A0H located in memory unit 21000H and 21001H to AX register.

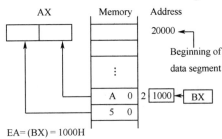

Figure 4-6 Example of register indirect addressing

Sometime, you can use the segment override prefix symbols if you want to access data in different segments. **For example**, as shown in the following instructions:

MOV AL, DS:[BX]   ; *transfer a byte to AL from DS:[BX]*
MOV DX, ES:[SI]   ; *transfer a word to DX from ES:[SI]*

The coding of "ES:" indicates an override operator that means "Replace the normal use of the DS segment register with that of the ES". The SI contains an offset address within ES segment. In this way, you can process the data in the other segment.

How many bytes the instruction will fetch from memory is determined by the width of the destination register, so the first instruction fetches one byte while the second instruction fetches two.

4.2.5 Register Relative Addressing

As shown in **Figure 4-7**, in this addressing mode, the operand is in the memory location addressed by an index or base register plus a displacement.

The effective address is the sum of an 8 or 16 bits displacement and the contents of a base register or an index register. The indexed addressing modes use the following syntax:

MOV AX/AL, DISP [BX]
MOV AX/AL, DISP [BP]
MOV AX/AL, DISP [SI]
MOV AX/AL, DISP [DI]

The offsets generated by these addressing modes are the sum of the constant and the specified register. The addressing modes involving BX, SI and DI are all use the data segment, while the BP addressing modes referring to stack segment.

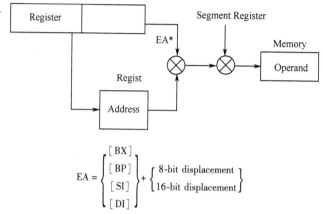

Figure 4-7 Register relative addressing

【Example 4-5】 Given that (DS) = 3000H, (SI) = 2000H, (AX) = 30AFH, COUNT = 3000H, for the instruction MOV AX, COUNT[SI], please determine the effective address, physical address and it's addressing mode.

The effective address is consisted of two parts, one is from register SI and another is from a constant COUNT. So, for the instruction MOV AX, COUNT[SI], there is :

EA = (SI) + COUNT = 2000H + 3000H = 5000H

Therefore the physical address is 35000H. The source operand is stored in 2 consecutive memory locations beginning with 35000H. It is a relative indirect addressing mode. As shown in **Figure 4-8**, the instruction will copy a word from these 2 memory locations to register AX. That is (AX) = 1234H.

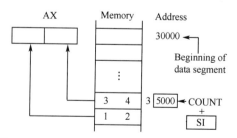

Figure 4-8 Example of register relative addressing

4.2.6 Base-plus-Index Addressing

The based plus index addressing modes are simply combinations of the register indirect addressing modes. The operand is in the memory location addressed by a base and an index register.

Figure 4-9 shows the formation of the effective address. The effective address is the sum of a base register and an index register, both of which are specified by the instruction.

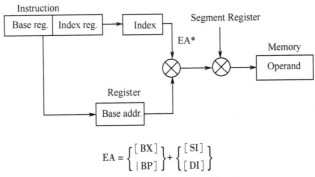

$$EA = \left\{ \begin{matrix} [BX] \\ [BP] \end{matrix} \right\} + \left\{ \begin{matrix} [SI] \\ [DI] \end{matrix} \right\}$$

Figure 4-9 Base-plus-Index addressing

Note that the addressing modes that involve BX, SI or DI use the data segment by default, except those that have BP as an operand use the stack segment by default.

[Example 4-6] If (SS) = 3000H, (DI) = 2000H, (BP) = 1000H, and (AX) = 1234H, for the instruction MOV AX, [BP][DI], please determine the effective address and it's addressing mode

As shown in **Figure 4-10**, the effective address is sum of value of register DI and BP. So, for this instruction, there is:

EA = (DI) + BP = 2000H + 1000H = 3000H

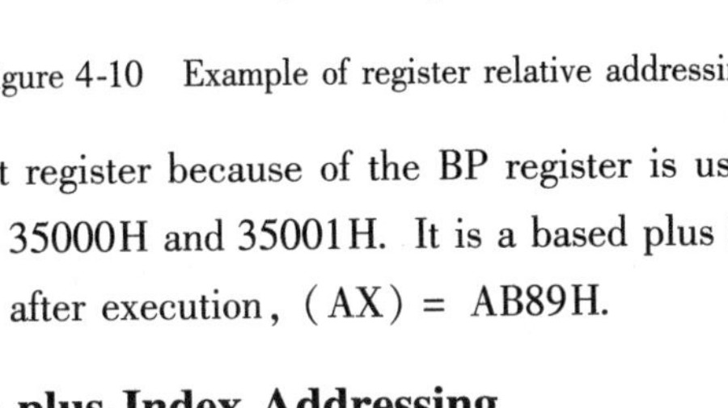

Figure 4-10 Example of register relative addressing

The SS is the segment register because of the BP register is used. The source operand is stored in memory locations 35000H and 35001H. It is a based plus index addressing mode. As shown in the **Figure 4-6**, after execution, (AX) = AB89H.

4.2.7 Base Relative-plus-Index Addressing

These addressing modes are a slight modification of the base plus index addressing modes with the addition of a 8-bit or 16-bit constant. As shown in **Figure 4-11**, the effective address is the sum of an 8-or 16-bit displacement and a based indexed address. For example, for the instruction.

MOV AX, DISP[BX][SI]

there are totally three terms in the source field: DISP, [BX] and [SI].

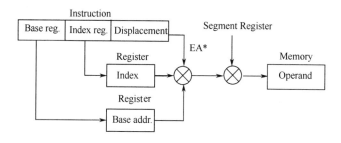

Figure 4-11 Base relative-plus-index addressing

【Example 4-7】 Suppose (AX) = AB89H, (BX) = 2000H, (SI) = 1000H, (DS) = 3000H, and DISP = 0250H, please indicate the addressing mode of the instruction MOV AX, DISP [BX][SI].

For MOV AX, DISP [BX][SI], it is equivalent to the following two instruction formats:

MOV AX, DISP[BX + SI]

or **MOV AX, [DISP + BX + SI]**

As shown in **Figure 4-12**, the effective address is sum of value of register SI, BX and constant DISP. So, for this instruction, there is :

EA = (BX) + (SI) + (DISP) = 3250H

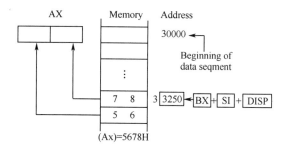

Figure 4-12 Example of base relative-plus-index addressing

Combination with the DS, the contents of the memory locations 33250H and 332501H are fetched, then (AX) = 5678H.

In base plus index addressing mode, you may combine operands in any sequence, but don't combine two base registers [BX + BP] or two index registers [DI + SI]. Only the index or base registers must be in square brackets so that the assembler knows to treat it as an index entry.

【Example 4-8】 Suppose (BX) = 0158H, (DI) = 10A5H, Displacement = 1B57H, (DS) = 2100H, please calculate the effective and physical address of the operand according to the addressing mode indicated in the left column.

Initial Status	(BX) = 0158H　(DI) = 10A5H　Displacement = 1B57H　(DS) = 2100H and DS is used as the segment register
Addressing Mode	**Effective and Physical Address of the Operand**
Direct	$EA = 1B57H$ $Physical\ address = 1B57H + 21000H = 22B57H$
Register	*No effective address - datum is in specified register*
Register indirect assuming register BX	$EA = 0158H$ $Physical\ address = 0158H + 21000H = 21158H$
Register relative assuming register BX	$EA = 0158H + 1B57H = 1CAFH$ $Physical\ address = 1CAFH + 21000H = 22CAFH$
Based indexed assuming registers BX and DI	$EA = 0158H + 10A5H = 11FDH$ $Physical\ address = 11FDH + 21000H = 221FDH$
Relative Based indexed assuming BX and DI	$EA = 0158H + 10A5H + 1B57H = 2D54H$ $Physical\ address = 2D54H + 21000H = 23D54H$

要点：

(1) 操作数。在微型计算机中，操作数可能在哪？不外乎三种可能：

① 操作数包含在指令中，即指令的操作数场包含着操作数本身。

② 操作数包含在 CPU 的某一寄存器中。

③ 操作数包含在内存的数据区中，且指令中的操作数场包含着此操作数的地址。

④ 操作数隐含。

(2) 寻址方式。　寻找操作数的方法称为寻址方式。通常可分为以下几类：

① 立即数寻址。8086CPU 指令系统中，有一部分指令所用的 8 位或 16 位操作数就在指令中提供，这种方式叫做立即数寻址方式。

② 寄存器寻址。如果操作数就在 CPU 的内部寄存器中，那么，寄存器名可在指令中指出，这种寻址方式就叫做寄存器寻址方式。

③ 直接寻址。数据存储单元的有效地址由指令直接指出的寻址方式。采用直接寻址方式时，如果指令前面没有用前缀指明操作数在哪一段，默认为段寄存器是数据段寄存器 DS。如果要对其他段寄存器所指出的存储区进行直接寻址，则本条指令前必须用前缀指出段寄存器名。

④ 寄存器间接寻址。采用寄存器间接寻址方式时，操作数一定在存储器中，存储单元的有效地址由寄存器指出，这些寄存器可以为 BX、BP、SI 和 DI 之一。即有效地址等于其中某一个寄存器的值：

$$EA = \begin{Bmatrix} [BX] \\ [BP] \\ [SI] \\ [DI] \end{Bmatrix}$$

和直接寻址的情况一样，如果指令前面没有用前缀指明操作数在哪一段，默认为段寄存器是数据段寄存器 DS。如寄存器为 BP 时，则对应的段寄存器为 SS。

采用寄存器间接寻址时,允许在指令中指定一个位移量,这样,有效地址通过将一个寄存器的内容加上一个位移量来得到。位移量可以是 8 位,也可以为 16 位,即

$$EA = \begin{Bmatrix} [BX] \\ [BP] \\ [SI] \\ [DI] \end{Bmatrix} + \begin{Bmatrix} 8\text{ 位位移量} \\ 16\text{ 位位移量} \end{Bmatrix}$$

细分起来,寄存器间接寻址可根据进行间接寻址的寄存器不同分为数据段基址寻址(以 BX 寄存器进行间接寻址)、堆栈段基址寻址(BP 寄存器)、变址寻址(以 SI 和 DI 寄存器)和基址加变址寻址(将 BX、BP 和 SI、DI 组合起来)等几种方式。

4.3 Data Movement Instructions

There are some basic 8086 instructions for transferring quantities to and/or from the registers and memory. They are the MOV, XCHG, XLAT, PUSH, POP, LEA, LDS, and LES instructions which are listed in **Table 4-1**.

Table 4-1 Data transfer instructions

Instruction form	Description	\multicolumn{6}{c}{Flag}	Remarks

Instruction form	Description	O	S	Z	A	P	C	Remarks
MOV DST, SRC	(DST)←(SRC)	–	–	–	–	–	–	SRC: register, memory, immediate; DST: register, memory
PUSH SRC	(Stack)←(SRC)	–	–	–	–	–	–	SRC: register, memory
POP DST	(DST)←(Stack)	–	–	–	–	–	–	DST: register(except CS), memory
XCHG DST, SRC	(OPR1)⟷(OPR2)	0	0	0	0	0	0	SRC: general register, memory; DST: general register, memory
XLAT	(AL)←Memory Table	–	–	–	–	–	–	
LEA DST, SRC	(REG)←SRC	–	–	–	–	–	–	SRC: memory operand; DST: 16-bit general register
LDS DST SRC	(REG)←(SRC) (DS)←(SRC+2)	–	–	–	–	–	–	SRC: memory operand; DST: 16-bit general register
LES DST, SRC	(REG)←(SRC) (ES)←(SRC+2)	–	–	–	–	–	–	SRC: memory operand; DST: 16-bit general register
LAHF	(AH)←(Low-order byte of PSW)	–	–	–	–	–	–	
SAHF	(Low-order byte of PSW)←(AH)	–	0	0	0	0	0	
PUSHF	(Stack)←(SRC)	–	–	–	–	–	–	
POPF	(DST)←(Stack)	0	0	0	0	0	0	
IN ACC port	(ACC)←(Port)	–	–	–	–	–	–	ACC: AL or AX (16-bit external bus use AX)
OUT port, ACC	(Port)←(ACC)	–	–	–	–	–	–	Port: Address 0-255 or indirect address register DX

4.3.1 MOV Instruction

MOV is the command that copy (move) the content of one memory to another memory location. Generally speaking, assembly language instructions manipulate data stored in memory and registers.

Consider the 8086 **MOV** instruction format

MOV < destination > , < source >

The instruction MOV copies the data from the source operand to the destination operand. The basic restriction of the MOV instruction is that both operands must be the same size.

The source operand can be a memory location or the register or an immediate, while the destination operand can be a memory location or the register. However, the best choice of the source or destination is using the registers, because it is shorter and faster than accessing the memory location. Generally speaking, the general purpose registers are usually one of the instruction operands in MOV instruction.

[**Example 4-9**] The examples of MOV instruction.

MOV AX, 0100H	;*The hexadecimal number 0100 immediately move to AX register*
MOV AX, BX	;*The content of the BX register is moved to the AX register*
MOV AL, DS:[8088H]	;*Loads the byte at memory location 8088H (in data segment) to the **AL** register*
MOV DS:[1234H], DL	;*Stores the value from the DL register to the memory location 1234H*
MOV AL, CS:[BX]	;*Refers the base address BX relative to the code segment*
MOV CL, 20H[BX]	;*If BX contains 1000H, the instruction will load CL from memory location DS:1020H*
MOV DH, 1000H[BP]	;*If BP contains 2020H, the instruction will load DH from location SS:3020H*
MOV AL, [BX][SI]	;*Suppose that BX contains 1000H and SI contains 880H, the instruction would load AL from the location DS:1880H*
MOV AX, [BP + DI]	;*If BP contains 1598H and DI contains 1004H, the instruction will load the 16-bit in AX from location SS:259CH and SS:259DH*
MOV AL, FFFFH[BX]	;*If BX contains 10H, the instruction, the instruction will load the AL register from location DS: 0FH, but not DS: 1000FH*

As shown in **Example 4-9**, we can notice that:

(1) At least one of the operands is always general purpose register.

(2) It is invalid in specifying CS as the destination operand.

(3) It is invalid that both of the operands are segment registers.

(4) The both operands must be the same size.

(5) No memory forms can both exist in destination and source operand.

(6) If the effective address calculation produces a value greater than FFFFH, the CPU ignores the overflow and the result wraps around back to zero.

(7) In register indirect addressing modes, the BX, SI and DI use the data segment (DS) address as default, but BP refers as the stack segment (SS) reference. Of cause, you can change the default segment register by adding the segment override prefix.

4.3.2 PUSH and POP

When a program is loaded into the memory, a stack is created. Stack segment register (**SS**) initialized the stack segment. Stack pointer (**SP**) will point to the top of the stack. The base register (**BP**) is used to access the elements inside the stack without popping out the top elements. When using **SP** and **BP**, the referencing pointer is **SS**.

There are two basic instructions for stack manipulation. They are PUSH and POP, and the syntax are:

        PUSH   { <register> | <memory> }
        POP   { <register> | <memory> }

The 8086-80286 PUSH instruction always transfer two bytes of data to the stack. The pop performs the inverse operation.

PUSH will decrease the stack pointer (**SP**) by 2 before placing the data in the stack. **POP** is an operation which copies the top element to the register before increase the stack pointer by 2. **PUSH** and **POP** are a word size (16 bits) operations.

Notice three items about the manipulation of the stack. First, it is always in the stack segment. Second, the stack grows down in memory, i.e. as you push the values onto the stack the CPU stores them into successively lower (smaller) memory locations. Finally, as shown in **Figure 4-13**, the SP (stack pointer) always contains the address of the value on the top of the stack.

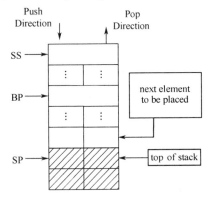

Figure 4-13 The structure of a stack

【Example 4-10】 Suppose (AX) = 1234H, (SP) = 1200H. After executing PUSH AX, the stack would be like:

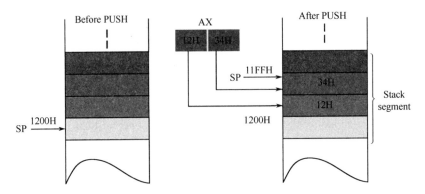

After executing the instruction POP AX, the stack would be like:

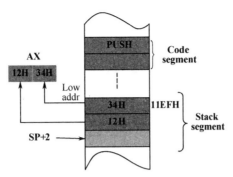

All pushes and pops are 16-bit operations. There is no way to push a single 8-bit value onto the stack. To push an 8-bit value you should load it into 16-bit register first, with the higher order byte being 0.

4.3.3 XCHG Instruction

The XCHG (exchange) instruction swaps two values. The general form is

    XCHG   < operand1 > , < operand2 >

There are two specific forms of this instruction on the 8086 machine:

    XCHG   < register > , < memory >
    XCHG   < register > , < register >

Since the 8086 often provides shorter and faster versions of instructions that use the **AX** register, you should try to arrange your computations so that the CPU can use the **AX** register as much as possible. However, both the operands must be the same size, and the **XCHG** instruction does not modify any status flags in the status register.

[Example 4-11] The examples of XCHG instruction.

XCHG   AL, BL   ;*will exchange the content of the register AL and BL*
XCHG BX,AX   ;*will exchange the two words of the BX and AX registers*
XCHG [2530H], CX   ;*will exchange the content of CX and the word at*
              ;*address 2530H and 2531H in memory*

Suppose (DS) = 2000H, (22530H) = 88H, (22532H) = 66H, for the third instruction in **Example 4-11**, after execution, (CX) = 6688H

4.3.4 XLAT Instruction

The XLAT (translate) instruction converts the contents of the AL register into a number stored in a memory table. This instruction performs the table lookup operation to convert one code to another. **Figure 4-14** illustrates the executing processing of the XLAT instruction. It adds the contents of AL to BX to form a memory address within the data segment. This is the only instruction that adds an 8-bit number to a 16-bit number.

The XLAT instruction adds and uses the resulting offset to point to an entry in an 8-bit translate table. This contains value referenced by BX + AL is fetched to substitute the original value in AL. The syntax is

XLAT

The flags are not affected in **XLAT** instruction operation. The table can have a maximum of 256 entries, the same range of values possible in the 8-bit AL register.

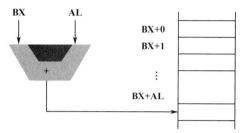

Figure 4-14 Function of the XLAT instruction

For example, store the characters representing all 16 hexadecimal digits in a table:

table DB   '0123456789ABCDEF'

The table contains the ASCII code of each hexadecimal digit. If we place 0AH in AL with the thought of converting it to ASCII, we can set BX to the table offset and invoke XLAT. The instruction adds BX and AL, generating an effective address that points to the eleventh entry in the table. It then moves the contents of this table entry to AL:

```
MOV   AL, 0AH            ;index value
MOV   BX, OFFSET table   ; point to table
XLAT                     ; (AL) = 41H, or 'A'
```

Suppose that a 7-segment LED display lookup table as shown in **Table 4-2** is stored in memory at address TABLE. The XLAT instruction then translates the BCD number in AL to a 7-segment code. The following example provides a short program that converts from a BCD code to a 7-segment code.

Table 4-2 7-Segment code lookup table

Decimal Number	Seven Segment Code						
	g	f	e	d	c	b	a
0	0	1	0	0	0	0	0
1	0	1	1	1	1	0	0	1
2	0	0	1	0	0	1	0	0
3	0	0	1	1	0	0	0	0
4	0	0	0	1	1	0	0	1
5	0	0	0	1	0	0	1	0
6	0	0	0	0	0	0	1	0
7	0	1	1	1	1	0	0	0
8	0	0	0	0	0	0	0	0
9	0	0	0	1	0	0	0	0

【**Example 4-12**】 Using an XLAT to convert from BCD to 7-segment code.

```
        DATA    SEGMENT                      ; define a DATA segment
           TABLE DB      40H, 79H, 24H       ; define 7-segment lookup table
              DB         30H ,19H, 12H, 02H
              DB         78H, 00H, 10H
           RESULT DB     ?                   ; reserve for result
        DATA    ENDS
        CODE    SEGMENT                      ; define a CODE segment
                ASSUME   DS:DATA,  CS:CODE   ; tell the assembler to associate segment names with
                                               CS, and DS segment registers
           START: MOV   AX,  DATA            ; start of program
                  MOV   DS,  AX              ; initialize DS register
                  MOV   AL, 2                ; load test data 2
                  MOV   BX,  OFFSET TABLE    ; address lookup table
                  XLAT                       ; convert data 2 to 7-segment, (AL) = 24H
                  MOV   RESULT,  AL          ; save 7-segment code 24H to RESULT
                  MOV AH,4CH
                  INT 21H                    ; end processing and exit to DOS
        CODE   ENDS
        END    START
```

要点:

(1) XLAT 是一条完成字节翻译的指令,称为换码指令。它的功能是使累加器中的一个值变换为内存表格中的某一个值,一般用来实现编码制的转换。

(2) 使用换码指令时,要求 BX 寄存器指向表的首地址, AL 中为表中某一项与表格首地址之间的偏移量,指令执行时,会将 BX 和 AL 中的值相加,把得到的值作为地址,然后将此地址所对应的单元中的值取到 AL 中去。

(3) XLAT 指令就是通过查表方式来完成翻译功能的,因此,在执行该指令之前,必须建立好一张翻译表,该表的最大容量为 256 字节。

4.3.5　LEA Instruction

LEA (load effective address) instruction is used to load effective address. It takes the form:

LEA　　　　< destination >, < source >

which takes the form

LEA　　　　< register >, < memory >

It initializes the specified general purpose register with an effective address (offset address).

【Example 4-13】 LEA instruction applications

LEA　　BX,　[BP + SI]　　　; After executing, the content of BX will be the value of BP + SI
LEA　　SP,　[**0428H**]　　　; Set SP to 0482H
LEA　　AX, 3[**BX**]　　　; loads the value of BX plus 3 into the AX register

An equivalent operation to LEA is MOV with the OFFSET operator, which generates slightly shorter machine code and is used like this:

MOV BX, OFFSET TABLE ; *load offset address*

For the string case, it will load the address of the first character (ASCII form) in the register (DX) to enable the string operation. Furthermore, it will display characters until it finds the " $ " sign.

4.3.6 LDS and LES

The LDS and LES instructions are the same except that the former loads the DS register from memory and the latter loads ES from memory. Both instructions also load a second non-segment register from memory and neither instruction affects the flags. Typical LDS and LES instructions are:

LDS   SI, STRINGA
LES   DI, TABLE [BX]

where STRINGA and TABLE are double-word variables.

【Example 4-14】

LDS   SI, [2100H]   ;*DS: 2100H&2101H - >SI*
                        ;*DS: 2102H&2103H - >DS*
LES   DI, [2100H]   ;*DS: 2100H&2101H - >DI*
                        ;*DS: 2102H&2103H - >ES*

The following is an example using the instructions we've learnt above, some of which may have no practical significance. Here we just show you how to use them.

【Example 4-15】 After executing the following instructions, (AX) = ?

LDS   DI, [1200H]
MOV   AX, [DI]

After the first instruction executing, the (DS) = 6000H and the register (DI) = 1234H. As shown in **Figure 4-15**, the DS register is reloaded. When the second instruction executing, the physical address of the accessed memory locations are 61234H and 61235H. So (AX) = 2233H.

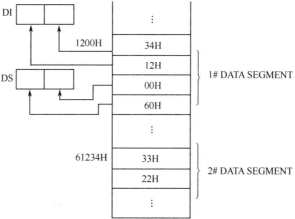

Figure 4-15   Data segment and data assignments

[Example 4-16] Suppose the square values look up table of 0-15 are stored in 16 consecutive memory locations, please develop a program to get the square value of number 4.

```
DATA1 SEGMENT                          ; define a data segment
    TABLE DB    0,1,4,9,16,25,36,49,64,81,100,
                121,144,169,196,225
                                       ; a look-up table definition
    DATA DB 4                          ; give the test data 4
    RESULT DB   ?                      ; reserve for result
DATA1 ENDS
CODE SEGMENT                           ; define a code segment
    ASSUME DS: DATA1, CS: CODE         ; associate the segment
                                         names with DS and CS
SR: MOV AX, DATA1                      ; start of program
    MOV DS, AX                         ; initialize DS register
    LEA BX, TABLE                      ; address lookup table
    MOV AL, DATA                       ; load test data 4
    XLAT                               ; table look-up 16→(AL)
    MOV RESULT, AL                     ; (AL) = 16→RESULT
    MOV AH, 4CH
    INT 21H                            ; end of processing and
                                         exit to DOS
    CODE ENDS
    END SR                             ; end of program
```

4.3.7 Flags Register Movement Instruction

1. LAHF and SAHF

The LAHF and SAHF instructions are seldom used because they are designed as bridge instructions. These instructions allowed 8085 software to be translated into 8086 software by a translation program.

LAHF loads the lower eight bits of the flags register into the AH register. On the contrary, for SAHF, it transfers the AH register into the rightmost eight bits of the flag register. **Figure 4-16** is the diagram of LAHF. Notice that SAHF only affects the lower eight bits of the flags register. These instructions do not deal with the overflow, direction, interrupt disable, or trace flags. The fact that these instructions do not deal with the overflow flag is an important limitation.

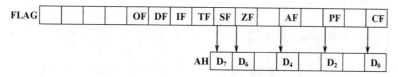

Figure 4-16 Operation of LAHF

2. POPF and PUSHF

The PUSHF and POPF instructions allow you to push/pop the processor status register (the flags). Note that these two instructions provide a mechanism to modify the 80x86's trace flag. Of course, you can set and clear the other flags in this fashion as well. However, most of the other flags you'll want to modify (specially, the condition codes) provide special instructions or other simple sequences for this purpose.

4.3.8 IN and OUT

IN and OUT perform the data transfer between accumulator register and I/O ports. An IN instruction transfers data from an external I/O device to AL or AX. An OUT instruction is inverse operation.

[Example 4-17] Examples of IN and OUT instructions.

```
IN   AX, 50H      ; 16-bit data are input to AX from ports 50H and 51H
IN   AX, DX       ; 16-bit data are input to AX from I/O port DX
OUT  44H, AL      ; 8-bit data are output from AL to port 44H
MOV  DX, 0345H
MOV  AL, 45H
OUT  DX, AL       ; 8-bit data 45H is output to I/O port 0345H
```

As shown in **Example 4-17**, two forms of I/O device (port) addressing exist for IN and OUT: fixed-port and variable-port. Fixed-port addressing allows data transfer between AL or AX using an 8-bit I/O port address. It is called fixed-port addressing because the port number follows the instruction's opcode. Variable-port addressing allows data transfers between AL, or AX and a 16-bit port address. It is called variable-port addressing because the I/O port number is stored in register DX, which can be changed (varied) during the execution of a program.

4.4 Arithmetic Instructions and Logic Instructions

The summary of the arithmetic operations that are directly implemented by 8086 instructions is shown as **Table 4-3**.

Table 4-3 Summary of the arithmetic operations on 8086

Function	Instruction form	Flag						Remarks
		O	S	Z	A	P	C	
Summation	ADD DST, SRC	0	0	0	0	0	0	SRC : register, memory, immediate; DST : register, memory
	ADC DST, SRC	0	0	0	0	0	0	SRC : register, memory, immediate; DST : register, memory
	INC DST	0	0	0	0	0	—	DST : general register, memory

(续)

Function	Instruction form	Flag						Remarks
		O	S	Z	A	P	C	
Subtract	SUB *DST* , *SRC*	0	0	0	0	0	0	*SRC*: register, memory, immediate; *DST* : register, memory
	SBB *DST* , *SRC*	0	0	0	0	0	0	*SRC* : register, memory, immediate; *DST*: register, memory
	DEC *DST*	0	0	0	0	0	—	*DST* : general register, memory
	NEC *DST*	0	0	0	0	0	1	*DST* : general register, memory
	CMP *DST* , *SRC*	0	0	0	0	0	0	*SRC* : register, memory, immediate; *DST* : register, memory
Multiplication	MUL *SRC*	*	*	*	*	*	0	*SRC* : general register, memory
	IMUL *SRC*	*	*	*	*	*	0	*SRC* : general register, memory
Division	DIV *SRC*	*	*	*	*	*	*	*SRC* : general register, memory
	IDIV *SRC*	*	*	*	*	*	*	*SRC* : general register, memory
	CBW	—	—	—	—	—	—	
	CWD	—	—	—	—	—	—	
Decimal System Adjust	AAA	*	*	*	0	*	1	
	DAA	*	0	0	0	0	0	
	AAS	*	*	*	0	*	0	
	DAS	0	0	0	0	0	0	
	AAM	*	0	0	*	0	*	
	AAD	*	0	0	—	0	*	

4.4.1 ADD and SUB Function

To begin with, we define ADD and SUB first.

ADD {destination} , {source}   ; *destination = destination + source*
SUB {destination} , {source}   ; *destination = destination - source*

which takes the form:

ADD { <register> | <memory> } , { <register> | <memory> | <immediate> }
SUB { <register> | <memory> } , { <register> | <memory> | <immediate> }

Note that no memory forms can both exist in destination and source operand. If you want to add both memory elements together, you must load one of the memories to the register, and perform the addition.

The **ADD** instruction adds the contents of the source operand to the destination operand. The **SUB** instruction subtracts the contents of the source operand from the destination operand.

For example, if (AL) = 60H and (AH) =00H

$$\text{ADD AL, 20H}$$

is executed, AX will become 80H. Why? Since it is overflow for AL, the value would store in the whole AX register.

【Example 4-18】 Examples of ADD and SUB instructions.

ADD	AL, 50H	;$AL \leftarrow (AL) + 50H$
ADD	CX, 1000H	;$CX \leftarrow (CX) + 1000H$
ADD	DI, SI	;$DI \leftarrow (DI) + (SI)$
ADD	[BX+DI], AX	;$[BX+DI+1][BX+DI] \leftarrow (AX) + ([BX+DI+1][BX+DI])$
ADD	AX, [BX+2000H]	;$AX \leftarrow (AX) + ([BX+2001H][BX+2000H])$
SUB	BX, CX	;$BX \leftarrow (BX) - (CX)$
SUB	[BP+2], CL	;$[BP+2] \leftarrow ([BP+2]) - (CL)$
SUB	AL, 20H	;$AL \leftarrow (AL) - 20H$
SUB	SI, 5010H	;$SI \leftarrow (SI) - 5010H$
SUB	WORD PTR[DI], 10H	;$[DI+1][DI] \leftarrow ([DI+1][DI]) - 10H$

In some cases, indirect addressing, like the last instruction in **Example 4-18**, requires specifying the size of the data with the special assemble directive BYTE PTR, WORD PTR or DWORD PTR. These directives indicate the size of memory data addressed by the memory pointer (PTR). For example, for the **SUB [DI], 10H** instruction, it is so difficult to determine this instruction address a byte, word or double word-size memory location. The instruction **SUB WORD PTR [DI], 10H** clearly designates the location addressed by DI as a word-sized memory location.

In order to deal with carry bit and borrow bit problem, we define ADC and SBB.

ADC {destination}, {source}   ; destination = destination + source + Carry
SBB {destination}, {source}   ; destination = destination - source - Carry

which takes the form:

ADC { <register> | <memory> }, { <register> | <memory> | <immediate> }
SBB { <register> | <memory> }, { <register> | <memory> | <immediate> }

ADC will add the three elements: destination, source operand and a carry bit (0 or 1). If the carry bit is clear before execution, **ADC** will behave exactly like the **ADD** instruction. **SBB** will subtract the contents of the source operand and a carry bit (0 or 1) from the destination operand. If the carry bit is clear before execution, **SBB** will behave exactly like the **SUB** instruction.

The following example will demonstrate the carry bit calculation.

【Example 4-19】 Examples of ADC and SBB instructions.

ADC	AX, SI	;$AX \leftarrow (AX) + (SI) + CF$
ADC	DX, [SI]	;$DX \leftarrow (DX) + [SI+1][SI] + CF$
ADC	BX, 3000H	;$BX \leftarrow (BX) + 3000H + CF$
ADC	AL, 5	;$AL \leftarrow (AL) + 5 + CF$
SBB	AX, 2030H	;$AX \leftarrow (AX) - 2030H - CF$
SBB	WORD PTR[DI+2], 1000H	;$[DI+3][DI+2] \leftarrow ([DI+3][DI+2]) - 1000H - CF$

【Example 4-20】 Please develop a program to calculate the sum of 0123BC62H and 0012553AH.

MOV	AX, W12	;Add rightmost

```
ADD     AX, W22
MOV     W32, AX
MOV     AX, W11         ;Add leftmost
ADC     AX, W21         ;with carry
MOV     W31, AX


; * * * * *    data definition * * * * *
W11     DW      0123H   ; here DW stands for defined word
W12     DW      BC62H
W21     DW      0012H
W22     DW      553AH
W31     DW      ?       ; not initialized
W32     DW      ?
```

It is clear that we cannot use one memory location to store the whole source and destination numbers. In order to deal with this, we break down each number into two words, defining W11 = 0123H, W12 = BC62H, W21 = 0012H, W22 = 553AH. The output will store in W31 and W32.

【**Example 4-21**】 Double word-sized data requires four bytes of memory because it is a 32-bit number. If a double word is stored in memory location beginning with 2000H, while another is stored in memory location with 3000H, please write a program to calculate the sum and difference of the two numbers and then stored the sum and difference to memory location 2000H and 3000H respectively.

```
CLC                     ; clear CF
MOV     SI, 2000H       ; address of the first number
MOV     AX, [SI]        ; load lower word to AX
MOV     BX, AX
MOV     DI, 3000H       ; address of the second number
ADD     AX, [DI]        ; addition of lower words
MOV     [SI], AX        ; store the lower sum to [2000H]
MOV     AX, [SI+2]      ; load higher word
MOV     CX, AX
ADC     AX, [DI+2]      ; addition of higher words with carry
MOV     [SI+2], AX      ; store the higher word sum to [2002H]
SUB     BX, [DI]        ; subtraction of lower words
MOV     [DI], B         ; store the lower word of the difference to [3000H]
SBB     CX, [DI+2]      ; subtraction of higher words with borrow
MOV     [DI+2], CX      ; store the higher word of the difference to [3002H]
```

During the ADD operation, although there is a carry bit, it would not be added into the result. However, the **carry flag** (**CF**) will be set to 1. When ADC operation is performed, it will add three elements, including AX contents, ([DI+2]) and **CF**, whatever **CF** is related to that calculation or not.

Note that the subtraction is not commutative.

The **ADD** and **SUB** instructions will affect the status registers. The overflow flag will be set to 1 if signed overflow/underflow occurs. Also the sign flag will be set if the result is negative. Set the zero flag if the result is zero, and set the carry flag if an unsigned overflow occurs.

4.4.2 INC and DEC Function

The **INC** instruction increases the memory or register content by 1. The **DEC** instruction decreases the memory or register content by 1. It will be used in the looping, which will be discussed in the next section. The syntax are:

INC { < memory > | < register > }
DEC { < memory > | < register > }

Note that INC and DEC does not affect the carry flag while add or subtract one to the memory or register content.

[Example 4-22] Examples of INC and DEC instructions.

```
INC    AL                       ;AL←(AL)+1
INC    CX                       ;CX←(CX)+1
INC    BYTE PTR[BX+DI+500]      ;[BX+DI+500]←([BX+DI+500])+1
DEC    AX                       ;AX←(AX)-1
DEC    BL                       ;BL←(BL)-1
DEC    BYTE PTR[DI+2]           ;[DI+2]←([DI+2])+1
```

4.4.3 NEG and CMP Function

1. NEG

The **NEG** instruction reverses the sign of a binary value. In effect, NEG reverses the bits and adds 1 to the number (as 2's complement operation). The syntax is

NEG { < memory > | < register > }

[Example 4-23]
If   (AX) = A5C0H
NEG AX   will change AX's value to 5A40H

If the operand is zero, its sign does not change, although this clears the carry flag. Negating any other value sets the carry flag. Negating a byte containing −128, a word containing −32,768, or a double word containing −2,147,483,648 does not change the operand, but will set the overflow flag. NEG always updates the A, S, P, and Z flags as though you were using the SUB instruction.

The **NOT** instruction inverts all bits of a byte, word, or double word. The NEG instruction two's complements a number, which means that the arithmetic sign of a signed number changed from positive to negative or from negative to positive. The **NOT** function is considered logical, and the **NEG** function is considered an arithmetic operation.

2. CMP

The comparison instruction (**CMP**) is a subtraction that changes only the flag bits, the destination operand never changes. A comparison is useful for checking the entire contents of a register or a memory location against another value. A CMP is normally followed by a conditional jump instruction, which tests the condition of the flag bits.

To compare the content of the two data fields before executing the conditional jump instruction, a compare instruction can be executed:

CMP　　　　{destination}, {source}　　　　; destination-source (set flags)

which takes the form:

CMP　　　　{ <register> | <memory> }, { <register> | <memory> | <immediate> }

Note that no both memories can exist in destination and source. In short, the only disallowed forms of compare are memory-to-memory and segment register compares. The **CMP** instruction updates the status flags according to the result of the subtraction operation.

【**Example 4-24**】CMP instruction examples.

CMP AX, 200H　　　　; compare the content of (AX) with 200H,
　　　　　　　　　　　　(AX) - 200H—≯ (AX), set flags
CMP AX, [BX + 1000H]　; compare the content of (AX) with the 16-bit data addressed by BX +
　　　　　　　　　　　　1000H and BX + 1001H
CMP AL, 10H　　　　　; compare (AL) with 10H
JAE AA　　　　　　　; jump if equal to 10H or above

It shows a comparison followed by a conditional jump instruction. In the last example, the contents of AL are compared with a 10H. Conditional jump instructions that often follow the comparison are JA (jump above) or JB (jump below). If the JA follows the comparison, the jump occurs if the value in AL is above 10H. If the JB follows the comparison, the jump occurs if the value in AL is below 10H. In this example, the JAE instruction follows the comparison. This instruction causes the program to continue at memory location AA if the value in AL is 10H or above. There is also a JBE (jump below or equal) instruction that could follow the comparison to jump if the outcome is below or equal to 10H. CMP Instruction do subtraction operation without storing the difference value between destination and source operand to the destination operand. The flag bits after execution are shown in **Table 4-4**.

Table 4-4　The flags status after the CMP Instruction execution

Relationship between DEST. SRC. Operand			CF	ZF	SF	OF
Signed Operand	DEST.	= SRC.	0	1	0	0
		<	—	0	1	0
		<	—	0	0	1
		>	—	0	0	0
		>	—	0	1	1
Unsigned Operand	DEST.	= SRC.	0	1	0	0
		<	1	0	—	—
		>	0	0	—	—

4.4.4 MUL and DIV Function

For multiplication, define the instructions as follows.
MUL   {source}   ; $(AX) \leftarrow (AL) \times source/(DX:AX) \leftarrow (AX) \times source$
IMUL {source}   ; $(AX) \leftarrow (AL) \times source/(DX:AX) \leftarrow (AX) \times source$
which takes the form:
MUL   { <register> | <memory> }
IMUL { <register> | <memory> }

MUL multiplies unsigned 8 or 16 bits data while **IMUL** multiples the signed (2's complement) 8 or 16 bits data. **Note that** when multiplying two n-bit values, the result may require as many as 2n bits. Therefore, the basic operations can be divided into two types: byte times byte or word times word. In the **byte times byte** operation, the multiplicand will be in the **AL** register and the result will put in **AX**. All contents in AH will be erased after multiplication. In the **word times word** operation, the multiplicand will be in the **AX** register. After multiplication, the most-significant word is in the **DX** while the least-significant word will be in **AX**. The original content in DX will be erased.

【**Example 4-25**】MUL and IMUL instruction example.
If the following program segment is executed,
MOV   AL, B1
MUL   B2
IMUL   B2
:   :
; * * * * * data definition * * * * *
B1   DB   80H
B2   DB   40H   ; *DB stands for define byte*

The first **MUL** B2 will treat 80H as +128, whereas **IMUL** will treat 80H as −128 (2's complement). After **MUL** executes, since $128 \times 64 = 8192$ which is 2000H, therefore AX = 2000H. However, when **IMUL** executes, since $-128 \times 64 = -8192$ which is E000H, therefore (AX) = E000H.

Division shares the same properties with multiplication. The syntax are:
DIV   {source}   ; $(AL) = (AX) / source \ldots (AH \text{ or } AX) = (DX:AX) / source$
IDIV {source}   ; $(AL) = (AX) / source \ldots (AH \text{ or } AX) = (DX:AX) / source$
which takes the form:
DIV { <register> | <memory> }
IDIV { <register> | <memory> }

DIV handles unsigned data while **IDIV** handles signed division. The basic operations can be divided into two types: byte into word or word into double word. In the **byte into word** operation, the dividend will be in the **AX** register. The remainder will put in **AH** while the quotient will put in **AL**. In the **word into double word** operation, the most significance dividend will be in the **DX** register while the least significance dividend will be in the **AX** register. The remainder will put in **DX** while the quotient will put in **AX**.

You cannot simply divide the 8-bit value by another 8-bit value. If the denominator is an eight-bit value, the numerator must be a 16-bit value. You can extend that 8-bit value into 16 bit and loaded it into a 16-bit register (e. g., AX), and perform the division.

Furthermore, if the dividend and divisor have the same sign, DIV and IDIV generate same results. However, if they are different in their sign, DIV generates a positive quotient while IDIV generates a negative quotient.

We summarized the MUL and DIV operation in the **Table 4-5** and **Table 4-6**.

Table 4-5 Summary of MUL Operation

Instruction	Multiplier	Multiplicand in	Product in
MUL	CLCL (byte)	AL	AX
MUL BX	BX (word)	AX	DX AX

Table 4-6 Summary of DIV Operation

Instruction	Divisor	Dividend in	Quotient in	Remainder in
DIV CL	CL (byte)	AX	AL	AH
DIV BX	BX (word)	DX AX	AX	DX

【Example 4-26】 DIV and IDIV instruction example.

```
MOV   AX, 0400H        ;(AX) =0400H =1024
MOV   BL, 0B4H         ;(BL) =B4H =180
DIV   BL               ;1024/180   (AH) =7CH =124, (AL) =05H =5
MOV   AX, 0400H        ;(AX) =0400H =1024
MOV   BL, 0B4H         ;(BL) =B4H = -76
IDIV  BL               ;1024/( -76)   (AH) =24H =36   (AL) =F3H = -13
```

4.4.5 Type Conversion Functions

1. CBW

With 8-bit division, the numbers are usually 8 bits wide. This means that one of them, the **dividend**, must be converted to a 16-bit wide number in AX. This is accomplished differently for signed and unsigned numbers. For the unsigned number, the most-significant 8 bits must be cleared to zero (**zero-extended**). For signed numbers, the least-significant 8 bits are sign-extended into the most-significant 8 bits. In the microprocessor, a special instruction **sign-extends** AL into AH, or converts an 8-bit signed number in AL into a 16-bit signed number in AX. The CBW (**convert byte to word**) instruction performs this conversion.

Example 4-27 illustrates two unsigned operands division. Here, the quotient is stored in location ANSQ and the remainder is stored in location ANSR. Notice the method how to extend 8-bit unsigned dividend to 16-bit.

【Example 4-27】

```
MOV   AL, 12        ; stores Dividend 12 into AL register
```

```
MOV     AH, 0           ; zero-extend, (AX) = 000CH
MOV     BL, 3           ; stores divisor 3 into BL register
DIV     BL              ; divide by BL
MOV     ANSQ, AL        ; save quotient 04H into ANSQ
MOV     ANSR, AH        ; save remainder 00H into ANSR
```

In **Example 4-27**, we can extend AL to AX by using the CBW instruction.

Example 4-28 shows the same basic program except that the numbers are signed numbers. This means that instead of zero-extending AL into AH, it is sign-extended with the CBW instruction.

[Example 4-28]

```
MOV     AL, 8EH         ; (AL) = 8EH
CBW                     ; sign-extend, (AX) = FF8EH
MOV     BL, 3
IDIV    BL              ; divide by number 3
MOV     ANSQ, AL        ; save quotient
MOV     ANSR, AH        ; save remainder
```

2. CWD

16-bit division is similar to 8-bit division, except that instead of dividing into AX, the 16-bit number is divided into DX – AX, a 32-bit dividend.

As with 8-bit division, numbers must often be converted to the proper form for the dividend. If a 16-bit unsigned number is placed in AX, DX must be cleared to 0. If AX is a 16-bit signed number, the CWD (**convert word to doubleword**) instruction **sign-extends** it into a signed 32-bit number.

Example 4-29 shows the division of two 16-bit signed numbers. Here, a – 100 in AX is divided by a + 9 in CX. The CWD instruction converts the – 100 in AX to a – 100 in DX – AX before the division. After the division, the results appear in DX – AX as a quotient of – 11 in AX and a remainder of – 1 in DX.

[Example 4-29]

```
MOV     AX, -100        ; load -100 to AX register
MOV     CX, 9           ; load +9 to CX register
CWD                     ; sign-extend (AX)→(AX:DX)
IDIV    CX              ; (DX) = -1, (AX) = -11
```

4.4.6 BCD Conversion Functions

Packed BCD values are decimal integer coded in binary form with one decimal digit (0...9) per nibble. Unpacked BCD values contain a single decimal digit per byte, the high nibble of the byte should contain zero.

The BCD operations occur in systems such as point-of-sales terminals (e.g., cash register) and others that seldom require arithmetic. The BCD operations are performed on BCD data used by many programs. The **Figure 4-17** is the BCD application illustration diagram.

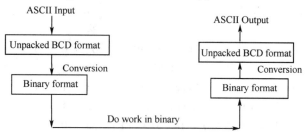

Figure 4-17   BCD conversion function

1. Packed BCD Arithmetic

Packed BCD numbers are stored two digits to a byte, in 4-bit groups referred to as **nibbles**. The ALU is capable of performing only binary addition and subtraction, but by adjusting the sum or difference the correct result in packed BCD format can be obtained. The **correction rule** for addition is:

If the addition of any two digits results in a binary number between 1010 and 1111, which are not valid BCD digits, or there is a carry into the next digit, then 6 (0110) is to be added to the current digit. Example 4-30 shows the adjustment process.

【Example 4-30】 BCD correction examples.

```
Carry from       → 0              0
previous digit      5           0101
                 + 8          + 1000
                   13           1101
                             +  110   ← Add 6 because 1101 is
Carry to next digit → 1      0011     not a valid BCD digit

Carry from       → 1              1
previous digit      7           0110
                 + 9          + 1001
                   17         1 0001
                             +  110   ← Add 6 because of carry
Carry to next digit → 1      0111     to next digit
```

Table 4-7 indicates Packed BCD adjust instruction. DAA is used to decimal adjust for addition and DAS is for subtraction.

Table 4-7   Packed BCD adjust instructions.

Name	Mnemonic and Format	Description
Decimal adjust for addition	DAA	(AL)←Sum in AL adjusted to packed BCD format
Decimal adjust for subtraction	DAS	(AL)←Difference in AL adjusted to packed BCD format

【Example 4-31】 Suppose two packed BCD numbers are 1234(number1) and 6789(number2) respectively, please develop a program to calculate the sum of these two numbers.

Instruction			Operation	Content of		
				AL	CF	AF
MOV	AL,	number1	(AL)←34	34	—	—
ADD	AL,	number2	(AL)←34+89	BD	0	0
DAA			ADJUST	23	1	*
MOV	sum,	AL	(sum)←23	23	1	*
MOV	AL,	number1+1	(AL)←12	12	1	*
ADC	AL,	number2+1	(AL)←18+67+(CF)	80	0	1
DAA			ADJUST	86	0	*
MOV	sum+1,	AL	(sum+1)←86	86	0	*

【**Example 4-32**】Suppose two packed BCD numbers are 1289(number1) and 6734(number2) respectively, please develop a program to calculate the difference of these two numbers.

Instruction			Operation	Contents of		
				AL	CF	AF
MOV	AL,	number1	(AL)←89	89	—	—
SUB	AL,	number2	(AL)←89-34	55	0	0
DAS			ADJUST	55	0	*
MOV	sub1,	AL	(sub1)←55	55	0	*
MOV	AL,	number1+1	(AL)←12	12	0	*
SBB	AL,	number2+1	(AL)←12-67-(CF)	AB	1	1
DAS			ADJUST	45	1	*
MOV	sub1+1,	AL	(sub1+1)←(AL)	45	1	*

Result = 4555 = -5445 in 10's complement

2. Unpacked BCD Arithmetic

In unpacked BCD there is only one digit per byte and because of this, unpacked multiplication and division can be done.

Table 4-8 indicates 4 instructions for unpacked BCD adjust.

Table 4-8 Unpacked BCD adjust instructions

Name	Mnemonic and Format	Description
Unpacked BCD adjust for addition	AAA	(AL) ← Sum in AL adjusted to unpacked BCD format (AH) ←(AH) + carry from adjustment
Unpacked BCD adjust for subtraction	AAS	(AL) ← Difference in AL adjusted to unpacked BCD format (AH) ← (AH) - borrow from
Unpacked BCD adjust for multiplication	AAM	(AX)←Product in AL adjusted to unpacked BCD format with AH containing high-order digit
Unpacked BCD adjust for division	AAD	(AL) 10 * (AH) + (AL) (AH)←0

【**Example 4-33**】Please develop a program to accomplish calculating the sum of 2658 and 3619, which stored in memory locations named STRING1 and STRING2 respectively, and put the result in location SUM. Suppose the addend and augend are stored in the memory in the type of ASCII as shown in the **Figure 4-18**.

```
              LEA  SI, STRING1
              LEA  DI, STRING2
              LEA  BX, SUM
              MOV  CX, 4
              CLC
     NEXT:    MOV  AL, [SI]
              ADC  AL, [DI]
              AAA
              MOV  [BX], AL
              INC  SI
              INC  DI
              INC  BX
              DEC  CX
              JNZ  NEXT
```

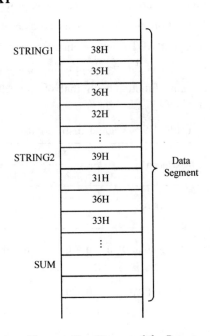

Figure 4-18 Diagram of the Data

[Example 4-34] Example involving unpacked BCD addition and subtraction. DAT1, DAT2, and DAT3 are defined as byte variables

```
     MOV  AL, DAT1       ; add low-order
     ADD  AL, DAT2       ; digits and
     AAA                 ; adjust
     MOV  DL, AL         ; put sum in DL
     MOV  AL, DAT1 + 1   ; add high-order
     ADC  AL, DAT2 + 1   ; digits with carry
```

```
    AAA                      ;and adjust
    XCHG    AL , DL          ;exchange AL and DL
    SUB     AL, DAT3         ;subtract low-order digit
    AAS                      ;from sum and adjust
    XCHG    AL, DL           ;exchange AL and DL
    SBB     SUB  AL, DAT3+1  ;subtract high-order digit
    AAS                      ;from sum and adjust
    MOV     DH , AL          ;move AL to DH
    ...                      ;output and other code
```

4.4.7 Boolean Operations

Logic function is important in circuitry design. The instructions for Boolean logic are **AND, OR, XOR, TEST** and **NOT**, which can be used to clear and set bits and to handle ASCII data for arithmetic purpose. The syntax and form are listed in **Table 4-9**.

As we learned before, the AND, OR and XOR operation are applied bit by bit.

Note that for above instructions, no memory location can appear in both destination and source. Of course, both the destination and source must be same size.

Under all the operations (except NOT), these instructions will clear the carry and overflow flag, and copy the high order bit of the result into the sign bit.

TEST instruction has the same function with that of the **AND**, except that it will not change the destination operand.

Table 4-9 Logical instructions

Instruction Function	Instruction Form	Description	Flag						Remarks
			O	S	Z	A	P	C	
Logical Operation	NOT *DST*	*DST* = NOT *DST*	–	–	–	–	–	–	*DST*: general register, memory
	AND *DST*, *SRC*	*DST* = *DST* AND *SRC*	0	0	0	*	0	0	*SRC*: register, memory, immediate; *DST*: register, memory
	OR *DST*, *SRC*	*DST* = *DST* OR *SRC*	0	0	0	*	0	0	*SRC*: register, memory, immediate; *DST*: register, memory
	XOR *DST*, *SRC*	*DST* = *DST* XOR *SRC*	0	0	0	*	0	0	*SRC*: register, memory, immediate; *DST*: register, memory
	TEST *DST*, *SRC*	*DST* = *DST* TEST *SRC*	0	0	0	*	0	0	*SRC*: 8-bit/ 16-bit immediate; *DST*: general register, memory

【Example 4-35】 If (AL) = 1100 0101 (C5H) and (BH) = 0101 1100 (5CH), please write down the contents of AL register after the following instruction operation.

```
    AND  AL, BH      ;(AL) =01000100 (44H)
    OR   AL, 02H     ;(AL) =11000111 (C7H)
    XOR  AL, AL      ;(AL) =00000000 (00H)
```

```
TEST AL, BH        ;(AL) = 11000101 (C5H)
```

As shown in **Example 4-35**, the **TEST** instruction performs the AND operation. The difference is that the AND instruction changes the destination operand, while the TEST instruction does not. A TEST only affects the condition of the flag register, which indicates the result of the test. The TEST instruction uses the same addressing modes as the AND instruction.

The TEST instruction functions in the same manner as a CMP instruction. The difference is that the TEST instruction normally tests a single bit (or occasionally multiple bits), while the CMP instruction tests the entire byte or word. The zero flag (ZF) is a logic 1 (indicating a zero result) if the bit under test is a zero, and ZF = 0 (indicating a non-zero result) if the bit under test is not zero.

Usually the TEST instruction is followed by either the JZ (jump if zero) or JNZ (jump if not zero) instruction. The destination operand is normally tested against immediate data. The value of immediate data is 1 to test the rightmost bit position, 2 to test the next bit, 4 for the next, and so on.

For example:
```
TEST   DL, 01H        ; Does the DL contain an odd number
TEST   CX, FFFFH      ; Does the CX contain a zero value
TEST   AH, F0H        ; Are any of the 4 leftmost bits in AH nonzero
```

[**Example 4-36**] Please develop a short program to accomplish a data output by port number 38EFH when the bit1 of the input data from the port 38F0H is one. Otherwise, there is no data output on the port 38EFH.

The flow chart of this program is given in **Figure 4-19**.

```
        LEA    SI, DATA
        MOV    DX, 38F0H
WATT:   IN     AL, DX
        TEST   AL, 02H
        JZ     WATT        ; ZF = 1 transfer
        MOV    DX, 38FEH
        MOV    AX, [SI]
        OUT    DX, AX
```

The NOT Instruction

The NOT instruction simply reverses the bits in a byte, word, or doubleword (80386 and later) in a register or memory so that 0s become 1s and 1s become 0s. Its general format is

[Label:]   NOT   Register/memory

For example, if the BL contains 00111010, the instruction NOT BL changes the BL to 11000101. (The effect is exactly the same as that of XOR BL, 0FFH) Flags are unaffected. NOT differs from NEG, which changes a binary value from positive to negative and vice versa by reversing the bits and adding 1.

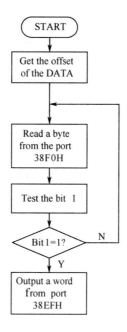

Figure 4-19 The flow chart of example 4-36

4.4.8 Shifting and Rotation

There are several shifting and rotation instructions, the main function of them are integer multiplication. Using shifting and rotation to perform such calculation on integer is faster than using multiplication. Destination operand must either be register or memory. Source operand is immediate provided that it is 1. The syntax is shown in **Table 4-10**.

Table 4-10 Shifting and rotation instructions

Instruction Function	Instruction Form	Description	Flag						Remarks
			O	S	Z	A	P	C	
Shift	SHL DST, counter	Shift left	0	0	0	*	0	0	DST: general register, memory; Counter: 1/shift count of CL
	SAL DST, counter	Shift arithmetic left	0	0	0	*	0	0	DST: general register, memory; Counter: 1/shift count of CL
	SHR DST, counter	Shift right	0	0	0	*	0	0	DST: general register, memory; Counter: 1/shift count of CL
	SAR DST, counter	Shift arithmetic right	0	0	0	*	0	0	DST: general register, memory; Counter: 1/shift count of CL
Circular Shift	ROL DST, counter	Rotation to left	0	–	–	*	–	0	DST: general register, memory; Counter: 1/shift count of CL
	ROR DST, counter	Rotation to right	0	–	–	*	–	0	DST: general register, memory; Counter: 1/shift count of CL
	RCL DST, counter	Rotation through carry bit left	0	–	–	*	–	0	DST: general register, memory; Counter: 1/shift count of CL
	RCR DST, counter	Rotation through carry bit right	0	–	–	*	–	0	DST: general register, memory; Counter: 1/shift count of CL

SAL would perform the same thing with **SHL**. Both operations shift bits in the designed register to the left. Each bit shifted off enters the carry flag. However, when the arithmetic right shifting is executed, as shown in **Figure 4-20**, the most significant bit will remain the original position.

Figure 4-20   SHL and SAL instruction

As shown in **Figure 4-21**, **SHR** will shift all the bits right by one bit, supply 0 to the missing bit and put the thrown bit to the carry bit C.

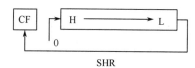

Figure 4-21   SHR instruction

For example, if AH = **1010110**1 and
  SHR   AH, 1

is executed, AH will become 0**1010110** and the carry bit C = 1 (thrown bit).

As shown in **Figure 4-22**, **SAR** instruction shifts all the bits in the operand to the right one bit, and replicating the high order bit. The main purpose is to perform a signed division by some power of two.

The **SHR** and **SAR** operations shift bits in the designated register to the right. Each bit shifted off enters the carry flag. SHR (Shift Logical Right) provides for logical (unsigned) data and SAR (Shift Arithmetic right) for arithmetic (signed) data. SAR differs from SHR in one important way: SAR uses the sign bit to fill leftmost vacated bits. In this way, positive and negative values retain their signs.

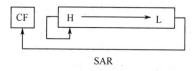

Figure 4-22   SAR instruction

Right shifts are especially useful for halving values and execute significantly faster than does a divide operation.

ROL does the similar thing with **SHL**. However, as shown in **Figure 4-23**, **ROL** & **ROR** will put the thrown bit to the missing part of the value.

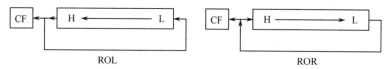

Figure 4-23   ROL and ROR instructions

For example, if (AH) = **1010110**1 and
    ROR   AH, 1

is executed, AH will become **11010110** and the carry bit CF = 1.

RCR does the similar thing with **ROR**. However, **RCR** will put the thrown bit to the carry bit and get the last result of thrown bit to the missing part, just as shown in **Figure 4-24**.

Notice that if you rotate through carries bit (n + 1) times, where n is the number of bits in the operand; you will get your original value. **For example**, if AH = **1010110**1 and carry bit

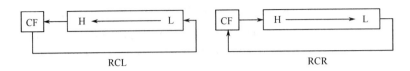

Figure 4-24   RCL and RCR instructions

CF = 0, and

RCR   AH, 1

is executed, AH will become **01010110** and the carry bit CF = 1.

Like ROL, RCL rotate the bits in the designed register to the left. RCL differs from ROL in the rotation way. Each bit rotated off on the left moves into the CF, and the CF bit moves into the vacated bit position on the right.

For 8088 and 8086 CPUs, the number of bits to be shifted or rotated is either 1 or in CL. If the number of bits shifted is larger than 1, the source operand must be initialized in **CL**.

[Example 4-37] Differentiate the following two instructions true or false.

SHL   AL, 3
SHR   AL, BL

The first instruction is invalid because the specified count that shifts to left in SHL instruction can only be equal to 1. If 3 bit right shifting is needed, the following must be followed.

MOV   CL, 3
SHR   AH, CL

The second instruction is also incorrect for only CL can be used as the count register.

Since shifting an integer value to the left one position is equivalent to multiplying that value by two (2H), use the shift left instruction for multiplication by powers of two.

[Example 4-38] Using shift instruction to accomplish calculation of $5_{10} \times 10_5$.

$5_{10} \times 10_{10} (= 50_{10})$ where $5_{10} = 00000101_2$

It is noticed that $5_{10} \times 10_{10} = 5 \times (8+2) = 5 \times (2^3 + 2^1) = 5 \times 2^3 + 5 \times 2^1$

Therefore, when we do the following procedure,

MOV	AH, 5H	; store the value of (AH) = 5d
MOV	AL, AH	; store the value of (AL) = 5d
MOV	CL, 3	
SHL	AH, CL	
SHL	AL, 1	; left shifting the (AL) by 1
ADD	AH, AL	; Add the value to perform the multiplication
		; AH will become 00110010 = 50_{10}.

Division by shifting: For division by a power of 2 (2, 4, 8, and so on), you may gain faster processing simply by shifting right the required number of bits. For the 8088/8086, a shift greater than 1 requires a shift value in the CL register. The following examples assume that the dividend is in the AX.

[Example 4-39] Using shift instruction to accomplish calculation of $40_{10} \div 8_{10}$.

```
MOV    AX,   40        ;Load AX with 40
MOV    CL,   03
SHR    AX,   CL        ; Divide by 8   (shift right 3)
```

Try more examples by yourself!

4.5 String Instructions

The string instructions are summarized in **Table 4-11**. Because the string instructions can operate on only a single byte or word unless they are used with the REP prefix, they are often referred to as string primitives, or simply primitives. All of the primitives are 1byte long, with bit 0 indicating whether a byte (bit = 0) or a word (bit 0 = 1) is being manipulated.

Table 4-11   String primitives

Instruction	Instruction Form	Flag O	S	Z	A	P	C	Instruction Function	Description
Basic String Operation	MOVSB	-	-	-	-	-	-	Move byte string	Byte operands $(SI) \leftarrow (SI) + 1$; $(DI) \leftarrow (DI) + 1$
	MOVSW	-	-	-	-	-	-	Move word string	Word operands $(SI) \leftarrow (SI) + 2$, $(DI) \leftarrow (DI) + 2$
	CMPSB	0	0	0	0	0	0	Compare byte string	Byte operand $((SI)) - ((DI))$, $(DI) \leftarrow (DI) + 1 (SI) \leftarrow (SI) + 1$
	CMPSW	0	0	0	0	0	0	Compare word string	Word operand $((SI)) - ((DI))$, $(DI) \leftarrow (DI) + 2 (SI) \leftarrow (SI) + 2$
	SCASB	0	0	0	0	0	0	Scan byte string	Byte operand $(AL) - ((DI))$, $(DI) \leftarrow (DI) + 1$
	SCASW	0	0	0	0	0	0	Scan word string	Word operand $(AX) - ((DI))$, $(DI) \leftarrow (DI) + 2$
	LODSB	-	-	-	-	-	-	Load byte string	Byte operand $(AL) \leftarrow ((SI))$, $(SI) \leftarrow (SI) + 1$
	LODSW	-	-	-	-	-	-	Load word string	Word operand $(AX) \leftarrow ((SI))$, $(SI) \leftarrow (SI) + 2$
	STOSB	-	-	-	-	-	-	Store byte string	Byte operand $((DI)) \leftarrow AL$, $(DI) \leftarrow (DI) + 1$
	STOSW	-	-	-	-	-	-	Store word string	Word operand $((DI)) \leftarrow AX$, $(DI) \leftarrow (DI) + 2$
Repeat Prefix	REP	-	-	-	-	-	-	Repeat unconditionally	
	REPE/REPZ	-	-	-	-	-	-	Repeat when equal/ zero	
	REPNE/REPNZ	-	-	-	-	-	-	Repeat when unequal/ non-zero	
* The B suffix indicates byte operands and the W suffix indicates word operands.									
* * Incrementing (+) is used if DF = 0 and decrementing (-) is used if DF = 1									
Flags: CMPS and SCAS affect all condition flags and MOVS, LODS and STOS affect no flags									
Addressing modes: Operands are implied									

There are three string data transfer instructions: LODS, STOS, MOVS. They can manipulate large blocks of data. Each string instruction allows data transfers that are either a single byte, word, or double word. SCAS and CMPS are string comparison instructions.

During the execution of a string instruction, memory accesses occur through either or both of the DI and SI registers. The DI offset address accesses data in the extra segment(ES) for all string instructions that use it. The SI offset address accesses data, by default, in the data segment (DS).

4.5.1 The Direction Flag

The direction flag (DF) (located in the flag register) selects the auto-increment (DF = 0) or the auto-decrement (DF = 1) operation for the DI and SI registers during string operations. The direction flag is used only with the string instructions. The CLD instruction clears the DF flag (DF = 0) and the STD instruction sets it (DF = 1). Therefore, the CLD instruction selects the auto-increment mode (DF = 0) and STD selects the auto-decrement mode (DF = 1).

Figure 4-25 is an example of data transfer between the data segment and the extra segment. The direction of data transfer is very important because of the two segments overlap. In **Figure 4-25 (a)**, the five data stored in memory beginning with DS:0000H will be transferred to the five memory locations started at ES:0000H. It is obviously that part of original data are modified when auto-increment (DF = 0) operation is selected, just like that shown in **Figure 4-25(b)**. The correct way is set DF = 1 to keep the auto-decrement operation for the DI and SI registers as shown in **Figure 4-25**(c) and (d).

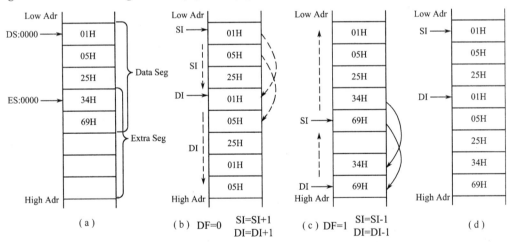

Figure 4-25 The direction control of data transfer

For the MOVS instruction, when a byte is transferred, the DI and/or SI increment or decrement by 1. If a word is transferred, the DI and/or SI increment or decrement by 2. But for the STOSB instruction, there is only DI increment or decrements without affecting SI. On the contrary, for the LODSB instruction, it only increases/ decreases SI without affecting DI.

4.5.2 String Data Transfers

1. MOVS

One of the more useful string data transfer instruction is MOVS because it transfers data from one memory location to another. This is the only memory-to-memory transfer allowed in the 8086-Pentium 4 microprocessors. The MOVS instruction transfers a byte, word, or doubleword from the data segment location addressed by SI to the extra segment location addressed by DI. Then the pointers will be increased or decreased as dictated by the direction flag DF. All the permissible forms of the MOVS instruction are listed as following:

MOVS DST, SRC
MOVSB   ;$(ES:(DI)) \leftarrow (DS:(SI))$
                ;$(DI) \leftarrow (DI) \pm 1$
                ;$(SI) \leftarrow (SI) \pm 1$
MOVSW   ;$ES:(DI) \leftarrow DS:(SI)$
                ;$(DI) \leftarrow (DI) \pm 2$
                ;$(SI) \leftarrow (SI) \pm 2$

Note that only the source operand (SI), located in the data segment, may be overridden so that another segment may be used. The destination operand (DI) must always be located in the extra segment.

【Example 4-40】 Please give the program sequence to transfer the 100 bytes beginning at memory location 2000H: 1200H to memory location 6000H: 0000H.

```
        MOV AX, 2000H
        MOV DS, AX       ; set data segment address
        MOV AX, 6000H
        MOV ES, AX       ; set extra segment address
        MOV SI, 1200H    ; offset of source string
        MOV DI, 0        ; offset of destination string
        MOV CX, 100      ; length of string
        CLD              ; auto-increment
NEXT:   MOVSB            ; transfer until CX = 0
        DEC CX
        JNZ NEXT
```

2. LODS

The LODS instruction loads AL, AX with data stored at the data segment offset address indexed by the SI register.

The form of LODS instruction is:

Assembly Language   **Operation**
LODS SRC
LODSB   ;$AL \leftarrow (SI)$, $(SI) \leftarrow (SI) \pm 1$
LODSW   ;$AX \leftarrow (SI/SI \pm 1)$, $(SI) \leftarrow (SI \pm 2)$

After loading AL with a byte, AX with a word, the contents of SI increment, if DF = 0 or decrement, if DF = 1. A 1 is added to or subtracted from SI for a byte-sized LODS, a 2 is selected for a word-sized LODS.

3. STOS

The STOS instruction stores AL, AX at the extra segment memory location addressed by the DI register. **Table 4-12** lists all forms of the STOS instruction. As with LODS, a STOS instruction may be appended with a B, W for byte, word transfers.

Table 4-12   Forms of the STOS instruction

Assembly Language	Operation
STOS DST	
STOSB	ES:[DI]→AL; (DI) = (DI) ±1
STOSW	ES:[DI/DI+1]→AX; (DI) = (DI) ±2

The **repeat prefix** (**REP**) is added to any string data transfer instruction, except the LODS instruction. It doesn't make any sense to perform a repeated LODS operation. The REP prefix causes CX to decrement by 1 each time the string instruction executes. After CX decrements, the string instruction repeats. If CX reaches a value of 0, the instruction terminates and the program continues with the next sequential instruction.

【Example 4-41】 Use STOS instruction to clear the 100 bytes beginning at memory location 6000H: 1200H.

```
    MOV   AX,  6000H
    MOV   ES,  AX          ; ES ← address of extra segment
    MOV   DI,  1200H       ; DI ← offset of the destination string
    MOV   CX,  100         ; CX ← length of string
    CLD                    ; auto-increment
    MOV   AL,  0           ; AL←0
    REP   STOSB            ; clear the 100 bytes
```

As shown in **Example 4-41**, when CX is loaded with a 100, the microprocessor automatically repeats the STOSB instruction 100 times. The DI register is automatically incremented or decremented after each datum is stored, this instruction stores the contents of AL(0) in a block of memory instead of a single byte of memory. REP can be used in **Example 4-41**. Thus, the last three instructions can be instead by one sentence. That is REP MOVSB.

4.5.3 String Comparisons

The string comparisons instructions include SCAS(string scan) and CMPS(string compare).

In this section, additional string instructions that allow a section of memory to be tested against a constant or against another section of memory are discussed. To accomplish these tasks, the SCAS or CMPS instruction is used.

1. SCAS

The forms of SCAS instruction are listed as following:

SCAS DST

SCASB   ;$(AL) - ((DI)), (DI) \pm 1$

SCASW   ;$(AX) - ((DI/DI+1)), (DI) \pm 2$

The SCAS (string scan instruction) compares the AL register with a byte block of memory, the AX register with a word block of memory. The SCAS instruction subtracts memory from AL, AX without affecting either the register or the memory location. The op-code used for byte comparison is SCASB, the op-code used for the word comparison is SCASW. In all cases, the contents of the extra segment memory location addressed by DI is compared with AL, AX. Recall that this default segment (ES) cannot be changed with a segment override prefix. SCAS instruction can accomplish repeat operation if prefixed by a conditional repeat prefix.

[Example 4-42] Suppose that there are 10 bytes data stored in a consecutive memory space which begins at location BLOCK, please develop a program to search this part of memory for a 01H by using SCASB instruction.

```
         MOV  DI, OFFSET BLOCK    ;address data
         CLD                      ;auto-increment
         MOV  CX, 10              ;load counter
         XOR  AL, AL              ;clear AL
         MOV  AL, 01H             ;load AL with 01H
    SER: SCASB                    ;(AL) - ((DI))
         DEC  CX                  ; counter decreasement
         JNZ  SER                 ; jump to SER on NOT zero
```

In this example, we also can use a repeat prefix to simplify the program.

$$\left. \begin{array}{l} \text{SER:SCASB} \\ \text{DEC CX} \\ \text{JNZ SER} \end{array} \right\} \Rightarrow \text{REPNZ SCASB}$$

The REPNE (repeat while not equal or non-zero) prefix causes the SCASB instruction to repeat until either the CX register reaches 0, or until an equal condition exists as the outcome of the SCASB instruction's comparison. Another conditional repeat prefix is **REPE/Z** (repeat while equal or zero). With either repeat prefix, the contents of CX decrements without affecting the flag bits. The SCASB instruction and the comparison it makes change the flag.

2. CMPS

The CMPS (compare strings instruction) always compares two sections of memory data as bytes (CMPSB), words (CMPSW).

The forms of CMPS are following:

CMP SRC, DST

CMPSB   ;$((SI)) - ((DI)); (SI) \leftarrow (SI) \pm 1, (DI) \leftarrow (DI) \pm 1$

CMPSW   ;$((SI)) - ((DI)); (SI) \leftarrow (SI) \pm 2, (DI) \leftarrow (DI) \pm 2$

The contents of the data segment memory location addressed by SI is compared with the contents of the extra segment memory location addressed by DI. The CMPS instruction increments or decrements both SI and DI. The CMP instruction is normally used with either the REPE or REPNE prefix.

Table 4-13 gives some repeat prefixs and their explanations. **Table 4-14** lists the repeat prefix that the string operation instructions can use.

Table 4-13 Repeat prefix

Repeat prefix	Explanation
REP prefix	Causes CX to decrement by 1 each time the string instruction executes. After CX decrements, the string instruction repeats. If CX reaches a value of 0, the instruction terminates and the program continues with the next sequential instruction
REPZ	Repeat while zero
REPNZ	Repeat while not zero
REPE	Repeat while equal
REPNE	Repeat while not equal

Table 4-14 String instruction and prefix

String Instruction	Prefix
MOVS	REP
CMPS	REPE/PEPZ REPNE/REPNZ
SCAS	REPE/PEPZ REPNE/REPNZ
LODS	—
STOS	REP

【Example 4-43】 Suppose that the 2 data sections in memory are both 256 bytes long and begin at location BUFFER1 and BUFFER2 respectively, please develop a short program to see whether the contents of these two sections are same or not.

```
SKIP    PROC    FAR
        LEA     DI, BUFFER1     ;load BUFFER1 effective address
        LEA     SI, BUFFER2     ;load BUFFER2 effective address
        CLD                     ;auto-increment
        MOV     CX, 256         ;counter
        REPE    CMPSB           ;search ((SI))-((DI))
        RET
SKIP    ENDP
```

Example 4-43 gives a short procedure that compares two sections of memory searching for a match.

【Example 4-44】 Compare two strings and find out the address of the first different character. Then put the character and its address in AL and BX. Assuming the length of the string is 200.

```
        LEA SI, String1         ; SI← address of string1
        LEA DI, String2         ; DI ←address of string2
```

```
            MOV CX, 200         ; CX ←length of string
            CLD                 ; auto-increment
            REPE CMPSB          ; compare if the same
            JZ SAME             ; if ZF =0, jump SAME
    DIFFER: DEC SI              ; else (SI)←(SI) - 1
            MOV BX, SI          ; BX← character address
            MOV AL, [SI]        ; AL← the character
    SAME:   MOV AH, 4CH
            INT 21H             ; end of processing
```

【**Example 4-45**】Please develop a program to accomplish transferring 17 numbers from BUFFER1 to BUFFER2.

```
    DATA  SEGMENT                        ; data segment definition
          BUFFER1  DB  Hello, everybody! $ '  ; define a byte variable  BUFFER1
    DATA  ENDS
    EXTRA SEGMENT                        ; define an extra segment
          BUFFER2  DB  17  DUP(?)        ; define a byte variable BUFFER2
    EXTRA ENDS
    CODE  SEGMENT                        ; define a code segment
    ASSUME  CS:CODE,DS:DATA,ES:EXTRA
      STA: MOV  AX,  DATA
           MOV  DS,  AX
           MOV  AX,  EXTRA
           MOV  ES,  AX
           LEA  SI,  BUFFER1
           LEA  DI,  BUFFER2
           MOV  CX,  17
           CLD
           REP  MOVSB           ; ((DI)) ←((SI)). (SI) +1→(SI). (DI) +1→(DI)
    CODE  ENDS
    END   STA
```

As shown in following, you can complete the program without using string operation instruction or repeat prefixes.

```
          KEA SI,SUFFER1                    LEA SI,SUFFER1
          KEA DI,SUFFER2                    LEA DI,SUFFER2
          MOV CX,17                         MOV CX,17
   AGAIN: MOV AL,[SI]        →              CLD
          MOV [DI],AL                AGAIN: MOV SB
          INC SI                            DEC CX
          INC DI                            JNZ AGAIN
          DEC CX
          JNZ AGAIN
```

```
                                            LEA SI,SUFFER1
                                            LEA DI,SUFFER2
                                            MOV CX,17
                                            CLD
                                            REP MOVSB
```

要点:

(1) 8086 串操作指令的特点:

① 通过加重复前缀来实现串操作。

② 可以对字节串进行操作,也可以对字串进行操作。

③ 源串一般存放在数据段 DS 中,偏移地址由 SI 指定。允许段重设。

④ 目标串必须在附加段 ES 中,偏移地址由 DI 指定。

⑤ 指令自动修改地址指针,修改方向由 DF 决定。DF = 0 时,SI 和 DI 作自动增量修改;DF = 1 时,SI 和 DI 作自动减量修改。

⑥ 数据块长度值由 CX 指定。

⑦ 可增加自动重复前缀以实现自动修改 CX 内容。

(2) 串操作指令可使用的重复前缀有以下三种形式:

① REP:无条件重复。

② REPZ/REPE:相等/为零则重复。

③ REPNZ/REPNE:不相等/不为零则重复。

4.6 Program Control Instructions

4.6.1 Program Flow Control Instructions

The program control instruction direct the flow of a program and allow the flow to charge. Some program control instructions are unconditional jump while some are conditional. **Table 4-15** lists some program flow control instructions.

1. Unconditional Jump

A commonly used instruction for transferring control is the unconditional jump instruction. The operation transfers control under all circumstances. The syntax is

JMP   { < address > | < label > | < register > }

You normally specify the target address by using a label. A statement label, as told before, is usually an identifier followed by a colon, usually on the same line as an executable machine instruction. The assembler determines the offset of the statement after the label and automatically computes the distance from the jump instruction to the statement label. Therefore, you do not have to worry about computing displacements manually.

Three types of unconditional jump instructions are available to the microprocessor: short jump, near jump, and far jump. The short jump is a two-byte instruction that allows jumps or branches to memory locations within + 127 and − 128 bytes from the address following the jump. The short jump instruction appears in **Figure 4-26**.

Table 4-15 Program flow control instructions

Instruction Function	Instruction Form	Instruction Functions
Unconditional Jump	JMP *Label* CALL *Procedure name* RET *Pop value*	*Unconditional transfer* *Procedure call* *Procedure return*

(续)

Instruction Function		Instruction Form		Instruction Functions
Conditional Jump	Unsigned Number	JA/JNBE	Label	Jump on above($CF=0$ and $ZF=0$)
		JAE/JNB	Label	Jump on above or equal($CF=0$ or $ZF=0$)
		JB/JNAE	Label	Jump on below($CF=1$ and $ZF=0$)
		JBE/JNA	Label	Jump on below or equal($CF=0$ or $ZF=1$)
Conditional Jump	Signed Number	JG/JNLE	Label	Jump on greater($ZF=0$ and $OF \oplus SF=0$)
		JGE/JNL	Label	Jump on greater or equal($SF \oplus OF=0$ or $ZF=1$)
		JL/JNGE	Label	Jump on less($SF \oplus OF=1$ or $ZF=0$)
		JLE/JNG	Label	Jump on less or equal($SF \oplus OF=1$ or $ZF=1$)
	Single Flag	JC	Label	Jump on $CF=1$
		JNC	Label	Jump on $CF=0$
		JE/JZ	Label	Jump on equal($ZF=1$)
		JNE/JNZ	Label	Jump on not equal($ZF=0$)
	Bit	JO	Label	Jump on overflow ($OF=1$)
		JNO	Label	Jump on not overflow($OF=1$)
		JNP/JNO	Label	Jump on $PF=0$
		JP/JO	Label	Jump on $PF=1$
		JNS	Label	Jump on $SF=0$
		JS	Label	Jump on $SF=1$
Loop Control		LOOP	Label	Loop
		LOOPE/LOOPZ	Label	Loop on equal or zero
		LOOPNE/LOOPNZ	Label	Loop on unequal or no-zero
		JCXZ	Label	Jump on $CX=0$
Interrupt		INT	Interrupt type	Interrupt
		INTO		Interrupt on overflow
		IRET		Interrupt and return

The three-byte near jump allows a branch or jump within ±32K bytes (or anywhere in the current code segment) from the instruction in the current code segment. **Figure 4-27** illustrates the operation of the real mode near jump instruction.

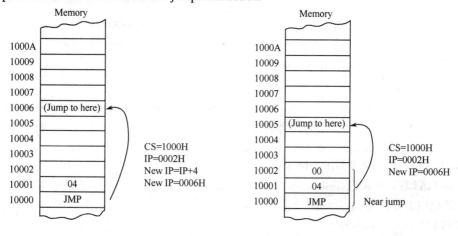

Figure 4-26　Short jump　　　　　　Figure 4-27　Near jump

The five-byte far jump allows a jump to any memory location within the real memory system. It's obviously that the near jump is very similar to short jump except that the distance is farther. The short and near jumps are often called intrasegment jumps, while the far jump is often called intersegment jumps. **Figure 4-28** illustrates the operation of the real mode far jump instruction.

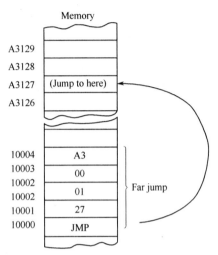

Figure 4-28   Far jump

Table 4-16 lists all formats for Jump instruction.

Table 4-16   Machine code formats for unconditional branch instructions.

Name	Mnemonic and Format	Description
Intrasegment direct short branch	JMP SHORT OPR	$(IP) \leftarrow (IP) + $ sign extended D8 determined by OPR
Intrasegment direct near branch	JMP NEAR PTR OPR	$(IP) \leftarrow (IP) + $ 16-bit displacement determined by OPR
Intrasegment indirect branch	JMP OPR *	$(IP) \leftarrow (EA)$ where EA is determined by OPR
Intersegment direct (far) branch	JMP FAR PTR OPR	$(IP) \leftarrow$ Offset of OPR within segment $(CS) \leftarrow$ Segment address of segment containing OPR
Intersegment indirect branch	JMP OPR *	$(IP) \leftarrow (EA)$ where EA is determined by OPR $(CS) \leftarrow (EA+2)$ where EA is determined by OPR
* Type of branch determined by type of operand		
Flags: No flags are affected		
Addressing modes: For intrasegment direct branches the mode is relative and for intersegment direct branches the mode is direct. Indirect branches cannot involve immediate modes and a memory addressing mode must be used in intersegment indirect branches		

2. CALL and RET Operations

The **CALL** and **RET** instruction enable us to call and return any procedure we like. The forms of **CALL** and **RET** are listed as following

```
CALL        < label >
RET         < immediate >
```

The CALL instruction transfers the flow of the program to the procedure. The CALL instruction differs from the jump instruction because a CALL saves a return address on the stack.

The **CALL** instructions take the same forms as the **JMP** instructions except there is no short intrasegment call.

Now, let's begin with **Example 4-46**.

〔Example 4-46〕 Simple procedure call.

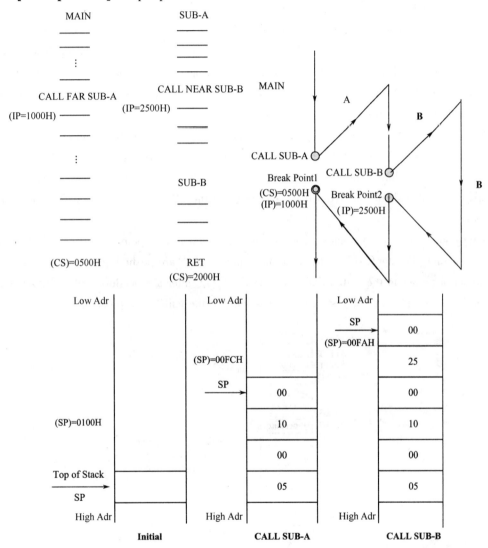

From the example, we got that there are two types of procedure calling, the **FAR CALL** and **NEAR CALL**.

The **FAR CALL** is also called as Intersegment CALL. For example, the SUB-A calling is a far call that pushes the CS:IP (0500H:1000H) of the break point 1 onto the stack. Then copies the 32-bit called procedure(SUB-A) address into the CS:IP register. Finally, executes the subroutine.

The **NEAR CALL** is an intrasegment CALL. Unlike a far call, it does not care the CS

147

register, for it is an intrasegment calling. So it just pushes the 16-bit offset of the break point 2 ((IP) = 2500H) onto the stack. Then copies the 16-bit effective address into the **IP** register.

The return address returns control to the instruction that immediately follows the CALL in a program when a RET instruction executes. The RET instruction, effectively the counterpart of CALL, return from the called procedure to the original calling procedure. RET should be the last instruction in the called procedure.

Near CALL. The near CALL instruction is three bytes long; the first byte contains the opcode, and the second and third bytes contain the displacement, or distance of ±32K in the 8086 through the 80286 processors. This is identical to the form of the near jump instruction. The 80386 and above use a 32-bit displacement, when operating in the protected mode, to allow a distance of ±2G bytes. A near call is a call to a procedure within the same segment. When the near CALL executes, it first pushes the offset address of the next instruction on the stack. The offset address of the next instruction appears in the instruction pointer (IP). After saving this return address, it then adds the displacement from bytes 2 and 3 to the IP to transfer control to the procedure. There is no short CALL instruction. A variation on the opcode exists as CALL, but this should be avoided in favor of using the PROC statement to define the CALL as near.

Why save the IP on the stack? The instruction pointer always points to the next instruction in the program. For the CALL instruction, the contents of IP are pushed onto the stack, so program control passes to the instruction following the CALL after a procedure ends. **Figure 4-29** shows the return address (IP) stored on the stack and the call to the procedure.

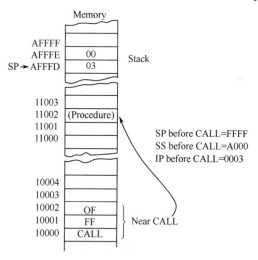

Figure 4-29   The effect of a near call

Far CALL. The far CALL instruction is like a far jump because it can call a procedure stored in any memory location in the system. The far CALL is a five-byte instruction that contains an opcode followed by the next value for the IP and CS registers. Bytes 2 and 3 contain the new contents of the IP, and bytes 4 and 5 contain the new contents for CS.

The far CALL instruction places the contents of both IP and CS on the stack before jumping to the address indicated by bytes 2-5 of the instruction. This allows the far CALL to call a procedure located anywhere in the memory and return from that procedure.

Figure 4-30 shows how the far CALL instruction calls a far procedure. Here, the contents of IP and CS are pushed onto the stack. Next, the program branches to the procedure.

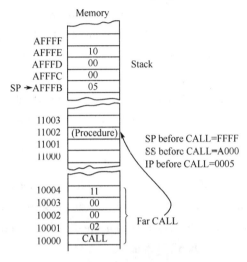

Figure 4-30   The effect of a far call

The return **RET** instruction returns control to the caller of a subroutine. It does so by popping the return address off the stack and transferring control to the instruction at the return address. Near call returns pop a 16-bit return address off the stack into the **IP** register. A far call returns pop a 16-bit offset into the **IP** register and a 16-bit segment value into the CS register.

The other form of the **RET** instruction is adding a displacement number after the **RET**. It is identical to those **RET** instruction, except the **CPU** adds the displacement value to the stack pointer immediately after popping the return address from the stack. This mechanism removes parameters pushed onto the stack before returning to the caller.

3. INT Instruction

The form of INT is:

INT   n   ; n is the interrupt vector number

On execution, an INT instruction interrupts processing and accesses the **interrupt vector table** in low memory to determine the address of the requested routine. The operation then transfers to DOS or to BIOS for specified action and returns to your program to resume processing. Most often, the interrupt has to perform the complex steps of an input or output operation. Interrupts require a trail that facilitates exiting a program and, on successful completion, returning to it. For this purpose, INT performs the following items one by one.

(1) Pushes the contents of the flags register onto the stack. (Push first decrements the stack pointer by 2.)

(2) Clears the interrupt and trap flags.

(3) Pushes the CS register onto the stack.

(4) Pushes the instruction pointer (containing the address of the next instruction) onto the stack.

(5) Times n by 4, the obtains the starting the interrupt address in interrupt vector table.

(6) Fetches the address of ISR, then the required operation to be performed.

To return from an interrupt, the routine issues an IRET (Interrupt Return), which pops the registers off the stack and thus causes a return to the instruction immediately following the INT.

Because the preceding process is entirely automatic, your only concerns are to define a stack large enough for the necessary pushing and popping and to use the appropriate INT operations.

4. Conditional Jump

Unlike **JMP**, **CALL** and **RET** instructions providing transfer of control, without making any decision before the jump, the conditional jump instructions are usually make a decision and a charge of flow occurs. They are the basic tool for creating loops and other conditionally executable statements like the "if... then" statement.

On a 8086 machine, the conditional jump instructions are all two bytes long, which takes the form:

JXX   { <address> | <label> }

where **XX** is the mnemonic representing the condition of branching. The first byte is a one byte opcode followed by a one byte displacement. Although this leads to a very compact instruction, a single byte displacement only allows a range of −128 bytes to +127 bytes.

The conditional jumps test one or more bits in the status register to see if they match some particular pattern. If the pattern matches, control transfers to the target location. If the match fails, the CPU ignores the conditional jump and execution continues with the next instruction. As shown in **Table 4-17**, some conditional jump instructions, test the conditions of the sign, carry, overflow, parity and zero flags.

Table 4-17 Conditional jump instructions

Definition	Description	Condition
Jump Based on Unsigned Data		
JE / JZ	Jump equal or jump zero	ZF = 1
JNE / JNZ	Jump not equal or jump not zero	ZF = 0
JA / JNBE	Jump above or jump not below/ equal	CF = 0 & ZF = 0
JAE / JNB	Jump above/ equal or jump not below	CF = 0
JB / JNAE	Jump below or jump not above/ equal	CF = 1
JBE / JNA	Jump below/ equal or jump not above	CF = 1 or ZF = 1

(续)

Definition	Description	Condition
Jump Based on Signed Data		
JE / JZ	Jump equal or jump zero	ZF = 1
JNE / JNZ	Jump not equal or jump not zero	ZF = 0
JG / JNLE	Jump greater or jump not less/ equal	SF$\oplus$OF = 0 & ZF = 0
JGE / JNL	Jump greater/ equal or jump not less	SF$\oplus$OF = 0
JL / JNGE	Jump less or jump not greater/ equal	SF$\oplus$OF = 1
JLE / JNG	Jump less/ equal or jump not greater	SF$\oplus$OF = 1 or ZF = 1
Arithmetic Jump		
JS	Jump sign	SF$\oplus$OF = 1
JNS	Jump no sign	SF$\oplus$OF = 0
JC	Jump carry	CF = 1
JNC	Jump no carry	CF = 0
JO	Jump overflow	OF = 1
JNO	Jump not overflow	OF = 0
JP / JPE	Jump parity even	PF = 1
JNP / JPO	Jump parity odd	PF = 0

Conditional branching instructions test the corresponding status register. If the condition is **TRUE**, the program will flow to the label specified. The status register would not change according to the JXX condition. The JXX command will be summarized in the table above.

The conditional jump instruction gives you the ability to split program flow into one of the two paths depending upon some logical condition.

[Example 4-47] Suppose you want to increment the AX register if BX is equal to CX. You can accomplish this with the following codes:

```
CMP     BX, CX
JNE     NEXTSTAT
INC     AX
NEXTSTAT::
         :  :
```

As shown in the example, we execute a conditional jump after the comparison. We can test the result of the comparison by checking the appropriate flags in the status register. **Figure 4-31** show us how to use the opposite branch to skip over the instructions you want to execute if the condition is true. Always use the opposite branch rule given earlier to select the opposite branch, otherwise you may use more than one conditional jump to perform one jumping.

5. Looping Operation

The **JMP** instruction would cause an endless loop. However, a routine is more likely to loop a specified number of times until it reaches a particular condition. The **LOOP** instruction, which serves the above purpose, requires an initial value in the **CX** register. It will decrement

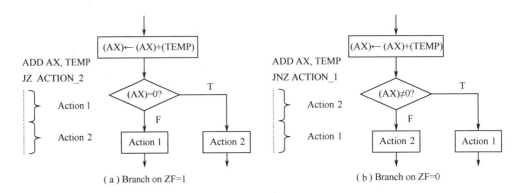

Figure 4-31 Conditional branches based on the ZF flag

the CX register and branches to the target location if the CX register does not contain 0. The syntax is:

LOOP　　　　< label >

Although the loop instruction's name suggests that you would normally create loops with it, keep in mind that all it is really doing is decreasing CX and branching to the target address if CX does not contain zero after the decrement.

【Example 4-48】 The application of LOOP instruction.
The following routines will loop 10 times.

```
        MOV     CX, 10
A:      INC     AX
         :   :
        LOOP    A
```

Figure 4-32 illustrates the loop instruction's execution. The loop instruction does not affect any flags.

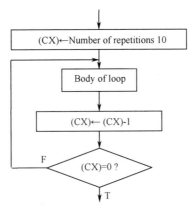

Figure 4-32 Typical structure of a post-test loop

The LOOP instruction does the same operation as following 2 sentences:

```
DEC  CX
JNZ  BEGIN
```

That is to say, the loop instructions are designed to simplify the decrementing, testing, and branching portion of the loop.

Table 4-18 are the forms of LOOP:

Table 4-18 Loop instructions

Name	Mnemonic and Format	Alternate Mnemonic	Test Condition *
Loop	LOOP OPR		$(CX) \neq 0$
Loop while zero, or equal	LOOPZ OPR	LOOPE	$ZF = 1$ and $(CX) \neq 0$
Loop while nonzero, or not equal	LOOPNZ OPR	LOOPNE	$ZF = 0$ and $(CX) \neq 0$
Branch on CX	JCXZ OPR		$(CX) = 0$
Except for JCXZ which leaves (CX) unchanged, $(CX) \leftarrow (CX) - 1$. Then if the test condition is met, $(IP) \leftarrow (IP) +$ sign extended D8; otherwise (IP) are unchanged and the program continues in sequence			
Addressing modes: Mode is relative to IP. OPR must represent a label that is within -128 to 127 bytes of the instruction following the loop instruction			
No flags are affected			

[Example 4-49] Develops a program to calculate the sum of 10 word data stored at the address of ARRAY, and then stores the result to memory location TOTAL.

```
        DATA SEGMENT
            ARRAY  DW,  20H, 50H, 35H, 46H, 6EH, 36H, 4DH, 52H, 2AH, 04H
            TOTAL  DW,  0
            TOTAL  DW,  10
        DATA ENDS
        CODE SEGMENT
            ASSUME  CS:CODE,  DS:DATA
                MOV    AX,  DATA
                MOV    DS,  AX
                MOV    CX,  COUNT        ;put count in CX
                MOV    AX,  0            ;zero AX
                MOV    SI,  AX           ;and SI
START_LOOP:     ADD    AX,  ARRAY[SI]    ;add next element to AX
                ADD    SI,  2            ;increment index by two
                LOOP   START_LOOP        ;repeat LOOP if CX nonzero
                MOV    TOTAL,  AX        ;store result in TOTAL
        Output and other code
        CODE ENDS
```

[Example 4-50] Search example using LOOPNE.

Suppose 10 byte data are stored at the memory location which begins at ASCII_STR, develop a program to search whether it contains a space or not.

Data related directives, input, and other code.

```
                MOV    CX,  10           ;put array size in CX
                MOV    SI,  -1           ;initialize index, and
```

```
              MOV     AL,  20H            ;put code for space in AL
NEXT:         INC     SI                  ;increment index
              CMP     AL,  ASCII_STR[SI]  ;test for space
              LOOPNE NEXT                 ;LOOP if not space and count
                                          ;is nonzero
              JNZ NOT_FOUND               ;if space not found, branch to NOT_FOUND;
NOT_FOUND:  :  :                          ; Output and other code
```

4.6.2 Machine Control and Miscellaneous Instructions

1. NOP and HLT Instructions

Table 4-19 is the manipulation illustration of NOP and HLT.

Table 4-19 NOP and HLT

Name	Mnemonic and Format	Description	Flag					
			O	S	Z	A	P	C
No operation	NOP	Cause no action	—	—	—	—	—	—
Halt	HLT	Cause the operation of the computer to cease	—	—	—	—	—	—
No flags are affected								
Addressing modes: None								

2. Flag Manipulation Instructions

As we have learned, some instructions set or clear the flags depending on their results. Flag Manipulation Instructions are used to set or clear a flag bit, you are requested to accomplish direct control to this flag bit.

Table 4-20 is the set of flag manipulation instructions.

Table 4-20 Flag manipulation instructions

Name	Mnemonic and Format	Description	Flag					
			O	S	Z	A	P	C
Clear Carry	CLC	(CF)←0	—	—	—	—	—	0
Complement Carry	CMC	(CF)←OT CF	—	—	—	—	—	C
Set Carry	STC	(CF)←1						
Clear Direction	CLD	(DF)←0	—	0	—	—	—	1
Set Direction	STD	(DF)←1	—	1	—	—	—	1
Clear Interrupt	CLI	(IF)←0	—	—	0	—	—	1
Set Interrupt	STI	(IF)←1	—	—	1	—	—	1
Load AH from Flags *	LAHF	(AH)←(Low−order byte of PSW)						
Store AH to Flags *	SAHF	(Low−order byte of PSW)←(AH)						
* Bits 0, 2, 4, 6, and 7 are transferred. Bits 1, 3, and 5 are indeterminate								
Addressing modes: None								

3. External Timing Instructions

Table 4-21 shows the external timing instructions for 8086/8088 system. The function of each instruction such as WAIT、ESC and LOCK are listed as following:

Table 4-21　External timing instructions

Name	Mnemonic and Format	Description	Flag					
			O	S	Z	A	P	C
External Timing	WAIT	Waiting	—	—	—	—	—	—
	ESC	Exit Windows	—	—	—	—	—	—
	LOCK	Lock the bus	—	—	—	—	—	—

要点：

（1）LOOP 指令的功能是先将 CX 的内容减 1，再判断 CX 中是否为 0，为 0 则退出，否则循环。其操作与下两条指令的操作的功能一致：

DEC　CX

JNZ　符号地址

（2）LOOPNZ/LOOPNE 指令的功能是先将 CX 的内容减 1，再判断 CX 中是否为 0，并且判断 ZF 是否为 1。如果 ZF = 0，并且 CX≠0，则继续循环；只有在 CX = 0，或 ZF = 1 情况下，则退出循环。该指令可用于查找不相等的情况。

（3）LOOPZ/LOOPE 指令的功能是先将 CX 的内容减 1，再判断 CX 中是否为 0，并且判断 ZF 是否为 1。如果 ZF = 0，或 CX = 0，则退出；只有在 ZF = 1 并且 CX≠0 情况下，才继续循环。该指令可用于查找相等的情况。

（4）JMP 指令是无条件转移指令，可以转移到内存中存放的任何程序段。指令必须指定转移的目标地址（或称转移地址）。无条件转移指令不影响标志位。JMP 指令可分为段内转移和段间转移。段内转移只须改变 IP 寄存器的内容，用新的转移目标地址代替原有的 IP 的值就可以。段间则转到另一段去执行程序，因此，不仅要修改 IP 寄存器的内容，而且还要修改 CS 寄存器的内容。

（5）子程序调用指令 CALL 和返回指令 RET。

子程序结构相当于高级语言中的过程。为了方便模块化程序设计，往往把程序中某些具有独立功能的部分编写成独立的程序模块，称为子程序。子程序调用和返回指令都不影响标志位。

子程序调用时会将调用指令的下一条指令的地址（断点）压入堆栈，获取子过程的入口地址（子过程第 1 条指令的偏移地址），执行子过程（含相应参数的保存及恢复），最后将断点地址由堆栈弹出，返回原程序。段内调用是指子过程与原调用程序在同一代码段，在调用之前只需保护断点的偏移地址；段间调用是指子过程与原调用程序不在同一代码段，在调用之前需保护断点的段基址和偏移地址。压栈时，先保存断点的 CS，再压入 IP。

（6）带参数的返回指令：RET n。

主程序为某个子程序提供一定的参数或参数地址，在进入子程序之前，主程序将这些参数或参数的地址压入堆栈中，通过堆栈供子程序调用，子程序使用后，这些参数已无使用价值，堆栈中不需要保存这些参数，因而在返回指令 RET 中提供参数 n 在返回同时堆栈指针自动增加几个字节，以去掉堆栈中的几个字节。n 可以表示为 0 ~ FFFFH 范围中的任何一个偶数。

4.7 The Symbolic Instruction Set

The following is a list of the symbolic instructions for the 8086 processor family, arranged by category. Although the list seems formidable, many of the instructions are rarely needed.

Arithmetic

ADC: Add with Carry   INC: Increment by 1
ADD: Add Binary Numbers   MUL: Unsigned Multiply
DEC: Decrement by 1   NEG: Negate
DIV: Unsigned Divide   SBB: Subtract with Borrow
IDIV: Signed (Integer) Divide   SUB: Subtract Binary Values
IMUL: Signed (Integer) Multiply

ASCII-BCD Conversion

AAA: ASCII Adjust After Addition
AAD: ASCII Adjust Before Division
AAM: ASCII Adjust After Multiplication
AAS: ASCII Adjust After Subtraction
DAA: Decimal Adjust After Addition
DAS: Decimal Adjust After Subtraction

Bit Shifting

RCL: Rotate Left Through Carry   SAR: Shift Algebraic Right
RCR: Rotate Right Through Carry   SHL: Shift Logical Left
ROL: Rotate Left   SHR: Shift Logical Right
ROR: Rotate Right   SAL: Shift Algebraic Left

Comparison

CMP: Compare
CMPS: Compare String
TEST: Test Bits

Data Transfer

LDS: Load Data Segment Register   LEA: Load Effective Address
LES: Load Extra Segment Register   LODS: Load String
MOV: Move Data   MOVS: Move String
STOS: Store String   XCHG: Exchange
XLAT: Translate

Flag Operations

CLC: Clear Carry Flag
CLI: Clear Interrupt Flag
LAHF: Load AH from Flags
PUSHF: Push Flags onto Stack
STC: Set Carry Flag
STI: Set Interrupt Flag

CLD: Clear Direction Flag
CMC: Complement Carry Flag
POPF: Pop Flag off Stack
SAHF: Store AH in Flags
STD: Set Direction Flag

Input/Output

IN: Input Byte or Word

OUT: Output Byte or Word

Logical Operations

AND: Logical AND
NOT: Logical NOT

OR: Logical OR
XOR: Exclusive OR

Looping

LOOP: Loop Until Complete
LOOPZ: Loop While Zero
LOOPNZ: Loop While Not Zero

LOOPE: Loop While Equal
LOOPNE: Loop While Not Equal

Processor Control

ESC: Escape
HLT: Enter Halt State
LOCK: Lock Bus

NOP: No Operation
WAIT: Put Processor in Wait State

Stack Operation

POP: Pop Word off Stack

PUSH: Push Word onto Stack

String Operation

CMPS: Compare String
LODS: Load String
MOVS: Move String
REP: Repeat String
REPE: Repeat While Equal

REPZ: Repeat While Zero
REPNE: Repeat While Not Equal
REPNZ: Repeat While Not Zero
SCAS: Scan String
STOS: Store String

Transfer (Conditional)

INTO: Interrupt on Overflow
JA: Jump If Above
JAE: Jump If Above/Equal
JB: Jump If Below

JNC: Jump If No Carry
JNE: Jump If Not Equal
JNG: Jump If Not Greater
JNGE: Jump If Not Greater/Equal

JBE: Jump If Below/Equal   JNL: Jump If Not Less
JC: Jump If Carry   JNLE: Jump If Not Less/Equal
JCXZ: Jump If CX is Zero   JNO: Jump If No Overflow
JE: Jump If Equal   JNP: Jump If No Parity
JG: Jump If Greater   JNS: Jump If No Sign
JGE: Jump If Greater/Equal   JNZ: Jump If Not Zero
JL: Jump If Less   JO: Jump If Overflow
JLE: Jump If Less/Equal   JP: Jump If Parity Odd
JNA: Jump If Not Above   JPE: Jump If Parity Even
JNAE: Jump If Not Above/Equal   JPO: Jump If Parity Odd
JNB: Jump If Not Below   JS: Jump If Sign
JNBE: Jump If Not Below/Equal   JZ: Jump If Zero

Transfer (Unconditional)

CALL: Call a Procedure   JMP: Unconditional Jump
INT: Interrupt   RET: Return
IRET: Interrupt Return

Type Conversion

CBW: Convert Byte to Word   CWD: Convert Word to Doubleword

Tips

ADC (add with carry flag) 带进位的加法指令
addressing mode 寻址方式
ambiguity 多义性
ampersand &(=and)的记号名称
assembly instruction 汇编指令
auto-decrement 自动减少
auto-increment 自动增加
base relative-plus-index addressing 相对的基址加变址寻址
base-plus-index addressing 基址加变址寻址
Boolean logic 布尔逻辑(布尔代数的基础是一整套定义了基本函数的逻辑运算,包括与、或、非等运算。)
byte operand 字节型操作数
CMPS (string comparison) 串比较指令
coercion operator 强制转换操作符
colon 冒号
comment 注释
commutative 可交换的
compiler 编译器
conditional jump 条件跳转

cyclic 循环的
DAA (decimal adjust for addition) 压缩BCD码的加法调整指令
DAS (decimal adjust for subtraction) 压缩BCD码的减法调整指令
default 缺省值,默认值
descriptor 描述符
destination operand 目的操作数
dictate 命令
direct addressing 直接寻址
displacement 偏移量
dividend 被除数
divisor 除数
flag manipulation instruction 标志位操作指令
generic 同类的
immediate addressing 立即数寻址
immediate operand 立即数
increment 增量
inter-segment jumps 段间跳转
intra-segment jumps 段内跳转

intrinsically 固有地	register indirect addressing 寄存器间接寻址
invoke 调用	register relative addressing 寄存器相对寻址
label 标号	repeat prefix 重复前缀
LAHF（load AH with flag） 读取标志指令	SAHF 设置标志指令
LEA（load effective address instruction）取有效地址指令	SBB（subtract with carry flag） 带借位的减法指令
	SCAS（string scan） 串扫描指令
loop 循环	segment override prefix 设置段前缀
machine control instruction 控制指令	semicolon 分号（即";"）
miscellaneous instruction 其他指令	skip over 跳过，略去
mnemonic 助记符	source operand 源操作数
negate 求反	string comparison instruction 串比较指令
nibble 四位	string instruction 串操作指令
op-code 操作码	suffix 后缀
operand 操作数	syntax 句法
permissible 可允许的	type coercion operator 类型强制操作数
primitives 基本类型	word operand 字型操作数
quotient 商	XCHG（exchange instruction） 交换指令
register addressing 寄存器寻址	XLAT（translate instruction） 换码指令

Exercise

1. Write instructions to:

   (a) Put 65 into the AH register.

   (b) Put 32 into the AL register.

   (c) Add AH to AL.

   (d) Swap the content of AH with AL.

   (e) Mov the content of AH to BL register.

   (f) Convert the content of AL to 16 bits long.

   (g) Divide the content of AX by that in BL register.

2. Assume that the CH contains 0111 1001 and that an item named TESTVAL contains 1110 0011. Determine the effect on the CH for the following unrelated operations:

   OR   CH, TESTVAL
   AND CH, TESTVAL
   XOR CH, TESTVAL
   AND CH, 0000 0000B
   XOR CH, 1111 1111B

3. Assume that the BX contains binary 1011 1001 1011 1001 and the CL contains 03. Determine the hex contents of the BX after execution of the following unrelated instructions:

   SHL BL, 1
   SHL BL, CL
   SHR BL, CL

```
SHR    BL,  1
SAL    BH,  1
ROR    BX,  CL
ROR    DL,  CL
```

4. Use shift, move, and add instructions to initialize the CX with 40H and to multiply it by 10.

5. For an instruction with two operands, which operand is the source and which one is the destination?

6. In what significant way do the following instructions differ in execution?

```
ADD    CX,  2548
ADD    CX,  [2548]
```

For the second ADD, one operand is in square brackets. What is the name of this feature?

7. In what significant way do the following ADD instructions differ in execution?

```
ADD    BX,  25
ADD    [BX], 25
```

For the second ADD what sort of addressing is involved with the first operand?

8. Please describe the addressing mode of the source operand in the following instructions.

(a) ADD   DX,   [BX + SI + 8]
(b) MOV   AX,   200
(c) MOV   AX,   BX
(d) MOV   AL,   [2000H]
(e) MOV   AL,   [SI]
(f) MOV   AL,   [BP]
(g) MOV   AL,   COUNT[BX]
(h) MOV   AL,   OFFSET DATA

9. Code the following as instructions with immediate operands:

(a) Add hex 48 to CX

(b) Substract hex 48 from DX

(c) Shift the DH one bit to the right

(d) Shift BYTEX one bit to the left

(e) Store 248 in the CX

(f) Compare BYTEX to zero

10. Code the instructions to divide(DIV) the following:

(a) the word VALUE1 by 36

(b) the doubleword beginning at VALUE1 by the word VALUE2

11. During execution of a program, the CS contains 6C3AH, the SS contains 6C62H, the IP contains 42H, and the SP contains 36H. (Values are shown in normal, not reversed-byte, sequence.) Calculate the addresses of :

(a) the instruction to execute

(b) the top (current location) of the stack

12. Suppose (DS) = 5000H, (ES) = 4000H, (SS) = 2500H, (BX) = 0100H, (BP) = 0050H, (SI) = 1250H, (DI) = 1A7CH, DISP = 10H, please calculate the memory address of the source operand for the following:

   (a) MOV   AL   DISP[BX]

   (b) MOV   AL   [BP]

   (c) INC   BYTE PTR DISP[SI]

   (d) CMP   DL   ES:[BX + SI]

   (e) MOV   AL   DISP[BX][DI]

13. Suppose that (DS) = 0200H, (BX) = 0300H and (DI) = 400H, Determine the memory address accessed by each of the following instructions, assuming real mode operation:

   (a) MOV   AL,   [1234H]

   (b) MOV   [DI],   AL

14. What is wrong with a MOV [BX], [DI] instruction?

15. Suppose that (DS) = 1000H, (SS) = 2000H, (BP) = 1000H and (DI) = 0100H, Determine the memory address accessed by each of the following instructions, assuming real mode operation:

   (a) MOV   AL,   [BP + DI]

   (b) MOV   CX,   [DI]

16. Suppose that (DS) = 1100H, (BX) = 0200H, LIST = 0250H and (SI) = 0500H, Determine the memory address accessed by each of the following instructions, assuming real mode operation:

   (a) MOV   CL,   LIST [BX + SI]

   (b) MOV   CH,   [BX + SI]

17. Assume that (SP) = 0100H and (SS) = 0200H, (AX) = 3000H, (BX) = 5000H. Explain what happens when the PUSH BX instruction executes. Make sure to show where BH and BL are stored. When POP AX instruction executes, what values will appear in SP, AX and BX registers.

18. What values appear in SP and SS if the stack is addressed at memory location 0200H?

19. Compare the operation of a MOV DI, NUMB instruction with an LEA DI, NUMB instruction.

20. Describe how the LDS BX, NUMB instruction operates.

21. Explain the operation of the LODSB instruction.

22. Explain how the LODSB instruction operates.

23. Select a SUB instruction that will:

(a) subtract BX from CX

(b) subtract 0EEH from DH

(c) subtract DI from SI

(d) subtract the data address by SI from CH

(e) subtract the data stored 10 words after the location addressed by SI from DX

(f) subtract Al from memory location FROG

24. Explain the difference between the IDIV and DIV instructions.

25. Differentiate the following instructions True or False. For the False one, please give the reason why it is wrong.

(a) MOV   AL, BX   (b) IN   BL, 0F5H
(c) SHL   AL, 2   (d) MOV   DS, 2000H
(e) INC   [BX]   (f) POP   CS
(g) MOV   5, AL   (h) MUL   5
(i) PUSH   CS   (g) MOV   [BX],[SI]
(k) MOV   AL, [BX][BP]   (l) XLAT   AL
(m) MOV   DS, SS

26. Explain how the OUT DX, AX instruction operates, and what the IN AL, 2500H instruction accomplishes.

Chapter 5 Directives and Macro Processing

When a program has development, all instructions are translated into machine language instructions and correspond to executable statements in high-level language programs. Just as high-level language programs must have non-executable statements to pre-assign values, reserve storage, assign names to constants, form data structures, and terminate a compilation, assembler language programs must contain directives to perform similar tasks. A directive is an artifact of the assembler MASM not the CPU. They are generally used to either instruct the assembler to do something or inform the assembler of something. They are not translated into machine code.

5.1 The Format of the Directives

Figure 5-1 is the general format for a directives statement, where square brackets indicate an optional entry:

| [identifier] | directive/operation | [operand(s)] | [;comment] |

Figure 5-1 The directive/operation format

An identifier (if any), operation, and operand (if any) are separated by at least one blank or tab character. There is a maximum of 132 characters on a line (512 since MASM 6.0), although most programmers prefer to stay within 80 characters because that is the maximum number most screens can accommodate.

Notice that the identifier, directive/operation, and operand may begin in any column. Now, let's illustrate the format of the Directives by **Example 5-1**.

[Example 5-1] Directive Examples.
```
DATA    DB    1           ; data definition, define a byte with 1 value
CODE1   SEGMENT           ; segment definition
```
From **Example 5-1**, we can get that:

Identifier: An identifier (or symbol) is a name that you apply to an item in your program that you expect to reference. The two types of identifier are name and label. In **Example 5-1**, the DATA is the name refers to the address of a data item.

CODE1 is a label that refers to the address of segment, sometimes it refers to the address of an instruction, procedure, or something like that.

The term name applies to the name of a defined item or directive, whereas the term label applies to the name of an instruction. The same rules apply to both names and labels. An identifier can use the following characters:

CATEGORY	ALLOWABLE CHARACTERS
Alphabetic letters:	A through Z and a though z
Digits:	0 through 9 (may not be the first character)
Special characters:	question mark (?)
	Underline (_)
	Dollar ($)
	At (@)
	Period (.) (may not be the first character)

The first character of an identifier must be an alphabetic letter or a special character, except for the period. Because the assembler uses some special words that begin with the @ symbol, you should avoid using it for your own definitions. By default, the assembler treats uppercase and lowercase letters the same. The maximum length of an identifier is 31 characters (247 since MASM 6.0).

Operation: the operation must be mnemonic. It is most commonly used for defining data areas and coding instructions. In **Example 5-1**, DB means "define byte". It defines a field work area, or constant. SEGMENT defines a segment. For an instruction, an operation such as MOV or ADD indicates an action to perform.

Operand: The operand (if any) provides information for the operation to act on. For a data item, the operand defines its initial value. In **Example 5-1**, the first directive defines a data item named DATA and the operand initializes its contents with a one value.

要点:

(1) 用汇编语言编写的程序叫做源程序。此程序不能由机器直接执行,必须翻译成由机器代码组成的目标程序,这个翻译过程称为汇编。用来把汇编语言编写的程序自动翻译成目的程序的软件叫做汇编程序。

(2) 伪指令与指令语句不同,它本身不产生对应的机器目标代码,仅仅是告诉汇编程序,对后面的指令语句和伪指令语句的操作数应该如何产生机器目标代码。每条伪指令语句共分4个字段,指令格式如图5-1所示。

① 标号名字段这是一个任选字段。标号名后面不能用冒号":",这是它与指令语句的突出区别。不同的伪指令,标号名可以是常量名、变量名、过程名、结构名、记录名等。

② 伪指令字段是伪指令语句不可省略的主要成分。伪指令种类很多,如定义数据伪指令 DB、DW、DD、段定义伪指令 SEGMENT、定义过程伪指令 PROC 等。它们是伪指令语句要求汇编程序完成的具体操作命令。

③ 操作数字段是否需要、需要几个、需要什么样的操作数等都由伪指令字段中伪指令来确定。

④ 注释字段是一个任选字段,它必须以分号为开始,它的作用与指令语句的注释字段相同。

5.2 Operators and Expression

The various assembly language features at first tend to be somewhat overwhelming. But once you have become familiar with the simpler and more common features described in earlier chapters, you should find the descriptions of the various types specifies operators, and directives in this chapter more easily understood and a handy reference.

For an expression, it includes operands and operator. And it forms a content value assembling by MASM.

An operator provides a facility for changing or analyzing operands during an assembly. Operators are divided into various categories:

(1) Calculation operators: Arithmetic, index, logical, shift, and structure field name.

(2) Record operators: MASK and WIDTH, covered later in this chapter under the RECORD directive.

(3) Comparison operators: EQ, GE, GT, LE, LT, and NE.

(4) Segment operators: OFFSET, SEG, and segment override.

(5) Type (or Attribute) operators: HIGH, HIGHWORD, LENGTH, LOW, LOWWORD, PTR, SHORT, SIZE, THIS, and TYPE.

The operators are introduced as following in alphabetic sequence.

1. Arithmetic Operators

The Arithmetic Operators include " + " " – " " * " " / " " MOD " " SHL " and " SHR ". The assembler automatically calculates the values of the arithmetic operators. Following is a list of the operators, together with examples and the effect obtained:

SIGN	TYPE	EXAMPLE	EFFECT
+	Addition	DAT + 25H	*Adds 25H to address of DAT*
–	Subtraction	EA2 – EA1	*Calculates difference between two offset addresses*
*	Multiplication	value * 5	*Multiplies value by 5*
/	Division	value/5	*Divides value by 5*
MOD	Remainder	32 MOD 5	*Delivers remainder 2*
SHL	Shift left	21H SHL 2	*84H*
SHR	Shift right	20H SHR 2	*08H*

Except for addition(+) and subtraction (–), all operators must be integer constants.

The operators **SHL** and **SHR** can also shift an expression during an assembly. The general formats are

Expression SHL/SHR count

【Example 5-2】 Please check the content of BL register after the instruction MOV BL, 01011101B SHR 3 executing.

For instruction MOV BL, 01011101B SHR 3, the SHR operator shifts the bit constant 3 bits to the right. *So after execution , the BL is load with 00001011B.*

Most likely, the expression would reference a symbolic name rather than a constant value.

2. Logical Operators

The logical operators perform logical operations on the bits in an expression:

OPERATOR	USED AS	EFFECT
AND	expression1 AND expression2	*ANDs the bits*
OR	expression1 OR expression2	*ORs the bits*
XOR	expression1 XOR expression2	*Exclusive ORs the bits*
NOT	NOT expression1	*Reverses the bits*

[Example 5-3] Following 2 instructions are examples for logical operators, please write down the content of destination register for each instruction.

```
MOV    CL,   00111100B  AND  01010101B   ; (CL) = 00010100B
MOV    DL,   NOT  01010101B              ; (DL) = 10101010B
```

[Example 5-4] Tell the difference between the following 2 AND mnemonics in the instruction:

AND　　AL, 38H　　AND　　05H

The first **AND** is an instruction executed by CPU, while the second one is performed by assembler. MASM for it is a operator.

3. Comparison Operators

Following is a list of the comparison operators, together with some examples and the effect obtained:

Operators	Functions	Examples (Suppose DAT = 25)
EQ	Equal	25 EQU DAT = 0FFFFH
GE	Greater than or equal	
GT	Greater than	
LE	Less than or equal	
LT	Less than	5 LE 6 = 0FFFFH
NE	Not equal	25 NE DAT = 0

4. SEG Operator

The **SEG** operator returns the address of the segment in which a specified variable or label is placed. Programs that combine separately assembled segments would most likely use this operator. The general format is

SEG　　variable or label

[Example 5-5] Suppose there are a data segment named DAT1 and a code segment named COD1. The following MOV instructions return the address of the data segment and code segment:

```
MOV    DX,  SEG  DAT1   ; send address of data segment to DX register
MOV    DX,  SEG  COD1   ; send address of code segment to DX register
```

5. OFFSET Operator

The **OFFSET** operator returns the offset address (that is, the relative address within the data segment or code segment) of a variable or label. The general format is

OFFSET   variable or label

〖Example 5-6〗 Suppose a data segment definition as following:
DATA SEGMENT
        DAT1 DB 5
        DAT2 DB ?
DATA ENDS

Please write down the content of DX register for following each instruction.
MOV   DX, OFFSET DAT2
LEA   DX, DAT2

In the first instruction, the MOV returns the offset address of DAT2, then (DX) = 1; LEA instruction loads DX with 1, although it doesn't require OFFSET. These two instructions are functional equivalent.

6. TYPE Specifiers

Type specifiers can provide the size of a data variable or the relative distance of an instruction label. Type specifiers that give the size of a data variable are BYTE, WORD, DWORD, FWORD, QWORD, and TBYTE. Those that give the distance of an instruction label are NEAR, FAR, and PROC. A near address, which is simply an offset, is assumed to be in the current segment; a far address, which consists of a segment: offset address, can be used to access data in another segment.

The PTR and THIS operators, as well as the LABEL, and PROC directives, use type specifiers. The general format of TYPE is

TYPE   variable or label

The **TYPE** operator returns the number of bytes, according to the definition of the referenced variable. The number of bytes for numeric variable as listed in **Table 5-1**:

Table 5-1   The number of bytes for numeric variable

Type	DB/BYTE	DW/WORD	DD/DWORD	DQ/QWORD	DT/TWORD	NEAR Label	FAR Label
Type Value	1	2	4	8	10	−1	−2

7. SIZE Operator

The **SIZE** operator returns the product of LENGTH times TYPE and is useful only if the referenced variable contains the DUP entry. The general format is

SIZE   variable

8. LENGTH Operator

The format is LENGTH variable.

The **LENGTH** operator returns the number of entries defined by a DUP operator, as shown by the following MOV instruction:
DAT1   DW   10   DUP(0)
        ...
   MOV   DX , LENGTH   DAT1   ; *Return length 10 to DX*

If the referenced operand does not contain a DUP entry, the operator returns the value 01 (a limit to its usefulness).

【Example 5-7】 The following examples illustrate the TYPE, LENGTH, and SIZE operators, please give the content of the destination register for each instruction:

```
DAT     DB      7                    ; Define one byte
DAT1    DW      10   DUP (?)         ; Define 10 words
        ...
    MOV    AX,   TYPE    DAT         ; (AX) = 0001H
    MOV    AX,   TYPE    DAT1        ; (AX) = 0002H(10)
    MOV    AX,   LENGTH  DAT         ; (AX) = 1
    MOV    AX,   LENGTH  DAT1        ; (AX) = 10
    MOV    AX,   SIZE    DAT         ; (AX) = 0001H
    MOV    DX,   SIZE    DAT1        ; (DX) = 0014H(20)
```

Because DAT1 is defined as DW, TYPE returns 0002H, LENGTH returns 000AH(10) based on the DUP entry, and SIZE returns type times length, or 14H(20).

9. Coercion Operator-PTR Operator

There are times when you would probably like to treat a byte variable as a word, or treat a word as a double word (addressing). Temporarily changing the type of a label for some particular occurrence is called coercion, as shown below:

Type   PTR   expression

Type is any of byte, word, dword, near, far, or other types. **Expression** is any general expression which can besa variable or constant. The coercion operator returns an expression with the same value as **expression**, but with the type specified by **Type**.

The **PTR** operator can be used on data variables and instruction labels. It uses the type specifiers BYTE, WORD, DWORD, QWORD, and TBYTE to specify a size in an ambiguous operand or to override the defined type (DB, DW, DD, DQ, or DT) for variables. It also uses the type specifiers NEAR, FAR, and PROC to override the implied distance of labels.

In some case, indirect addressing needs specifying the size of data with PTR.

【Example 5-8】 Suppose the data definition in a data segment and some instructions in code segment are given as following , please compare the data stored in memory after the following instructions executing.

```
    BYTEA       DB      22H
                DB      35H
    WORDA       DW      2672H       ; Data stored as 7226
                ...
    MOV   AH,   BYTE PTR WORDA      ; Move first byte (72) to AH
    ADD   BL,   BYTE PTR WORDA + 1  ; Add second byte (26)
    MOV   BYTE PTR WORDA,   05      ; Move 05 to first byte
    MOV   BX,   OFFSET BYTEA        ; Load the offset of BYTEA
                                    ;(BX) = 0
```

```
INC   BYTE  PTR [BX]              ; 22H→23H
MOV   AX,   WORD PTR BYTEA        ; Move two bytes (2335) to AX
CALL  FAR   PTR [BX]              ; Call far procedure
```

In **Example 5-8**, the WORDA is defined as a word variable and BYTEA is a byte variable. For the WORDA, we can use the PTR to temporarily change its type to byte, just like that in the first 3 instructions in the example. **Figure 5-2** is the data assignments for *example* 5-8.

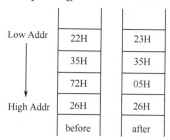

Figure 5-2   The assignment of data in memory locations

10. THIS Operator

The **THIS** operator creates an operand with segment and offset values that are equal to those of the current location counter. Its general format is

THIS   type

The type specifier can be BYTE, WORD, DWORD, QWORD, or TBYTE for variables and NEAR, FAR, or PROC for labels. THIS would typically be used with the EQU or equals sign (=) directives. The following example defines PARTREC:

```
PARTREC     EQU     THIS BYTE
```

The effect is the same as if you used the LABEL directive as

```
PARTREC     LABEL   BYTE
```

11. LOW Operator

The **LOW** operator returns the low (rightmost) byte of an expression.

12. HIGH Operator

The **HIGH** operator returns the high (leftmost) byte of an expression.

[Example 5-9]

```
EQUVAL      EQU     1234H
MOV   BL,   LOW EQUVAL            ; Load 34H in BL
MOV   CL,   HIGH EQUVAL           ; Load 12H in CL
```

13. MASK Operator

See "RECORD Directive" in the later section "Directives".

14. WIDTH Operator

See "RECORD Directive" in the following section.

15. Index Operators

The square bracket[ ] is an index operator.

[Example 5-10] Suppose a data segment is defined as:

```
DATA    SEGMENT
DAT     DB    25
DAT1    DB    10 DUP(05H)
DATA    ENDS
```

Learn how to use index operators in the following instructions.

(1) MOV   AL,   DAT
(2) MOV   BX,   OFFSET DAT1
    MOV   AL,   [BX]
(3) MOV   AL,   [BX+4]

The first instruction is a direct memory reference, its source operand is specified by the name DAT, so the processor locates the data item 25 by combining the DS with the offset value of DAT. For the second and the third group, they are indirect addressing of memory, the operands will be fetched from DS:0001H and DS:0005H respectively.

For indirect addressing of memory, an operand references a base or index register constants, offset variables, and variables. The index operator, which uses square brackets, acts like a plus (+) sign. A typical use of indexing is to reference data items in tables, you can use the [ ] to reference indexed memory. For example, load the fifth entry of Data_table into the CL (note that Data_table [0] is the first entry):

```
Data_table  DB   10   DUP ( 31H )      ; Defined table
...
MOV    CL,   Data_table [4]            ; Get fifth entry from Data_table
```

Here, an immediate number 4 is in square bracket.

16. Segment Override Operator

This operator, coded as a colon (:), calculates the address of a label or variable relative to a particular segment. Its general format is

Segment : expression

The named segment can be any of the segment registers or a segment name. The expression can be a constant, an expression, or a SEG expression.

[Example 5-11]

```
MOV    BH,    ES:10H       ; Override the default DS segment register, access from ES + 10H
MOV    CX,    SS:[BX]      ; Access from SS + offset in BX
```

An instruction may have a segment override operator apply to only one operand.

17. SHORT Operator

The purpose of the SHORT operator is to modify the NEAR attribute of a JMP destination that is within +127 to −128 bytes. The format is

JMP   SHORT   label

The assembler reduces the machine code operand from two bytes to one. This feature is useful for near jumps that branch forward, since the assembler initially doesn't know the distance of the jump address and may assume two bytes for a near jump.

5.3 Directives

After discussing the format of the directive, some details about the directives (**pseudo-operations**) that control the assembly process must be learned. **Table 5-2** lists part of assembly directives. Directives indicate how an operand or section of a program is to be processed by the assembler. Some directives generate and store information in the memory, while others do not. For example, the DB (**define byte**) directive stores bytes of data in the memory, while the **BYTE PTR** directive never stores data.

Table 5-2 Some common assembly directives

Directive	Function
EXIT	Exits to DOS
ALIGN	Starts the next data item or instruction on an address according to a given value which must be a power of 2
ASSUME	Informs the assembler of the name of each segment for full segment definitions
BYTE	Indicates byte-sized, as in BYTE PTR
DB	Defines byte(s) (8-bit)
DD	Defines doubleword(s) (32-bit)
DQ	Defines quadword(s) (64-bit)
DT	Defines ten byte(s) (80-bit)
DUP	Generates duplicates
DW	Defines word(s) (16-bit)
DWORD	Indicates doubleword-sized, as in DWORD PTR
END	Ends a program file
ENDM	Ends a macro sequence
ENDP	Ends a procedure
ENDS	Ends a segment or data structure
EQU	Equates data to a label
FAR	Defines a far pointer
MACRO	Designates the start of a macro sequence
NEAR	Defines a near pointer
OFFSET	Specifies an offset address
ORG	Sets the origin within a segment
PROC	Starts a procedure
PTR	Designates a pointer
SEGMENT	Starts a segment
STACK	Starts a stack segment
WORD	Indicates word-sized, as in WORD PTR
STRUC	Defines the start of a data structure

5.3.1 Data Definition and Storage Allocation

Statements that pre-assign data and reserve storage have the general format:

Variable   Mnemonic   Expression or data items; Comments

Variable: It is optional, but if it is presented, it is assigned the offset of the first byte that is reserved by the directive. Unlike the label field, a variable must be terminated by a blank, not a colon.

Mnemonic: it determines the length of each operand and is one of the following:

(1) **DB (Define Byte)**: Each operand datum occupies one byte.

(2) **DW (Define Word)**: Each operand datum occupies one word, with its low-order part being in the first byte and its high-order byte being in the second byte.

(3) **DD (Define Double Word)**: Each operand datum is two words long with the low-order word followed by the high-order word.

To pre-assign data the operand must be a constant, an expression or a string constant.

Example 5-12 is an example of data definition directive:

[Example 5-12] Suppose the data items are defined as following.

```
DATA            DB  ?
DATA _BYTE      DB  10, 4, 10H
DATA _WORD      DW  100, 100H, -16
DATA _DW        DD  3 * 20, 0FFFDH
DATA _STRING    DB  'ABCD'
```

In the example, all the variables have their own types. For example, the DATA, DATA _BYTE and DATA _STRING are defined as byte variables, DATA _WORD is a word variable while DATA _DW is a double word variable. They will cause a sequence of bytes to be pre-assigned as shown in **Figure 5-3**:

As shown in **Example 5-12**, we can see that the expression in an operand may define an uninitialized value or an initial constant. A question mark is used to, just like the definition of DATA in the example, indicate an uninitialized item. Also, the variables defined in the example have different offsets. The offset of DATA to the starting addressing of the data segment is 0, for the DATA_BYTE, it is one. For DATA_WORD, it is 4. For DATA _DW, it is 10, and for DATA _

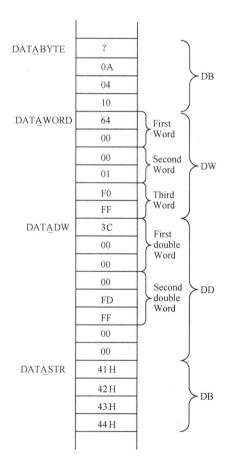

Figure 5-3 Typical pre-assignment of data using the DB, DW, and DD directives

STRING, it's 18.

The **ASCII** character string, DATA_STRING can be pre-assigned by using a string constant, which is defined within single quotes, such as PC', or within double quotes, such as "PC". The assembler stores character strings as object code in normal ASCII format, without the apostrophes.

[Example 5-13] The **ASCII** character string data definition and it's memory allocation. The data items are defined as following,

STRING1 DB 'ABCDEF'
STRING2 DW 'AB','CD','EF'
STRING3 DD 'AB''CD'

They will cause a sequence of bytes to be pre-assigned as shown in **Figure 5-4**:

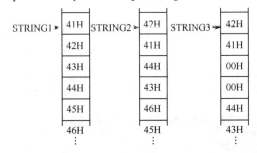

Figure 5-4 Memory allocation caused by defined character string

As shown in **Figure 5-4**, for the STRING1, microprocessor puts the **ASCII** codes for A (41H), B(42H), C(43H), D(44H), E(45H), and F(46H) in consecutive bytes beginning with the byte whose address is associated with the variable STRING1.

Note that in the STRING1, the first character in the string goes in the first byte, the second in the second byte, and so on. It's quite different from the STRING2 and STRING3,

The assembler uses the type attribute to determine whether the machine instruction is to operate on a byte or word.

Sometimes, we can use the duplication operator **DUP** to define data items. Several operands or operand patterns can be replaced with a form such as

$$\text{Exp DUP} \quad (\text{Operand}, \ldots, \text{Operand})$$

where **Exp** is an expression that evaluates to a positive integer, which causes the operand pattern to be repeated the number of times indicated by **Exp**.

[Example 5-14]

ARRAY1   DB   2 DUP(0,1,2,?)

;would cause the pre-assignment and allocation shown in **Figure 5-5** (**a**)

ARRAY2   DB   100 DUP(0, 2 DUP(1,2), 0, 3)

;would cause the pre-assignment and allocation shown in **Figure 5-5** (**b**)

The statements would cause the offsets of ARRAY1 and ARRAY2 to be stored as shown in **Figure 5-5(a)**. ARRAY1 and ARRAY2 may be variables or labels.

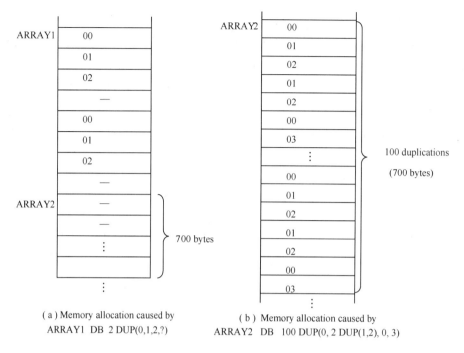

Figure 5-5 Application of the DUP operator.

5.3.2 EQU Directive

The EQU directive is used to redefine a data name or variable with another data name, variable, or immediate value. The directive should be defined in a program before it is referenced.

The formats of EQU directive is:

name EQU expression

It is very convenient to use EQU directive when an expression appears several times in a program, where the expression name may be any valid identifier and the expression may have the format of any valid operand, be any expression that evaluates to a constant (the expression name is then a constant name), or be any valid mnemonic.

The assembler replaces each occurrence of the name with the operand. Because EQU is used for simple replacement, it takes no additional storage in the generated object program.

【**Examples 5-15**】 Examples of the use of EQU with numeric and string data.

```
        DAT1    DB      4, 0AH
        COUNTER DW      1, 0BH
        SUM     EQU     COUNTER     ; Another name for COUNTER
        TEN     EQU     10          ; Numeric value
        ARRY    EQU     <'ABCD$'>
        MESSAGE EQU     ARRY        ; Replace with string
                ...
        INC     SUM                 ; Increment COUNTER(COUNTER=2)
        ADD     SUM, TEN            ; Add 10 to COUNTER(COUNTER=12)
```

```
            MOV      DX ,    OFFSET MESSAGE
            MOV      AH      09
            INT      21H                        ;Display the character string ABCD on the screen
```

5.3.3 Segment Definition Directive

The general format for a segment definition is
seg-name SEGMENT [**align**] [**combine**] ['class']
 ...
seg-name ENDS

As described earlier, a memory physical address is obtained by combining an offset with a segment address that is contained in a segment register. One of the tasks an assembler must perform is to assign the offsets of the labels and variables as it translates the instructions into machine language.

The assembler must also pass to the linker (via the object modules) all of the information that the linker will need in putting the various segments and modules together to form a program. Several directives are designed to instruct the assembler how to perform these functions.

To be able to assign the variable and label offsets the assembler must know the exact structure of each segment.

SEGMENT is a segment definition directive, it indicates starting of a segment. ENDS directive indicates the end of a segment (defined by SEGMENT) or a structure (defined by STRUC or STRUCT).

All operands are optional. The following subsections describe the entries for align, combine, and class.

1. Align

The align operand indicates the starting boundary for a segment. The following are the four orientations.

PARA is commonly used for all types of segments. **BYTE** and **WORD** can be used for segments that are to be combined within another segment, usually a data segment.

```
BYTE   Next address                          x x x x  x x x x  x x x x  x x x x  x x x x
WORD   Next even address (divisible by 2)    x x x x  x x x x  x x x x  x x x x  x x x 0
PARA   Next paragraph (divisible by 16, or 10H)  x x x x  x x x x  x x x x  x x x x  0000
PAGE   Next page address (divisible by 256, or100H)  x x x x  x x x x  x x x x  0000 0000
```

2. Combine

The combine operands **NONE, PUBLIC, STACK,** and **COMMON** indicate the way the linker to handle a segment:

(1) **NONE (default)**: the segment is to be logically separate from other segments, although it may end up physically adjacent to them. the segment is presumed to have its own base address.

(2) **PUBLIC**: Link loads PUBLIC segments of the same name and class adjacent to one another. One base address is presumed for all such PUBLIC segments.

(3) **STACK**: Link treats STACK the same as PUBLIC. There must be at least one STACK defined in a linked .EXE program. If there is more than one stack, the SP is set to the start of the first stack.

(4) **COMMON**: If COMMON segments have the same name and class, the linker gives them the same base address. During execution, the second segment overlays the first one. The largest segment, even if overlaid, determines the length of the common area.

(5) **AT paragraph-address**: The paragraph must be defined previously. The entry facilitates defining labels and variables at fixed offsets within fixed areas of memory.

3. Class

The class entry can help the linker associate segments with different names, identify segments, and control their order. Class may contain any valid name, contained in single quotes. The linker uses the name to relate segments that have the same name and class.

[Examples 5-16] Examples of the segment definition.

```
VAR_DATA SEGMENT
        DATA1   DB   12H
        DATA2   DB   34H
        DATA3   DW   9876H
VAR_DATA ENDS
```

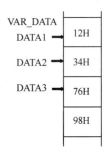

Figure 5-6 The memory allocations and offsets of the three variables

In **Examples 5-16**, the variable DATA1, DATA2, DATA3 are defined in the same Segment VAR_DATA. DATA1 and DATA2 occupy one byte respectively, while the DATA3 is 2 bytes long. The memory allocation and offset of these three variables are shown in **Figure 5-6**.

5.3.4 Assume Directive

The assignments of the segments to the segment registers are made with directives which are written :

ASSUME   Segment register name: Segment name, Segment register name: Segment name,...

For example, the statement

ASSUME   CS: CODE, DS:DATA, ES:EXTRA

would inform the assembler that it is to assume that the segment address of CODE_SEG is in CS, of DATA is in DS, and of EXTRA is in ES. An assignment is not made for SS, presumably because either the stack is not used or the assignment for SS is in a separate ASSUME statement.

It is important to note that the ASSUME directive does not load the segment addresses into the corresponding segment registers. The starting address of a segment has to load to segment register by transfer instruction.

[Example 5-17]

CODE_SEG   SEGMENT

```
            ASSUME    CS:CODE, DS:DATA, ES: EXTRA
    START:  MOV       AX,   DATA
            MOV       DS,   AX          ; load starting address of DATA segment to DS
            MOV       AX,   EXTRA
            MOV       ES,   AX          ; load starting address of EXTRA segment to ES
            ⋮
    CODE_SEG  ENDS
```

5.3.5 PROC Directive

A procedure is a block of code that begins with the PROC directive and terminates with ENDP. The format of a procedure definition is:

<procedure name> **PROC**　　{**FAR** | **NEAR**}
　　⋮　　　; * * * your (main/ sub) program here * * *
<procedure name> **ENDP**

In the procedure definition, **FAR** allows us to call outside the code segment (inter-segment call) while **NEAR** only allows us to call inside the same code segment (intra-segment call).

Although technically you may enter a procedure inline or by a JMP instruction, the normal practice is to use CALL to enter and RETN or RETF to exit. The CALL operand may be a NEAR or FAR type specifier.

An omitted operand defaults to NEAR. If a called procedure is external to the calling segment, it must be declared as PUBLIC, and you should use CALL to enter it.

For an .EXE program, the main PROC that is the entry point for execution must be FAR.

Also, a called procedure under a different ASSUME CS value must have the FAR attribute. A far label may be in another segment, which CALL accesses by a segment address and offset.

[**Example 5-18**] Simple Procedure Call.

```
CODE_SEG    SEGMENT
BEGIN   PROC    FAR
⋮
CALL    A
CALL    B
⋮
BEGIN   ENDP
A    PROC    NEAR
⋮
RET
A    ENDP
CODE_SEG    ENDS
CODE_SEG1   SEGMENT
B    PROC    FAR
```

```
    ⋮
    RET
B       ENDP
CODE SEG1       ENDS
```

From the example, we notice that if the subroutine A to be called within the code segment, **PROC NEAR** can be defined for the procedure. However, when we want to call the procedure outside the code segment, e. g. B, **PROC FAR** must be defined. After finishing all the procedure and returning to the program after **CALL** function, **RET** tells the assembler to return back to the original program. Otherwise, the program would continue to execute with unpredictable results.

5.3.6 END Directive

The END directive is needed to signal the end of a set of assembler language code, the form of END directive is:

END   Label

【Example 5-19】 Read the following program to learn how to use the directives.

```
DATA_SEG    SEGMENT                         ;Data segment beginning
    OPER1   DW      12                      ;Variables definition
    OPER2   DW      230
    RESULT  DW      ?
DATA_SEG    ENDS                            ;Data segment ending
CODE_SEG    SEGMENT                         ;Code segment beginning
    ASSUME  CS:CODE_SEG, DS:DATA_SEG        ;Indicates the segment address of CODE_SEG is
                                             in the CS and DATA_SEG is in the DS register
    START:  MOV     AX,     DATA_SEG
            MOV     DS,     AX
            MOV     AX,     OPER1
            ADD     AX,     OPER2
            JGE     STORE
            NEG     AX
    STORE:  MOV     RESULT,AX
            MOV     AH,     4CH
            INT     21H                     ;Return DOS
CODE_SEG    ENDS                            ;Code segment ending
            END     START                   ;The codes ending
```

5.3.7 ORG Directive

The ORG directive is used to indicate the starting offset of the data in the data segment. Its general format is:

ORG   expression

The expression must be from a 2-byte absolute number and must not be a symbolic name.

【Example 5-20】 Suppose 2 data segments are defined as following:

```
    DAT    SEGMENT                    DAT    SEGMENT
        DATA1   DW    1234H               ORG    20H
        DATA2   DB    56H                 DATA1  DW    1234H
        DATA3   DW    'ABCD$'             DATA2  DB    56H
    DAT    ENDS                           DATA3  DW    'ABCD$'
                                          ORG    $+9
                                      DAT    ENDS
```

Please diagram the data memory location of these 2 segments:

As shown in Figure 5-7, we can get that initially, the assembler stores the data 1234H beginning at 00. Because DATA1 is 2-byte long, the location counter is incremented to 02 for the location of the next item. Because DATA2 is 1-byte long, the location counter is incremented to 03, and so forth. But when the ORG directive is used, it will redefine the memory locations originally. The DATA1 will be located beginning with 20H.

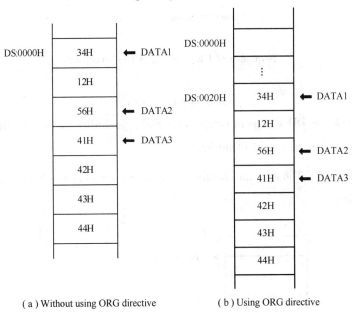

(a) Without using ORG directive   (b) Using ORG directive

Figure 5-7   The pre-assignment of data in DAT segment

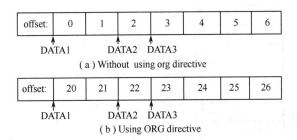

Figure 5-8   The offset of each variable in DAT segment

An operand containing a dollar symbol ($), as in the last ORG, refers to the current value in the location counter. The operand $ +5 therefore sets the location counter to 26 +5, or 1FH.

When you use ORG to redefine memory locations, be sure to reset the location counter to the correct value and that you account for all redefined memory locations. Also, the redefined variables should not contain defined constants, these would overlay constants on top of the original ones. ORG cannot appear within a STRUC definition.

5.3.8 Structures

All elements allocated by a single storage definition statement must be of the same type (bytes, words, or double-word). It is desirable, especially in business data processing applications, for a variable to have several fields, with each field having its own type.

A structure definition gives the pattern of the structure and may have the simplified form

        Structure name   STRUC

        ...

        Sequence of DB, DW, and DD directives

        ...

        Structure name ENDS

If a DB, DW, or DD statement includes a variable identifier, it denotes the beginning of a field and is referred to as a field identifier.

[Example 5-21] The structure for the personnel record shown in **Figure 5-9** could be defined as following:

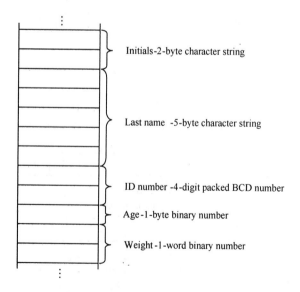

Figure 5-9 Fields in a typical personnel record data structure

```
PERSONNEL_DATA    STRUC
    INITIALS      DB    'XX'
    LAST_NAME     DB    5 DUP(?)
    ID            DB    0,0
    AGE           DB    ?
    WEIGHT        DW    ?
PERSONNEL_DATA    ENDS
```

The structure definition does not reserve storage or directly preassign values; it merely defines a pattern. Therefore, to reserve the necessary space it must be accompanied by a statement for invoking the structure.

【**Example 5-22**】Please read the program given below, then give the assignment of data in diagram after its execution.

```
DATA SEGMENT
        DD1 DB 30H,40H,60,71,78,30, 91,83,64,100
        DD2 DB 10 DUP(?)
        TAN1 STRUC
                D1 DB 2,25,10H
                D2 DB 'WTAN'
                SYY DB 0FFH
        TAN1 ENDS
        SUN TAN1 < ,'DOG1',2 >
        CAT TAN1 < ,'be', >
DATA ENDS
CODE SEGMENT
        ASSUME CS:CODE , DS:DATA
START:
        MOV AX , DATA
        MOV DS , AX
        MOV DH , SUN.SYY
        MOV AL , CAT.SYY
        MOV AH , 4CH
        INT 21H
CODE ENDS
END START
```

Example 5-22 is an example application of a structure directive. The program execution will cause a sequence of bytes to be assigned as shown in **Figure 5-10**.

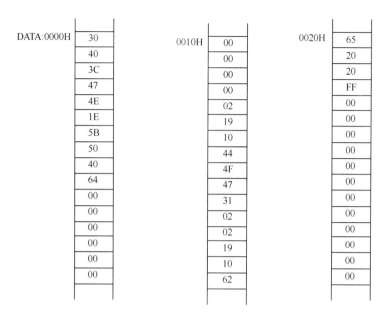

Figure 5-10 The data assignment using the structure directive

5.3.9 Records

1. RECORD Definition

The RECORD directive is for defining a bit pattern within a word or byte. It has the form

Record name **RECORD** Field specification, ... , Field specification where each field specification is of the form

Field name: Length = Pre-assignment

with the pre-assignment being optional.

【Example 5-23】 RECORD example

TAN　　　RECORD　　X:6,　　Y:4,　　Z:6
AP1　　　TAN　　　　<5 , 10, >
AP2　　　TAN　　　　<12, , 20>

TAN would break a word into 3 fields and give them the names X, Y, and Z. The lengths of the fields in bits would be 6, 4, and 6 respectively.

The statement AP1 would actually reserve the word, associate it with the variable AP1, and pre-assign X to 5 and Y to 10 as shown in **Figure 5-11**.

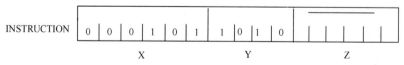

Figure 5-11 Typical use of RECORD to subdivide a word

For AP2, it is the same as that for AP1, except it pre-assign X to 12 and Z to 20.

2. WIDTH Operator

The WIDTH operator returns a width as the number of bits in a RECORD or in a RECORD field.

3. MASK Operator

The MASK operator returns a mask of 1-bit representing the specified field and, in effect, defines the bit positions that the field occupies.

【**Example 5-24**】 Using the RECORD Directive.

The record definition and its initial are given as these:

    BITREC RECORD BIT1:3, BIT2:7, BIT3:6   ;*Define record*
    DEFBITS BITREC <101B,0110110B,011010B>   ;*Init record*

PA1:
    MOV BH, WIDTH BITREC   ; *Width of record (16)*
    MOV AL, WIDTH BIT2   ; *Width of field (07)*

PA2:
    MOV CL, BIT1   ; *Hex 0D*
    MOV CL, BIT2   ; *06*
    MOV CL, BIT3   ; *00*

PA3:
    MOV AX, MASK BIT1   ; *Hex E000*
    MOV BX, MASK BIT2   ; *1FC0*
    MOV CX, MASK BIT3   ; *003F*

PA4:
    MOV AX, DEFBITS   ; *Get record*
    AND AX, MASK BIT2   ; *Get record*
    MOV CL, BIT2   ; *Clear BIT1 & 3*
    SHR AX, CL   ; *Shift right*

PA5:
    MOV AX, DEFBITS   ; *Get record*
    MOV CL, BIT1   ; *Get shift 13*
    SHR AX, CL   ; *Shift right*
    MOV AX, 4C00H   ; *End processing*
    INT 21H

In **Example 5-24**, the three instructions following PA3 are the examples of the MASK. The MASK for each of the field defined in BITREC is:

FIELD	BINARY	HEX
BIT1	1110000000000000	E000
BIT2	0001111111000000	1FC0
BIT3	0000000000111111	003F

The three instructions following PA3 return the MASK values for BIT1, BIT2, and BIT3. The instructions following PA4 and PA5 isolate BIT2 and BIT1, respectively, from BITREC. PA4 gets the record into the **AX** register and use a MASK of BIT2 to AND it:

Record:	101 0110110 011010
AND MASK BIT2:	000 1111111 000000
Result:	000 0110110 000000

The effect is to clear all bits except those of BIT2. The next two instructions cause the AX to shift 6 bits so that BIT2 is right-adjusted:

$$0000000000110110 \ (0036H)$$

The example following PA5 gets the record into the AX, and because BIT1 is the left most field, the routine simply uses its shift factor to shift right 13 bits:

$$0000000000000101 \ (0005H)$$

【Example 5-25】 Please read the program given below, then give the assignment of data after its execution.

```
DATA SEGMENT
        ORG 0050H
        TAN RECORD X:6,Y:4,Z:6
        AP TAN <5,10,>
        BB TAN <12,,20>
DATA ENDS
CODE SEGMENT
        ASSUME CS:CODE , DS:DATA
START:
        MOV AX , DATA
        MOV DS , AX
        MOV AX , WIDTH TAN
        MOV BH , WIDTH Y
        MOV AL , MASK Z
        MOV CX , BYTE PTR MASK  X
        MOV AH , 4CH
        INT 21H
CODE ENDS
END START
```

Example 5-25 is an example application of a record directive. The program execution will cause a sequence of bytes to be assigned as shown in **Figure 5-12**.

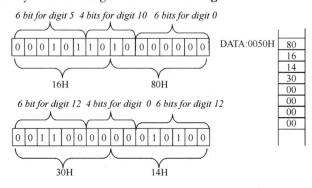

Figure 5-12 The data assignment using the structure directive

5.3.10 The PAGE and TITLE Listing Directives

The PAGE and TITLE directives help to control the format of a listing of an assembled program. This is their only purpose, and they have no effect on subsequent execution of the program.

PAGE. At the start of a program, the PAGE directive designates the maximum number of lines to list on a page and the maximum number of characters on a line. Its general format is

PAGE [length] [, width]

For example, the directive PAGE 60, 132 provides 60 lines per page and 132 characters per line.

TITLE. You can use the TITLE directive to cause a title for a program to print on line 2 of each page of the program listing. You may code TITLE once, at the start of the program. It is general format is

TITLE text [comment]

For the next operand, a recommended technique is to use the name of the program, as cataloged on disk. **For example,** if you named the program ASMSORT, code that name plus an optional descriptive comment (a leading ';' is not required), all up to 60 characters in length, like this:

TITLE ASMSORT   *Assembly program to sort customer names*

SUBTTL/SUBTITLE Directive

The SUBTTL directive (SUBTITLE since MASM 6.0 and % SUBTTL in TASM Ideal mode) causes a subtitle of up to 60 characters to print on line 3 of each page of an assembly source listing. You may code this directive any number of times. The general format is:

SUBTTL/SUBTITLE   text

5.3.11 EXTRN/EXTERN Directive

The EXTRN (or EXTERN since MASM 6.0) directive informs the assembler and linker about data variables and labels that the current assembly references, but that another module (linked to the current one) defines. The general format is

EXTRN/EXTERN name: type [,...]

The name is an item defined in another assembly and declared in it as PUBLIC. The type specifier can refer to either of the following:

(1) **Data items:**   ABS (a constant), BYTE, WORD, DWORD.
(2) **Distance:**   NEAR or FAR.

5.3.12 GROUP Directive

A program may contain several segments of the same type (code, data, or stack). The purpose of the GROUP directive is to collect segments of the same type under one name, so that they reside within one segment, usually a data segment. The general format is

Name GROUP seg-name [, seg-name],...

The following GROUP directive combines DSEG1 and DSEG2 in the same assembly module:

```
GROUPX   GROUP      DSEG1,   DSEG2
  DSEG1    SEGMENT  PARA     DATA'
           ASSUME   DS: GROUPX
           ...
           ...
  DSEG1    ENDS
                    ;
  DSEG2    SEGMENT  PARA     'DATA'
           ASSUME   DS: GROUPX
           ...
           ...
  DSEG2    ENDS
```

The effect of using GROUP is similar to giving the segments the same name and the PUBLIC attribute.

5.3.13 INCLUDE Directive

The include directive, when encountered in the source file, switches program input from the current file to the file specified in the parameter list of the include. This allows you to construct the text files containing macros, source codes and other assembler items. The syntax for the include directive is:

                INCLUDE   < filename >

Filename must be a valid DOS filename. The assembler will merge the specified file into the assembly at the point of the include directive. Note that you can nest INCLUDE the statements inside files you include.

Using the include directive by itself does not provide separate compilation. You could use the include directive to break up a large source file into the separate modules and join these modules together when you assemble your file. The following example would include the PRINTF. ASM file during the assemble of the program:

                INCLUDE   PRINTF. ASM

Now the program can benefit from the modularity gained by this approach. The INCLUDE directive inserts the source file at the point of the include during assembly, exactly as though you had typed that code in the program.

5.3.14 LABEL Directive

The LABEL directive enables you to redefine the attribute of a data variable or instruction label. Its general format is

                Name   LABEL   type-specifier

For labels, you may use LABEL to redefine executable code as NEAR, FAR, or PROC, such as for a secondary entry point into a procedure. For variables, you can use the type specifiers BYTE, WORD, DWORD, FWORD, QWORD, or TBYTE, or a structure name, to redefine data items and the names of structures, respectively. **For example**, LABEL enable you define a field as both DB and DW.

要点：

伪指令没有对应的机器指令，它不是由 8086/8088 CPU 来执行，而是由 MASM - 86 识别，并完成相应的功能。

(1) 数据定义伪指令 DB、DW、DD 等：主要为数据项分配存储单元并预置初值。在数据定义的指令中，表达式是给变量赋予的初值，它可以是数值表达式、? 表达式（不带引号的?，表示可以预置任何内容）、字符串表达式和带 DUP 的表达式。DUP 是定义重复数据操作符。在表达式中，使用 DUP 操作符格式时，DUP 后括号里的内容是重复的内容，其前是重复的次数。

(2) 段定义伪指令 SEGMENT/ENDS：存储器在逻辑上是分段的，各段的定义由伪指令实现。段定义伪指令为程序的汇编和连接说明了段名、分段的各种属性以及分段的开始和结束。段名是自定义符，开始的段名与结束的段名必须相同。段的长度不超过 64KB。SEGMENT 后面的参数是可选项。

(3) 段寄存器说明伪指令 ASSUME 是告诉汇编程序在汇编时，段寄存器"CS:"、"DS:"、"SS:"和"ES:"应具有的符号段基址，以便汇编指令时确定段和建立错误信息。但是段寄存器的实际值（CS 除外）还要由传送指令在执行程序时赋值。

(4) 过程定义伪指令 PROC/ENDP 是为了实现过程调用而设的，过程起始名和终止名必须相同。类型有 NEAR 和 FAR 两种，默认时系统约定是近过程。调用格式为 CALL 〈过程名〉。当选 NEAR 时，过程是段内调用，过程中的 RET 是段内返回。当选 FAR 时，过程是段间调用，过程中的 RET 是段间返回。

(5) 定位伪指令 ORG 把以下语句定义的内存数据或程序，从表达式指定的起点（偏移地址）开始连续存放，直至遇到新的 ORG 指令。表达式的值是一个无符号数。

(6) 结构伪指令 STRUC/ENDS 是把多个数据定义语句组织成一个结构，而把每一个数据定义语句称为一个结构字段，内存变量名称为字段名。

结构定义不真正为结构分配存储空间，它是在汇编时进行结构说明的。

对结构存储单元分配及赋值时，应采用如下格式：[变量名]　结构名〈字段值表〉。

变量名是自定义符，字段值表的各字段值赋给结构的各字段中，其顺序与结构定义的顺序相同，若字段值表中某些项缺省，则保留初值。应用时，尖括号"〈〉"不能缺省。

结构引用采用的格式为："结构变量名．结构字段名"。

(7) 记录伪指令主要有以下三种类型：

① 记录定义伪指令 RECORD：完成对内存单元二进制位的定义，这在实际应用中可以实现按位开关量或按位组合信息的处理。

记录名和字段名是自定义符，字段宽度是 1~16 的常数。记录是把 1~16 个二进制位分为段并赋予一个字段名。记录定义伪指令不真正为记录分配内存单元，它只是在汇编时进行记录名、字段名以及记录长度的说明。

② 记录存储单元分配及赋值:格式为"[变量名]　记录名〈字段值表〉"。变量名是自定义符,字段值表中的各字段值赋给记录的各字段中,其顺序与记录定义的顺序相同,若字段值表中的某些项缺省,则缺省值为 0。应用时尖括号"〈〉"不能缺省。

③ 记录操作符有两个,其中 WIDTH 可用于求出记录或字段所占的位数;MASK 用于返回记录字段位或使用情况,它是 8 位或 16 位二进制数,1 表示是该字段位,0 表示不是该字段位。

5.4 Macro Processing

5.4.1 Macro Definition

For each symbolic instruction, the assembler generates one machine language instruction. However, if you want to repeat the same instruction(s) with a calling statement, there will be overhead in the calling procedure. In this regard, a macro statement can help to solve the problem. First of all, macro can simplify and reduce the amount of coding. Furthermore, it is easy to be read. Finally, errors can be reduced in the program.

A macro is like a procedure that inserts a block of statements at various points in the assembly program during assembly. Unlike the assembly instructions you write, the conditional assembly (talked later) and macro language constructs execution during assembly. The conditional directives and macro statements do not exist when your assembly program is running. The purpose of these statements is to control which statements the assembler assembles into your final execution file. The macro directives let you emit repetitive sequences of instructions to an assembly language file like high level language procedures and loops, but without any running overheads.

The assembler has facilities that programmers use to define the macros. A specific macro name is defined for the macro along with the assembly instructions that the macro is to generate. Next, the instructions are defined within the macro.

The syntax of a macro definition is:

```
<Macro name>     MACRO      (arg1,arg2,...)
                 : : : : :       ;  * * * your macro here * * *
                 ENDM
```

The **MACRO** directive tells the assembler that the following instructions up to **ENDM** are to be part of the macro definition. The **ENDM** tells the assembler the end of the macro definition.

【Example 5-26】Example of Macro definition.

```
AD1     MACRO     Y, X
        MOV       CL,X
        RO&Y      AX,CL
        ENDM
```

AD1 is the name of the macro that you defined before the **MARCO** directive. It should be a valid and unique symbol in the source file. You can use this identifier to expand the macro (as the calling subroutine). For example, AD1, R, 5. When you expand the macro AD1, the arguments Y and X are instead by R and 5 respectively. (Just like the formal and actual parameters in the calling subroutine).

During the expansion, the assembler will expand every occurrence of the macro in the code segment. It will replace the macro name by the macro body at the destination address, and the formal arguments inside macro body by the supplied arguments. In this case, no run-time overhead is needed. The only thing the assembler done is a simple substitution followed by normal assembling.

Note that the assembler does not immediately assemble the instructions between the MACRO and ENDM directives when the assembler encounters the macro. Instead, the assembler stores the text corresponding to the macro into a special table (called the symbol table). The assembler inserts these instructions into your program when the assembler expands the macro.

[Example 5-27] Macro definition and its expansion.

```
DATA SEGMENT
        BB2 MACRO M , N
                MOV    CL , N
                MOV    AX , 01H
                RO&M   AX , CL
        ENDM
DATA ENDS
CODE SEGMENT
        ASSUME CS:CODE , DS:DATA , ES:DATA
START:
        MOV AX , DATA
        MOV DS , AX
        MOV ES , AX
        BB2  R ,  5         ; Shift AX to right 5 bits
        BB2  L ,  5         ; Shift AX to left 5 bits
        MOV AH , 4CH
        INT  21H
CODE ENDS
END START
```

As shown in **Example 5-27**, in the code segment, use the symbol BB2 to call the macro procedure. Just write BB2. When you do this, the assembler will insert the statements between the **MACRO** and **ENDM** directives into your code at the point of the macro invocation.

If arguments are needed, formal parameters and actual parameters (arguments) are defined in the macro definition and the calling macro respectively. For example, the macro BB2 has 2 parameters M and N. They will be replaced by R and 5 respectively when the BB2 is first called. The assembler will substitute the actual parameters appearing as operands for the formal parameters appearing in the macro definition. The assembler does a straight textual substitution only when expanding the macro.

In **Example 5-27**, the '&' symbol is used before parameter M like &M. It means M will be substituted by R for the first BB2 expanding. The instruction "RO&M AX, CL" will be changed as "ROR AX, CL". For the second BB2 expanding, this instruction alters rotation direction. The left move operation carries out.

【**Example 5-28**】 Example of using parameters in macro.

```
           DATA_SEG    SEGMENT    PARA
           MSG1        DB         "Hello$"
           MSG2        DB         "Everybody$"
           DATA_SEG    ENDS
PRTMSG     MACRO       MSG            ;/ * * * printing message * * */
           MOV         AH, 09H
           LEA         DX, MSG
           INT         21H
           ENDM
CODE_SEG   SEGMENT
     ASSUME   CS: CODE_SEG, DS: DATA_SEG
     MOV    AX,   DATA_SEG
     MOV    DS,   AX
        ⋮
       PRTMSG    MSG1
       PRTMSG    MSG2
        ⋮
     MOV    AH,   4CH
     INT    21H
CODE_SEG   ENDS
```

The macro definition named PRTMSG provides for the use of INT 21H function 09H to display message; when using PRTMSG, you have to supply the name of the message, which references a data area terminated by a dollar sign.

After the assembling, the program will become:

```
CODE_SEG  SEGMENT    PARA
           ⋮
          MOV   AH, 09H
          LEA   DX, MSG1
          INT   21H
          MOV   AH, 09H
          LEA   DX, MSG2
          INT   21H
           ⋮
```

```
CODE_SEG    ENDS
```

As shown in **Example 5-28**, you can output different message in different case. It can save a considerable amount of typing in the program to avoid a large duplication of the same statements. Writing the macro can simplify your program, and easy to write.

If we call the macro procedure as:

```
MSG      < A * 5 >,
```

the assembler will automatically convert the text object A * 5 to the macro. If you use

```
MSG      %A * 5,
```

the assembler will evaluate the expression "A * 5" and convert the resulting numeric value to a text value consisting of the digits that represent the value before the expansion.

5.4.2 Macro Sequence and Procedure Calling

The macros are very similar with procedure calling except that there is no call or return. The procedure definition generates code when the assembler encounters the PROC directive.

When the CALL instruction executes, it will push the return address onto the stack and jump to the procedure. Then it will execute the code in the procedure one by one. When RET instruction executes, it will pop the return address off the stack and return the calling code.

The macro sequence, on the other hand, does not emit any code when processing the statements between the MACRO and ENDM directives. However, upon encountering macro in the mnemonic field, the assembler will assemble every statement between the MACRO and ENDM directives. Then it will emit the code to the output execution file (* .exe).

At the running time, the CPU executes these instructions without the procedure overhead.

The execution of a macro expansion is usually faster than that of a procedure calling when they have same code implemented. Furthermore, to call a macro, you simply specify the macro name as though it were an instruction or directive. To call a procedure, you need to use the CALL instruction.

Macros are useful when you need to simplify and reduce the amount of repetitive coding, and to reduce errors caused by repetitive code typing. It makes an assembly program more readable.

5.4.3 Macro Directives

1. Local Directives

The syntax of Local is:

LOCAL   (**label_1**, **label_2**, ...)

Since the assembler copies the macro text directly, if the same label is redefined many times, the assembler will generate a multiple definition error. To overcome this problem, the LOCAL directives can be used to define the local label within a macro.

[Example 5-29]

```
CLS      MACRO
LOCAL    A
```

```
              MOV       AH, 06H
              MOV       AL, 00H
          A:    MOV       BH, 07H
                 ⋮          ⋮
              LOOP      A
              ENDM
```
The local label definition should be defined after the macro directives. During the expansion, the assembler will assign the different labels in the program.

2. Repetition Directives
The syntax of the REPT directives is:
REPT   < expression >
                      < statements >
ENDM

Another macro format is the repeat macro. A repeat macro is nothing but a loop that repeats the statements within the loop at specified value. There are three types of repetition directives, REPT, IRP and IRPC. These three directives cause the assembler to repeat a block of statements, terminated by **ENDM**. These directives do not have to be contained in a **MACRO** definition, but if they are, one **ENDM** is required to terminate the repetition and the second **ENDM** to terminate the **MACRO** definition. Expression must be a numeric expression that evaluates to an unsigned constant. The repeat directive duplicates all the statements between REPT and ENDM with the repeat value indicated in the expression.

3. IRP directive
Another form of the repeat macro is the IRP macro. The IRP (Indefinite Repeat) operation will cause a repeat of block of instructions up to the ENDM. The syntax is:
IRP   < < parameter >, < arguments > >
                < statements >
ENDM

The " < > " brackets are required around the items in the parameter and arguments. The IRP directive replicates the instructions between IRP and ENDM once for each item appearing in the argument. Furthermore, for each iteration, the first symbol in the parameter is assigned the value of the successive items from the second arguments.

[Example 5-30] Indefinite Repeat Example
```
    IRP      N, <1,2,3,4>
    DB       N
    ENDM
```
The loop emits 4 DB instructions, generating DB 1, DB 2, DB 3 and DB 4. The arguments can be any number of legal symbols, string, numeric, or arithmetic constants. Remember, the IRP loop, like the REPT loop, executes at assembly time, not at run time.

4. IRPC directive
The third form of the repeat macro is the IRPC macro. It differs from the IRP macro in that

it repeats a loop the number of times specified by the length of a character string rather than by the number of the operands present. Here is the general syntax:

```
IRPC       < <parameter> , <string argument> >
           <statements>
ENDM
```

The statements in the loop repeat once for each character in the string operand. The angle brackets " < > " must appear around the string.

[Example 5-31] Indefinite Repeat Example

```
IRPC       N, <1234>
DB         N
ENDM
```

The assembler will generate a block of the code for each character in the string argument. After the expansion, the assembler will generate DB 1, DB 2, DB 3 and DB 4. The arguments can be any number of legal symbols, string, numeric, or arithmetic constants.

5.5 DOS Function Calls

MS-DOS and the PC's BIOS provide a software connection between your application program and the underlying hardware. It's very convenient for you to program the hardware by using these system software (MS-DOS and the BIOS). You can access the IBM PC system hardware by programming directly or using ROM BIOS routines or making MS-DOS calls. But it's much easier for you to simply call a routine built into your system than to write the routine yourself.

This section is addressed on providing a brief introduction to the various BIOS and DOS services available to you. Instead of describing all of the routines or the options available to each routine, we will focus on some function routines which used most commonly in the DOS Function Calls.

5.5.1 The IBM PC BIOS

Rather than placing the BIOS routines at fixed memory locations in ROM, IBM used a much more flexible approach in the BIOS design. To call a BIOS routine, you use one of the 80x86's INT software interrupt instructions. The INT instruction uses the following syntax:

INT value

Value is some number in the range 0-255. Execution of the INT instruction will cause the 80x86 to transfer control to one of 256 different interrupt handlers.

If all you're doing is calling BIOS routines (as opposed to writing them), you can view the INT instruction as nothing more than a special call instruction.

The IBM PC BIOS uses software interrupts 5 and 10H..1AH to accomplish various operations. Therefore, the INT 5, and INT 10H..INT 1AH instructions provide the interface to BI-

OS. The following table summarizes the BIOS services:

INT Function

5H	Print screen operation.
10H	Video display services.
11H	Equipment determination.
12H	Memory size determination.
13H	Diskette and hard disk services.
14H	Serial I/O services.
15H	Miscellaneous services.
16H	Keyboard services.
17H	Printer services.
18H	BASIC.
19H	Reboot.
1AH	Real time clock services.

Most of these routines require various parameters in the 80x86's registers. Some require additional parameters in certain memory locations. The following sections describe the exact operation of many of the BIOS routine.

5.5.2 An Introduction to MS-DOS' Services

MS-DOS provides all of the basic file manager and device manager functions required by most application programs running on an IBM PC. MS-DOS handles file I/O, character I/O, memory management, and other miscellaneous functions in a (relatively) consistent manner. And that's exactly what it means.

We're only going to concentrate on a small subset of the available DOS commands in this section, and totally ignore those obsolete commands that have been augmented by newer, better, commands in later versions of DOS. Furthermore, we're going to skip over a description of those calls that have very little use in day to day programming.

Interrupts 20H through 27H are served for DOS operation, as described in **Table 5-3**.

Table 5-3 Interrupt for DOS operations

INT Instruction	Function Description
INT 20H	Terminate program
INT 21H	Dos function request. Requires a function code in the ah
INT 22H	Terminate address
INT 23H	Ctrl + Break Address. Designed to transfer control to a Dos routine when you press < ctrl > + < break > or < ctrl > + < c >.
INT 24H	Critical_error handler
INT 25H	Absolute disk read
INT 26H	Absolute disk write
INT 27H	Terminate but stay resident

5.5.3 MS-DOS Calling Sequence

MS-DOS is called via the INT 21H instruction. It contains more than 80 subroutines. To select an appropriate DOS function, you load the AH register with a function number before issuing the INT 21H instruction. Most DOS calls require other parameters as well. Generally, these other parameters are passed in the CPU's register set. The specific parameters will be discussed along with each call. Unless MS-DOS returns some specific value in a register, all of the CPU's registers are preserved across a call to DOS.

5.5.4 Frequently Used MS-DOS Functions

Generally, we can use the INT (interrupt) instruction, which exits from our program, enters a MS-DOS routine, performs the requested function, and returns to our program. There are different types of INT operations, some of which require a function code in the AH register to request a specific actions. **Table 5-4** lists some DOS function codes and their function descriptions in detail.

Table 5-4 Some MS-DoS fumctions

The parts of the INT 21H services			
Function # (AH)	Format	Output Parameters	Description
1	MOV AH, 1 INT 21H	AL-char read	Console Input w/Echo

2	MOV DL, 'A'
	MOV AH, 2
	INT 21H
			Console Output: Writes a single character to the display

3	MOV AH, 3 INT 21H	AL-char	Read Auxiliary Input: Reads a single character from the serial port

4	MOV DL, '$'
	MOV AH, 4
	INT 21H
			Auxiliary Output: Writes a single character to the output port

6	MOV DL, 0FFH
	MOV AH, 6
	INT 21H
	MOV DL, 'A'
	MOV AH, 6
	INT 21H
		AL-char read (if input DL = 0FFH)	Direct Console I/O: On input, if DL contains 0FFH, this function attempts to read a character from the keyboard. If a character is available, it returns the zero flag clear and the character in AL. If no character is available, it returns the zero flag set. On input, if DL contains a value other than 0FFH, this routine sends the character to the display. This routine does not do Ctrl-C checking

7	MOV AH, 7 INT 21H	AL-char read	Direct Console Input: Reads a character from the keyboard. Does not echo the character to the display. This call does not check for Ctrl-C

(续)

		The parts of the INT 21H services	
Function # (AH)	Format	Output Parameters	Description
8	MOV AH, 8 INT 21H	AL-char read	Read Keyboard w/o Echo: Just like function 7 above, except this call checks for Ctrl-C
9	BUF DB ' good bye $ ' ... MOV DX, OFFSET BUF MOV AH, 9 INT 21H		Display String: This function displays the characters from location DS:DX up to (but not including) a terminating ' $ ' character
0AH	BUF DB 20 DB ? DB 20 DUP (?) ... MOV DX, OFFSET BUF MOV AH, 0AH INT 21H		Buffered Keyboard Input: This function reads a line of text from the keyboard and stores it into the input buffer pointed at by DS:DX. This first byte of the buffer must contain a count between one and 255 that contains the maximum number of allowable characters in the input buffer. This routine stores the actual number of characters read in the second byte. The actual input characters begin at the third byte of the buffer
2AH	MOV AH, 2AH INT 21H	AL-day (0 = Sun, 1 = Mon, etc.) CX-year DH-month (1 = Jan, 2 = Feb, etc.) DL-Day of month (1 – 31)	Get Date: Returns the current MS-DOS date

2BH	MOV CX, 2003H
	MOV DH, 08H
	MOV DL, 18H
	MOV AH, 2BH
	INT 21H
			Set Date:Sets the current MS-DOS date

2CH	MOV AH, 2CH INT 21	HCH-hour(24hr format) CL-minutes DH-seconds DL-hundredths	Get Time:Reads the current MS-DOS time. Note that the hundredths of a second field has a resolution of 1/18 second

As shown in **Table 5-4**, in order to use DOS function calls, we need always to place the function number(code) into register AH and load all other pertinent information into registers, as described in the table as entry data. Once this is accomplished, an INT 21H is executed. Note that some function calls require a segment and offset address. Sometimes, DOS function calls could change some registers in certain case. To prevent these problems, it is advisable to save registers where problems occur.

要点:

(1) 宏指令。为了简化汇编语言源程序的书写,把一些频繁出现的程序段定义为"宏指令",当程序中遇到这个程序段时,只需用一条宏调用语句,这样有效地缩短了源程序的长度,使源程序易读,也减少了由于重复书写而引起的错误。

宏指令与过程的区别:

① 宏调用语句由宏汇编程序中的宏处理程序来识别,并完成相应的处理,而调用过程中的 CALL 语句由 CPU 来执行。

② 汇编语言源程序在汇编过程中要将宏指令所代替的程序段汇编成相应的机器代码,每次调用均需插入到源程序的目标代码中,因此使用宏调用并不能缩短目标代码的长度。在过程调用中,由于其目标代码与主程序是分开独立存放且在存储器中只保留一份,因此能有效地缩短目标代码的长度。

(2) 系统功能调用。为给编写汇编语言源程序提供方便,MS-DOS 系统中设置了几十个内部子程序,它们可完成 I/O 设备管理、存储管理、文件管理和作业管理等功能。其入口地址已由系统置入中断入口地址表中,在汇编语言源程序中可用软件中断指令调用它们。

一般常用的软中断指令有 8 条,系统规定它们的中断类型码为 20H – 27H,其中"INT 21H"是系统功能调用,它本身包含 80 多个子程序,每个子程序对应一个功能号,其编号从 0 到 57H。

系统功能调用中的几十个子程序是汇编语言程序员的重要工具,程序员不必了解所使用设备的物理特性、接口方式及内存分配等,不必编写繁琐的控制程序。调用它们时采用统一的格式,只需要使用以下 3 个语句即可,

① 传送入口参数到指定寄存器中;

② 功能号送入 AH 寄存器中;

③ INT 21H。

有的子程序无入口参数,则只需安排后两个语句,调用结束后,系统将出口参数送到指定寄存器中或从屏幕显示出来。

5.6 Assembling, Linking and Executing a Program

5.6.1 The Assembler and Linker

When you finish a program by assembling language, you have already created a source file. Remember that the file should have an extension of ASM. The **Figure 5-13** indicates the assembly process.

The assembler program MASM converts a symbolic source file (module) into a hexadecimal object file (.OBJ). Meanwhile, a listing file named ××.LST is created. The source listing file (.LST) contains the assembler version of the source file and its hexadecimal machine language equivalent. The cross-reference file (.CRF) lists all labels and pertinent information required for cross-referencing. An .OBJ file, which is required for linking a program into an

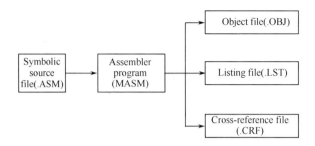

Figure 5-13　An assembly process of assemble program

executable form. An .LST file contains error diagnostics which can be used to examine the generated machine code. A .CRF file is useful for a large program where you want to see which instructions reference which data items. Also, requesting a .CRF file causes the assembler to generate statement numbers for in the .LST file to which the .CRF file refers.

The linker program reads the object files that are created by assembler program and link them together in to a single execution file, which is generated with the file name extension EXE.

Now, let's learn how to assemble, link and execute a program with an example.

【Example 5-32】 Develop a program to accomplish the string characters transfer.

According to what we have learned, the program can be written as following:

```
DATA    SEGMENT
        BUFFER1   DB   'Hello, everybody!'
DATA    ENDS
EXTRA   SEGMENT
        BUFFER2   DB   17   DUP(?)
EXTRA   ENDS
CODE    SEGMENT
    ASSUME   CS:CODE,DS:DATA,ES:EXTRA
    STA: MOV   AX,   DATA
         MOV   DS,   AX
         MOV   AX,   EXTRA
         MOV   ES,   AX
         LEA   SI,   BUFFER1
         LEA   DI,   BUFFER2
         MOV   CX,   17
         CLD
         REP   MOVSB
         MOV AH, 4CH
         INT 21H
CODE    ENDS
        END   STA
```

And we named the program as trychar5.asm.

5.6.2 Assembling a Source Program

The Microsoft assembler program (up to version 5.x) is MASM.EXE. As of version 6.0, the Microsoft assembler normally uses the ML command, but also accepts MASM for compatibility with earlier versions.

The general format for a command line to assemble a program is

MASM/TASM [options] source [,object] [,listing] [,crossref]

For example, the following information is displayed when **Example 5-32** is assembled:

D:\MASM > masm trychar5

Microsoft(r) Macro Assembler Version 5.00
Copyright(c) Microsoft Corp 1981 – 1985, 1987. All rights reserved.

Object filename [trychar5.OBJ]:
Source listing [NUL.LST]:
Cross-reference [NUL.CRF]:

48894 + 397426 Byte symol space free

0 Warning Errors
0 Severe Errors

(1) Options provide for such features as setting levels of warning messages. The assembler's defaults are adequate for our purposes at this time.

(2) Source identifies the name of the source program, try chars in **Example 5-32**. The assembler assumes the extension .ASM, so you need not enter it.

(3) Object provides for a generated .OBJ file.

(4) Listing provides for a generated .LST file that contains both the source and object code.

(5) Crossref generates a cross-reference file containing the symbols used in the program, which you can use for a cross-reference listing. The extension is .CRF for MASM. For .OBJ, .LST and .CRF file, the drive, subdirectory, and filename may be the same as or different from those in the source.

5.6.3 Linking an Object Program

When your program is free of error messages, your next step is to link the object module that was produced by the assembler and that contains only machine code. The linker version for Microsoft is LINK. The command line for linking is

> **LINK objfile,exefile**[,**mapfile**][,**libraryfile**]

For example, you key a command "link trychar5", you will see the following information on the screen:

    D:\MASM > link trychar5

    Microsoft(r) Macro Assembler Version 3.60
    Copyright(c) Microsoft Corp 1981 – 1987. All rights reserved.
    Run file [TRYCHAT5.EXE]:
    List files [NUL.MAP]:
    Libraries [.LIB]:
    LINK: warning L4021: no stack segment

(1) Objfile identifies the object file generated by the assembler. The linker assumes the extension.OBJ, so you need not enter it.

(2) Exefile provides for generating an .EXE file.

(3) Mapfile provides for generating a file with an extension .MAP that indicates the relative location and size of each segment and any errors that LINK has found. A typical error is the failure to define a stack segment.

(4) Libraryfile provides for the libraries option, which you don't need at this early stage of assembly programming.

If the filename is to be the same as that of the source, you need not repeat it: the reference to drive number is sufficient to indicate a request for the file. The linker performs the following functions:

(1) Combines, if requested, more than one separately assembled module into one executable program, such as two or more assembly programs or an assembly program with a C program.

(2) Generates an .EXE module and initializes it with special instructions to facilitate its subsequent loading for execution.

Once you have linked one or more .OBJ modules into an .EXE module, you may execute the .EXE module any number of times. But whenever you need to make a change in the program, you must assemble and link it again.

5.6.4 Executing a Program

Having assembled and linked a program, you can now (at last!) execute it. If the .EXE file is in the default drive, you could cause the loader to read it into memory for execution by typing

    "filename .exe" or "file name" (without the .EXE extension)

Because some programs have no visible output, it is suggested that you run it under DEBUG and use trace commands to step through its execution.

5.6.5 Using the DEBUG Program

The DOS system comes with a program named DEBUG that is used for testing and debugging executable programs. It can allow you to view memory, to enter programs in memory, and to trace their execution. A feature of DEBUG is that it displays all program code and data in hexadecimal format and any data that you enter into memory must also be in hex format. DEGUB also provides a single-step mode, which allows you to execute a program one instruction at a time, so that you can view the effect of each instruction on memory locations and registers.

5.6.6 DEBUG Commands Exercise

In this section, we still take **Example 5-32** as the example to introduce some DEBUG commands.

1. The T Command

The T command used to trace the program step by step. You can observe execution of each instruction. As shown above, key in T (Trace) and press < Enter >, the instruction will be executed. The machine code is B8 (B8 (move to AX register) followed by 0B632301. The operation moves the 23 63 to the low half (AL) of the AX register and the 01 0B to the high half (AH) of the AX register:

$$\begin{array}{ccc} & AH & AL \\ AX: & 01 & 23 \end{array}$$

DEBUG display the offset of the operation on the registers. The IP register now contains 0103H 0008 (the original 0100H plus 3 bytes for the first machine code instruction). The value indicates the offset location in the code segment of the next instruction to be executed, namely:

0B65:0008 8EC0   MOV   ES   AX

To execute this add MOV instruction, enter another T.

You can key in successive T commands to step through the remaining instructions.

2. The R Command

The R command used to view the contents of the registers and flags. To view the initial contents of the registers and flags, key in the R command, followed by < Enter >, as shown below:

R (followed by < Enter >)

DEBUG will display the contents of the registers and flags in hexadecimal format as following:

D:\MASM > debug trychar5.exe
  -t
  -t
AX = 0B61 BX = 0000 CX = 005C DX = 0000 SP = 0000 BP = 0000 SI = 0000 DI = 0000
DS = 0B51 ES = 0B51 SS = 0B61 CS = 0B65 IP = 0003 NV UP EI PL NZ NA PO NC
0B65:0003 8ED8   MOV   DS, AX
  -t
AX = 0B61 BX = 0000 CX = 005C DX = 0000 SP = 0000 BP = 0000 SI = 0000 DI = 0000
DS = 0B61 ES = 0B51 SS = 0B61 CS = 0B65 IP = 0005   NV UP EI PL NZ NA PO NC

```
0B65:0005 B8630B          MOV     AX,0B63
-r
AX=0B61 BX=0000 CX=005C DX=0000 SP=0000 BP=0000 SI=0000 DI=0000
DS=0B61 ES=0B51 SS=0B61 CS=0B65 IP=0005   NV UP EI PL NZ NA PO NC
0B65:0005 B8630B          MOV     AX,0B63
-t
AX=0B63 BX=0000 CX=005C DX=0000 SP=0000 BP=0000 SI=0000 DI=0000
DS=0B61 ES=0B51 SS=0B61 CS=0B65 IP=0008 NV UP EI PL NZ NA PO NC
0B65:0008 8EC0            MOV     ES,AX
-t
AX=0B63 BX=0000 CX=005C DX=0000 SP=0000 BP=0000 SI=0000 DI=0000
DS=0B61 ES=0B63 SS=0B61 CS=0B65 IP=000A NV UP EI PL NZ NA PO NC
0B65:000A 8D36000 LEA SI,[0000]                DS:0000=6548
```

The IP register should be IP=0005, indicating that instruction execution is to begin 0005 bytes past the start of the code segment.

The flag register shows the following settings for the overflow, direction, interrupt, sign, zero, auxiliary carry, parity, and carry flags:

NV UP EI PL NZ NA PO NC

These settings means no overflow, up (or right) direction, enable interrupt, plus sign, nonzero, no auxiliary carry, parity odd, and no carry, respectively. At this time none of these settings is important to us.

As shown above the CS register contains 0B65H. And we can see the last line as:

0B65: 0005   B8630B

(1) 0B65 indicates the start of the code segment as 0B65 [0]. The value 0B65:0005 means offset 5H bytes following the CS segment address 0B65 [0].

(2) B8630B is the machine code that you entered at CS:0005.

(3) MOV AX,0B63 is the symbolic assembly instruction that DEBUG determines from the machine code. This instruction means, in effect, moving the immediate value 0B63 into the AX register.

3. The D Command

The D command displays the contents of a requested data area on the screen. The display consists of three parts:

(1) To the left is the hex address of the leftmost display byte, in **segment: offset** format.

(2) The wide area in the center is the hex representation of the displayed area.

(3) To the right is the ASCII representation of bytes that contain displayable characters, which can help you interpret the hex area.

The operation displays 8 lines of data, each containing 16 bytes (32 hex, digits), for 128 bytes in all, beginning with the address that you specify in the D command. Diagrammatically, we have:

Address	|← Hexadecimal representation →|	← ASCII →|
××××:××10	×× .... ×× - ×× .... ××	× .... ×
××××:××20	×× .... ×× - ×× .... ××	× .... ×
××××:××30	×× .... ×× - ×× .... ××	× .... ×
......		
××××:××80	×× .... ×× - ×× .... ××	× .... ×

```
- dds:0
0B61:0000   48 65 6C 6C 6F 2C 20 65 - 76 65 72 79 62 6F 64 79   Hello, everybody
0B61:0010   21 00 00 00 00 00 00 00 - 00 00 00 00 00 00 00 00   !...............
0B61:0020   00 00 00 00 00 00 00 00 - 00 00 00 00 00 00 00 00   ................
0B61:0030   00 00 00 00 00 00 00 00 - 00 00 00 00 00 00 00 00   ................
0B61:0040   B8 61 0B 8E D8 B8 63 0B - 8E C0 8D 36 00 00 8D 3E   .a....c....6...>
0B61:0050   00 00 B9 11 00 FC F3 A4 - B4 4C CD 21 D1 E3 D1 E3   .........L.!....
0B61:0060   A1 3A 21 8B 16 3C 21 89 - 87 BE 22 89 97 C0 22 80   .:!..<!..."..."."
0B61:0070   3E 45 07 00 74 0A FF 36 - 56 07 E8 21 FC 83 C4 02   >E..t..6V..!....
-
```

The address to the left refers only to the leftmost (beginning) byte, in segment:offset format; you can count across the line to determine the position of each other byte. The hex representation area shows two hex characters for each byte, followed by a space for readability. Also, a hyphen separates the second 8 bytes from the first 8, again for readability. Thus if you want to locate the byte at offset 000AH, start with 0B61H, and count ten bytes successively to the right.

要点:

(1) 汇编程序的功能。汇编程序的主要功能是将由汇编语言(助记符)编写的源程序翻译成用机器语言(二进制代码)编写的目标程序。下图可用以说明这个功能。

汇编程序功能

可以看出,汇编语言源程序是汇编程序(MASM 或 ASM)这个"翻译"的输入,而这个"翻译"的输出有以下 3 个文件:

① 目标代码文件。其中的地址数据还是浮动的(相对的),不能直接运行。

② 列表文件。包含程序的逻辑地址、代码程序及源程序对照清单,附有引用符号表,可用 DOS 的 TYPE 命令输出。

③ 交叉索引文件。包含了符号定义行号和引用行号,不能用 TYPE 命令输出,需要执行 CREF 文件后生成索引列表文件(.REF)才能输出。

(2) 程序的编辑、汇编及连接过程。

汇编语言的程序一般要经过的步骤如下：
① 建立源程序文件。
② 汇编(MASM 或 ASM)源程序。
③ 程序连接(LINK)。
④ 执行程序。
⑤ 调试程序(DEBUG)。

Tips

(NONE)　空缺表示本段不与任何段连接
align　定位方式
alignment directive　定位伪指令
ASSUME　段寄存器说明伪指令
AT paragraph-address　表示本段定位在表达式值指定的段地址处
auxiliary　辅助的, 补助的
CALL & RET　过程调用与返回指令
call procedure　过程调用
class　类别名
combine　连接方式
COMMON　表示本段与同名同类别的段共用同一段起始地址, 即同名同段相重叠, 段的长度是最长段的长度
console　控制台
data definition　数据定义
descriptive　描述, 说明
directive　伪指令
diskette　磁盘
dummy　虚的
duplication operator　重复操作符
echo　回送
eject　弹出
far call　远程调用
INT function　中断功能
INT instruction　中断指令
interrupt handler　中断处理
intransigent　非妥协性的, 不妥协的人
IRP directive　重复定义语句1
IRPC directive　重复定义语句2
iteration　重复操作
local directive　局域符号定义语句
machine language　机器码, 目标码

macro definition　宏定义
macro expansion　宏扩展
macro invocation　宏调用
macro processing　宏处理
MASK operator　返回记录字段位或使用情况
MEMORY　表示本段在连接时定位在所有段之上, 即高地址处
merge　合并
module　模块
near call　近程调用
obsolete　废弃的
omission　省略
ORG　定位伪指令
overhead　在头上的, 高架的
overlay　覆盖
PAGE　起始地址的低8位为0, 也称为页边界
PARA　起始地址的低4位为0, 也称为段边界, 为默认定位方式
parameter　参数
pre-assign　预赋值
PUBLIC　告诉连接程序与本段其他同名同类别的段连接起来, 公用一个段的起始地址, 形成一个物理段
quote　引用, 引证, 提供
reboot　重新启动
RECORD　记录伪指令定义
repetition directive　重复定义语句
return address　返回地址
segment definition　段定义
STACK　表示本段是堆栈段, 连接方式同PUBLIC, 连接后的起始地址在SS寄存器中。连接程序要求源程序至少要有一个堆栈段, 否则提示警告, 此时取默认的堆栈段值

storage allocation 存储单元分配
STRUCTURE 结构伪指令定义
termination 终止
textual substitution 原文代替
variant 变体(型)
WIDTH operator 返回记录或字段所占的位数
WORD 起始地址的最低位为0,也称为字边界

Exercise

1. Code assemble, link, and use DEBUG to test the following program:

(a) Define byte items named BYTEX and BYTEY (containing any values) and a word item named WORDZ (containing zero).

(b) Move the contents of BYTEX to the AL.

(c) Add the contents of BYTEX to the AL.

(d) Move the immediate value 34H to the CL.

(e) Exchange the contents of AL and CL.

(f) Multiply the contents of AL by the CL (MUL CL).

(g) Transfer the product from the AX to WORDZ.

2. The assembler converts a source file to a(n) _____.

3. What files are generated from the source file trychar.asm if it is processed by MASM?

4. The linker program links object file and _____ files to create an execution file.

5. Explain the purpose of the stack.

6. What happens if AH = 02H and DL = 43H when the INT 21H instruction is executed? What does the INT 21H accomplish if AH contains a 4CH?

7. Identify the following INT 21H functions:

(a) 01H (b) 02H (c) 06H (d) 09H (e) 0AH (f) 4CH

8. What's a macro sequence?

9. How are parameters transferred to a macro sequence?

10. Develop a macro that sums a list of byte-sized data invoked by the macro ADDM LIST, LENGTH. The label LIST is the starting address of the data block and LENGTH is the number of data added. The result must be a 16_bit sum found in AX at the end of the macro sequence.

11. What is an assemble language directive?

12. Describe the purpose of the following assembly language directives: DB, DW, and DD.

13. Select an assembly language directive that reserves 30 bytes of memory for array BUF.

14. Suppose a data segment BUF is defined as following

BUF SEGMENT
    BUFDATA1 DB ?
    BUFDATA2 DW 0100H
    BUFDATA3 DB 2 DUP('AB')
BUF ENDS

Please diagram how the data items are stored in the memory locations.

15. Describe the purpose of the ORG directive. If the ORG 50H is inserted before the BUFDATA1 in BUF Segment, please diagram the data items stored in memory locations.

16. Explain the purpose of each of the following DEBUG commands:

(a) D (b) E (c) R (d) Q (e) T (f) A (g) U (h) P

17. Provide the DEBUG commands for the following unrelated requirements.

(a) Display the memory beginning at offset 1A5H in the data segment.

(b) Display the memory beginning at location B40H. (Note: Separate this address into its segment and offset values.)

(c) Key in the hex value 444E41 into the data segment beginning at location 18AH.

(d) Display the content of all registers.

(e) Display the content of the IP registers.

(f) Unassemble the symbolic code in location 100H through 11AH.

18. Provide the machine code instructions for the following operations:

(a) Move the hex value 324B to the AX register;

(b) Add the immediate hex value 024B to the AX.

19. Assume that you have used DEBUG to enter the following E command:

E CS:100 B8 36 01 05 25 00

The hex value was supposed to be 54. Code another E command to correct only the one byte that is incorrect; that is, change the 36 to 54 directly.

Chapter 6　Programming with Assembly Language

6.1　Design of Assembly Program

To be a good programmer, you should be familiar with the instruction sets and the computer architecture. Apart from these requirements, you should have a clear idea of the problem to be solved. Then sketch your ideas in general teams, and plan the overall logic. For example, if a problem is to perform multi-byte move operations, start by defining the fields to be moved. Then plan the strategy for the instructions: routines for initialization, for using a conditional jump, and for using a loop. The following, which shows the main logic, is pseudo code that many programmers use to plan a program:

(1) Initialize segment registers;
(2) Call the jump routine;
(3) Call the loop routine;
(4) End processing.

The jump routine could be planned as

(1) Initialize registers for count, addresses of names
(2) Jump1:Move one character of name
　　　　　Increment for next characters of names
　　　　　Decrement count:If nonzero, Jump1
　　　　　　　　　If zero, Return

The loop routine could be handled in a similar way.

You should organize the program into logical units such that related routines follow one another. Procedures that are about 25 lines (the size of the screen) are easier to debug than procedures that are longer.

Also, you can use other programs as guides. Attempts to memorize all the technical material and code "off the top of the head" often result in even more program bugs.

It's recommended that the comments are used to clarify what a procedure is supposed to accomplish, what arithmetic and comparison operations are performing, and what a seldom-used instruction is doing. (An example of the latter is LOOPNE: Does it loop **while** not equal or **until** not equal?)

To facilitate keying in the program, use a saved skeleton program that you can copy into a newly named file.

The remaining programs in this text make considerable use of JMP, LOOP, conditional jumps, CALL, and called procedures. Having covered the basics of assembly language, you are now in a position for more advanced and realistic programming.

6.2 Simple Procedures Designing

Simple Procedure is a kind of program which executes the instructions one by one from the first statement to the last one.

【Example 6-1】 Suppose the content of the memory location addressed 10010H is 9CH, please develop a program to split the 9CH into two parts of 09H and 0CH and then store them into the location 10011H and 10012H respectively.

The instruction sequences can be written as following:

```
MOV AX, 1000H
MOV DS,   AX         ;Set the starting address of data segment to DS register,
                     ;(DS) = 1000H
MOV SI,   10H        ;Send the offset address 10H to SI register
MOV AL,   [SI]       ;Send the content of the memory location 10010H to AL         10011100
AND AL,   0FH        ;Clear the high four bits of AL; (AL) = 0CH              AND  00001111
MOV [SI+1], AL       ;Send 0AH to the memory location of (10011H)                   00001100
MOV AL,   [SI]       ;Send the content of (10010H) to AL
MOV CL,   4          ;(AL) = 9CH, (CC) = 4
SHR AL,   CL         ;Shift the content of AL register to right 4 times
MOV [SI+2], AL       ;Put the message into the address of (10012H)
```

Note that different programmer has different programming habit. So the instruction sequences to accomplish the splitting data can be different. It depends on the programmer.

【Example 6-2】 Please develop a program to accomplish the addition, subtraction, multiplication and division of two 8-bit data which you defined in the data segment.

```
       DATA SEGMENT                  ;Data segment definition
           ORG 0050H                 ;Assignment the offset of data item
           DB 0AAH,66H,77H,55H       ;Pre-assign data and reserve storage
           DB ?,?
       DATA ENDS                     ;DATA segment ends
       CODE SEGMENT                  ;Code segment definition
           ASSUME CS:CODE,DS:DATA    ;Assignments of the segments to the segment registers

       START: MOV AX,DATA
              MOV DS,AX              ;Load the data segment address into the DS registers
              MOV SI,50H             ;Load the offset of data item to SI, DI and BX
              MOV DI,52H
              MOV BX,54H
```

XOR AX,AX	;*Clear AX register*
MOV AX,[SI]	;*66AAH→AX*
ADD AX,[DI]	;*66AAH + 5577H→AX*
MOV [BX],AX	;*(AX) →DS:0054H and DS:0055H*
MOV AX,[SI]	;*66AAH→AX*
SUB AX,[DI]	;*66AAH-5577H→AX*
MOV [BX+2],AX	;*(AX) →DS:0056H and DS:0057H*
XOR DX,DX	;*Clear DX register*
MOV AX,[SI]	;*66AAH→AX*
MUL WORD PTR [DI]	;*66AAH × 5577H→DX and AX*
MOV [BX+4],AX	;*(AX) →DS:0058H and DS:0059H*
MOV [BX+6],DX	;*(AX) →DS:005AH and DS:005BH*
MOV AX,[SI]	;*66AAH→AX*
CWD	;*000066AAH→DX , AX*
DIV WORD PTR [DI]	
MOV [BX+8],AX	;*Store the quotient in AX*
MOV [BX+10],DX	;*Store remainder in DX*
MOV AH,4CH	;*End the processing and return to DOS*
INT 21H	
CODE ENDS	;*CODE segment ends*
END START	

6.3 Branch and Looping Procedures

For assembly language, it has its own version of these flow control statements, such as conditional-jumping/decision, looping/iteration and sub-procedure calls statements. With these necessary language constructs, the programmer is able to control the logical direction of the program.

Different from other high-level programming languages such as C and C + + , programmer working with assembly language must have some instructions working together to perform a flow control statement. The CMP instruction and the jumping instructions can perform a conditional/decision controls that allow your program to perform differently at a proverbial "fork in the road." Assembly language provides a few different looping constructs, which allow you to perform the same code over and over again.

In a branch program, the decision is some sort of branch within the code that switches between two possible execution paths based on some conditions. Normally (though not always), conditional instruction sequences are implemented with the conditional jump instructions. The most commonly used conditional statement is the if... then or if... then... else statement. These two statements take the forms as shown in **Figure 6-1**:

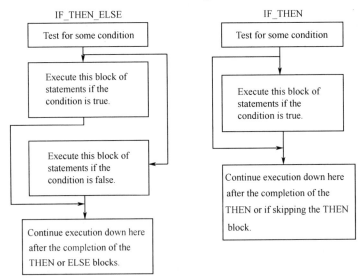

Figure 6-1   The structure of a branch program

【Example 6-3】 Develop a sequence of instructions to convert the number stored in DA1 into an even number.

        TEST   DA1, 01H   ; Test the last significant bit of DA1
        JE   NEXT   ; Jump to NEXT if ZF = 1 (even number)
        INC   DA1   ; If odd, increase DA1 by 1
NEXT: : : :

【Example 6-4】 Write a program to get the absolute value of (AX-BX), then put the result into memory location 2800H.

To sketch the branch clearly and logically, we draw a flow chart as shown in **Figure 6-2**.

    DATA SEGMENT   ; Define data segment
      ORG   2800H   ; Redefine the memory locations at DS:2800H
      DAT1   DW   ?
      ....
    DATA ENDS
    PROGRAM SEGMENT   ; Define code segment
      ASSUME CS:PROGRAM, DS:DATA   ; Assign the segments to the segment registers
    START: MOV   AX, DATA   ; Starting execution
          MOV   DS, AX   ; Load the data segment address to DS
          MOV   CX, AX   ; (CX) = (AX)
          CLC   ; Clear CF
          SUB   AX, BX   ; (AX) - (BX) → (AX)
          JC   AA   ; Jump AA on CF = 1 (with a borrow flag)

          MOV   DI, 2800H   ; CF = 0 2800H → DI;

```
        MOV    [DI], AX              ;(AX)→DS: 2800H
        JMP    OVER                  ;Jump OVER unconditional
AA:     SUB    BX, CX                ;CF=1,(Bx)-(Cx)→(Bx)
        MOV    DI, 2800H
        MOV    [DI], BX
OVER:
        MOV    AH, 4CH               ;End of processing and return DOS
        INT    21H

                                     ;End of main part of prog.
PROGRAM ENDS                         ;End of segment
END START                            ;End of assembly
```

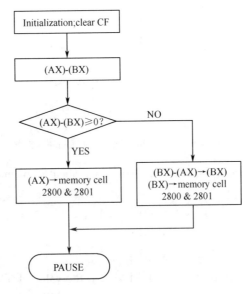

Figure 6-2 The flow chart of Example 6-4

【**Example 6-5**】Design a sequence of instructions to judge whether the value of AL is between 20H and 30H (include 20H,30H), if so, store the content of AL into memory location 40H. Otherwise, put it into 41H.

To ensure the sequences of instruction to meet the topic requirments, we draw a flow chart as shown in **Figure 6-3.**

```
        CMP    AL,    20H            ;Compare(AL) with 20H   (AL)-20H
        JB     L1                    ;Jump L1 if (AL)<20H
        CMP    AL,    30H            ;Compare(AL) with 30H   (AL)-30H
        JA     L1                    ;Jump L1 if (AL)>30H
        MOV    [40H], AL             ;Otherwise (AL)→DS:0040H
        JMP    L2                    ;Jump L2 unconditional
L1:     MOV    [41H], AL             ;(AL)→DS:0041H
L2:     MOV    AH,    4CH            ;End of processing and return DOS
        INT    21H
```

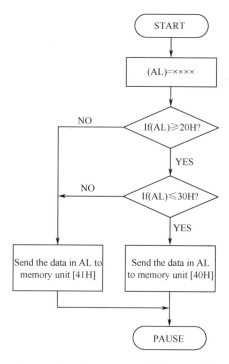

Figure 6-3 The flow chart of Example 6-5

【Example 6-6】 Design a sequence of instructions about sign function which indicates as following.

$$Y = \begin{cases} 1 & ,X>0 \\ 0 & ,X=0 \\ -1 & ,X<0 \end{cases} \quad (-128 < X < 127)$$

Suppose if the value of X is stored in DATA1 and the value of Y is stored in DATA2.

DAT SEGMENT			;*Define a segment*
	DATA1 DB X		;*Define Byte variable DATA1 and pre-assign data X*
	DATA2 DB Y		;*Define Byte variable DATA2 and pre-assign data Y*
DAT ENDS			
COD SEGMENT			
	ASSUME CS:COD,DS:DAT		;*Assign the segment COD to CS and DAT to DS*
ST:	MOV	AX, DAT	;*Load the starting address of COD segment to DS*
	MOV	DS, AX	
	MOV	AL, DATA1	;$X \rightarrow AL$
	CMP	AL, 0	;*Compare (AL) with 0, (AL) − 0*
	JGE	BIG	;*Jump BIG if (AL) ⩾0*
	MOV	AL, 0FFH	;*Otherwise send −1 to DATA2(Y) while (AL) <0*
	MOV	DATA2, AL	
	JMP	AA	;*Jump AA unconditional*
BIG:	JE	EQ	;*Jump EQ if ZF = 1*
	MOV	AL, 1	;*Otherwise send 1 to DATA2(Y) while (AL) >0*

		MOV	DATA2, AL	
		JMP	AA	;*Jump AA unconditional*
EQ:		MOV	AL, 0	;*0→Y while (AL) =0*
		MOV	DATA2, AL	
AA:		MOV	AH, 4CH	;*End of the processing and return DOS*
		INT	21H	

COD ENDS
END ST

Unlike branch programs, the loops represent the final basic control structures (sequences, decisions, and loops) which make up a typical program. Like so many other structures in assembly language, you'll find yourself using loops in places you've never dreamed of using loops. Program loops consist of three components:

(1) an optional initialization component,

(2) a loop termination test,

(3) and the body of the loop.

The order with which these components are assembled can dramatically change the way the loop operates. Three permutations of these components appear over and over again.

[**Example 6-7**] Write a procedure to find the maximum and minimum in a series of numbers, then put them into assigned locations.

```
DATA    SEGMENT                    ;Define data segment DATA
        BUFFER  DW   X₁,X₂,…,Xₙ    ;Define WORD variable buffer and pre-assign data items
                                    X₁,X₂,…,Xₙ
        COUNT   EQU  $-BUFFER       ;Define the byte numbers of BUFFER with a name COUNT

        MAX     DW   ?
        MIN     DW   ?
DATA    ENDS
PROG    SEGMENT
        SEARCH  PROC FAR            ;Define a procedure
        ASSUME CS:PROG, DS:DATA     ;Assign the segment PROG to CS and DATA to DS
        START:  MOV  AX, DATA
                MOV  DS, AX         ;Load the starting address of DATA to DS
                MOV  CX, COUNT      ;Load the bytes numbers of BUFFER to CX
                SHR  CX, 1          ;To get the number of the data items
                LEA  BX, BUFFER     ;Load the effect address (offset in the data segment) of
                                    BUFFER to the BX
                MOV  AX, [BX]       ;Xᵢ→AX(AX is used to store the max value of the BUFFER)
                MOV  DX, [BX]       ;Xᵢ→DX(DX is for min value of the BUFFER)
                DEC  CX             ;(CX) -1→(CX)
        LP:     INC  BX             ;Modify the address pointer to point the next data item
```

213

```
                INC    BX
                CMP    AX, [BX]       ;(AX) - X_i (i = 2,3,···,n)
                JGE    NEXT1          ;Jump NEXT1 if (AX) ≥ X_i
                MOV    AX, [BX]       ;X_i →AX if (AX) < X_i
        NEXT1:  CMP    DX, [BX]       ;(DX) - X_i (i = 2,3,···,n)
                JL     NEXT2          ;Jump NEXT2 if (DX) < X_i
                MOV    DX, [BX]       ;X_i →DX(MIN)
        NEXT2:  LOOP   LP             ;Jump LP if (CX) ≠ 0_i
                MOV    MAX, AX        ;Store the maximum value to MAX
                MOV    MIN, DX        ;Store the minimum value to MIN
                MOV    AH, 4CH        ;End of the processing and return to DOS
                INT    21H
        SEARCH  ENDP
    PROG    ENDS
        END  START
```

In this program, the data items pre-assignment is given as **Figure 6-4**, and the major logical flow is shown in **Figure 6-5**.

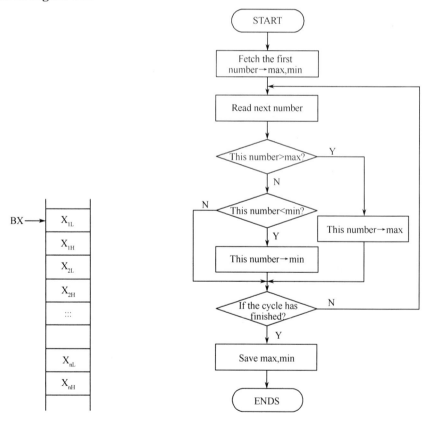

Figure 6-4 Data pre-assignment of Example 6-7   Figure 6-5 Logic flow chart of Example 6-7

【**Example 6-8**】 Suppose the scores of ten students are 56, 69, 84, 82, 73, 88, 98, 65, 100 and 81. Please develop a program to accomplish the number statistics which are less than

60, 60-69, 70-79, 80-89, 90-99 and 100 respectively.

```
        DATA SEGMENT                              ;Define a DATA segment
        GRADE  DW                                 ;Define word variable GRADE
               56,69,84,82,73,88,99,63,100,80     ;And pre-assign data
            S5     DW     0
            S6     DW     0
            S7     DW     0
            S8     DW     0
            S9     DW     0
            S10    DW     0
        DATA ENDS                                 ;DATA segment ends
        PROGNAM SEGMENT                           ;Define PROGRAM segment
        MAIN PROC FAR                             ;Procedure define
        ASSUME CS:PROGNAM, DS:DATA                ;Assign segment PROGRAM to CS and data to DS
START:
        PUSH   DS
        XOR    AX,    AX
        PUSH   AX
        MOV    AX,    DATA
        MOV    DS,    AX                          ;Load the starting address of DATA to DS
        MOV    S5,    0                           ;Load S5 S10 with 0
        MOV    S6,    0
        MOV    S7,    0
        MOV    S8,    0
        MOV    S9,    0
        MOV    S10,   0
        MOV    CX,    10                          ;Send COUNT 10 to CX
        MOV    BX,    OFFSET GRADE                ;Load the offset of GRADE to BX
COMPARE:                                          ;Label for loop
        MOV    AX,    [BX]                        ;Cope the data item of GRADE to AX
        CMP    AX,    60                          ;Compare (AX) with 60
        JL     FIVE                               ;Jump FIVE if (AL) <60
        CMP    AX,    70                          ;Otherwise compare (AX) with 70
        JL     SIX                                ;Jump SIX if (AX) <70
        CMP    AX,    80                          ;Otherwise compare (ax) with 80
        JL     SEVEN                              ;Jump seven if (ax) <80
        CMP    AX,    90                          ;Compare with 90 if (AX) ⩾80
        JL     EIGHT                              ;Jump EIGHT if (AX) <90
        CMP    AX,    100                         ;Compare with 100 if (AX) ⩾90
        JNE    NINE                               ;Jump NINE if (AX) ≠100
        INC    S10                                ;Otherwise S10 + 1→S10
        JMP    SHORT CHANGE_ADDR                  ;Jump to the label indicated unconditional
```

```
NINE:   INC       S9                        ;S9 + 1→S9
        JMP       SHORT CHANGE_ADDR         ;Jump to the label indicated unconditional
EIGHT:  INC       S8                        ;S8 + 1→S8
        JMP       SHORT CHANGE_ADDR
SEVEN:  INC       S7                        ;S7 + 1→S7
        JMP       SHORT CHANGE_ADDR
SIX:    INC       S6                        ;S6 + 1→S6
        JMP       SHORT CHANGE_ADDR
FIVE:   INC       S5                        ;S5 + 1→S5
CHANGE_ADDR:
        ADD    BX, 2                        ;Modify BX by adding 2
        LOOP   COMPARE                      ;( CX ) - 1→( CX ) ,jump compare if ( CX ) ≠0
        RET
MAIN ENDP                                   ;Procedure ends
PROGNAM ENDS                                ;PROGRAM segment ends
END START
```

6.4 Procedure Call and Return

Assembly language provides CALL instruction to call a subroutine and use RET instruction to return to the main procedure.

The 80x86 supports near and far subroutines. Near calls and returns transfer control between procedures in the same code segment. Far calls and returns pass control between different segments. The two mechanisms push and pop different return addresses. You generally do not use a near call instruction to call a far procedure or a far call instruction to call a near procedure. Given this little rule, the next question is "how do you control the emission of a near or far call or ret?".

Most of the time, the call instruction uses the following syntax:

CALL   ProcName

and the ret instruction is either:

RET

or

RET   disp

Unfortunately, these instructions do not tell MASM if you are calling a near or far procedure or if you are returning from a near or far procedure. The proc directive handles that chore. The proc directive has an optional operand that is either near or far. Near is the default if the operand field is empty. The assembler assigns the procedure type (near or far) to the symbol. Whenever MASM assembles a call instruction, it emits a near or far call depending on operand. Therefore, declaring a symbol with proc or proc near, forces a near call. Likewise, using proc far, forces a far call.

Besides controlling the generation of a near or far call, proc's operand also controls ret code generation. If a procedure has the near operand, then all return instructions inside that procedure will be near. MASM emits far returns inside far procedures.

【**Example 6-9**】Please design a sum subroutine to get the sum of each data item and then calculate the sum of two data arrays named ARY1 and ARY2 respectively by sum subroutine calling.

The major logical flow of **Example 6-9** is given in **Figure 6-6**, and the program can be written as following:

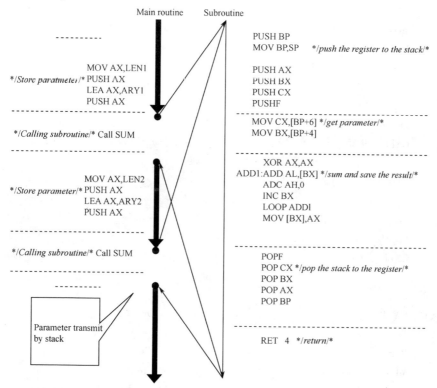

Figure 6-6 The major logical flow chart of Example 6-9

```
DATA    SEGMENT                          ;Define data segment
    ARY1   DB    03H, 07H, 50H, 06H,     ;Define byte variable named ARY1 and pre-assign the
                 23H, 45H, 0F6H, 0DFH    data
    LEN1   EQU   $ - ARY1                ;The byte numbers occupied by ARY1 is given to LEN1
    SUM1   DW    ?                       ;Define a word variable SUM1 to store the sum of ARY1
    ARY2   DB    33H, 44H, 55H, 12H,     ;Define byte variable named ARY1 and pre-assign the
                 78H, 89H, 0FFH, 0CDH    data
    LEN2   EQU   $ - ARY2                ;The byte numbers occupied by ARY2 is given to LEN2
    SUM2   DW    ?                       ;Define a word variable SUM2 to store the sum of ARY2
DATA    ENDS
STACK   SEGMENT STACK 'STACK'            ;Identify the start of the stack segment
    DB    100   DUP ('S')                ;Pre-assign 100 bytes s'
```

```
        STACK   ENDS                            ;Identify the end of the stack segment
        CODE  SEGMENT  PARA     'CODE'
            ASSUME  CS : CODE, DS : DATA,       ;Assign segment code to CS, data to DS and stack to SS
                    SS : STACK
        STA   PROC   FAR                        ;Define a procedure STA
        STB: PUSH   DS
             XOR    AX,   AX
             PUSH   AX
             MOV    AX,   DATA
             MOV    DS,   AX                    ;Load DS with the starting address of segment data
             MOV    AX,   STACK
             MOV    SS,   AX
             MOV    AX,   LEN1
             PUSH   AX                          ;Push the data count of ARY1 to stack
             LEA    AX,   ARY1                  ;Push the offset address of ARY1 to stack
             PUSH   AX
             CALL   SUM                         ;Sum subroutine calling, save a return address on the
                                                 stack

             MOV    AX,   LEN2
             PUSH   AX                          ;Push the data count of ARY2 to stack
             LEA    AX,   ARY2                  ;Push the offset address of ARY2 to stack
             PUSH   AX
             CALL   SUM                         ;Sum subroutine calling, save a return address on the
                                                 stack

             RET
        STA   ENDP                              ;STA procedure end
        SUM   PROC                              ;Define a SUM procedure
             PUSH   BP                          ;Push the (BP) to stack
             MOV    BP,   SP                    ;Copy (SP) to BP
             PUSH   AX                          ;Store (AX), (BX) (CX) to the stack
             PUSH   BX
             PUSH   CX
             PUSHF                              ;Store the flag register to stack
             MOV    CX,   [BP+6]                ;Copy the counter of ARY1 or ARY2 to CX
             MOV    BX,   [BP+4]                ;Copy the counter of ARY1 or ARY2 to BX
             XOR    AX,   AX                    ;Clear AX
        ADD1: ADD    AL,   [BX]                 ;Addition
             ADC    AH,   0                     ;Addition with carry flag
             INC    BX                          ;Modify the pointer to the next location
             LOOP   ADD1                        ;(CX)-1→CX, jump to add1 if (CX)≠0
             MOV    [BX], AX                    ;(CX) =0, store the sum of ARY1 or ARY2
             POPF                               ;Pop the flags to flag register
```

```
        POP   CX           ;Pop the values to CX,BX,AX,BP respectively from the
        POP   BX           stack
        POP   AX
        POP   BP
        RET   4            ;Remove a 16-bit number(near return) from the stack
        SUM   ENDP         and place it into IP
CODE    ENDS
        END   STA
```

要点：

(1) 子程序。子程序相对主程序而定，是一个子的程序段，确切地说，它是被父程序调用的程序。一般子程序具有公用性、重复性或有相对独立性的特点。

(2) 子程序的调用与返回

子程序的调用及返回由 CALL 和 RET 指令实现。子程序的调用实际上是程序的转移，但它与转移指令有所不同，转子指令 CALL 执行时要保护返回地址，而转移指令不考虑返回问题。每个子程序都有 RET 指令负责把压入栈后的返回地址弹出送 IP 或 CS 和 IP(段间返回)实现程序返回。

(3) 子程序设计与应用应该注意的问题。

① 现场保护与恢复

子程序使用时，要注意寄存器内容的保护。由于 CPU 的寄存器的数量有限，子程序使用的寄存器往往会和调用程序的寄存器发生冲突，破坏了调用程序中寄存器的内容，影响了子程序返回后主程序的继续使用。为了避免这种现象，在子程序入口处把所用寄存器的内容压入堆栈，保护起来，而在退出子程序前恢复寄存器的内容是必要的。恢复现场是保护现场的逆操作。当用栈区保护现场时，应注意恢复现场的顺序。

② 参数传递是指主程序与子程序之间相关信息或数据的传递。参数传递方式有寄存器传递、用内存单元传递或用栈区传递(寄存器、变量、地址表、堆栈等)。

③ 子程序说明

由于子程序有共享性，可被其他程序调用，因此，每个子程序应有必要的使用注释。主要包括子程序名，功能、技术指标，占用寄存器和存储单元，入口、出口参数，嵌套哪些子程序等。

(4) 这个转移地址是什么？

```
PUSH   DS
XOR    AX, AX
PUSH   AX
        ⋮
RET
```

当我们用编辑程序把原程序输入至机器中，用汇编语言把它转变为目标程序，用连接程序对其进行连接和定位后，连接程序为每一个用户程序建立了一个程序段前缀，共占用 256 个字节，在程序段前缀的开始处(0000H)安排了一条程序运行返回 DOS 的指令。给 DS 所赋的值(在未执行用户程序中的指令 MOV DS,AX 之前)就是程序段前缀的段地址。所以上面所提的 3 条指令就能在用户程序结束以后，利用 RET 指令返回到程序段前缀开始处,通过执行在程序段前辍中安放的这一指令,控制返回 DOS。

6.5 Programming Examples

【Example 6-10】 Use a table to convert the HEX(0-9,A-F) number into ASCII-code.

It's a table translation program. we need to write down the ASCII-code needed to convert to form the table at first.

```
    DATA SEGMENT
       TABLE  DB  30H, 31H, 32H, 33H, 34H, 35H    ;Form the table
              DB  36H, 37H, 38H, 39H, 41H, 42H
              DB  43H, 44H, 45H, 46H
       HEX    DB  4                                ;The data to convert
       ASCI   DB  ?
    DATA ENDS
    CODE SEGMENT
       ASSUME CS: CODE, DS: DATA
       STA: MOV  AX,   DATA
            MOV  DS,   AX
            LEA  BX,   TABLE      ;Load base addr of translation table
            XOR  AH,   AH         ;Clear AH register
            MOV  AL,   HEX        ;Copy the data needed convert to AL
            ADD  BX,   AX         ;Calculate the offset of the data
            MOV  AL,   [BX]       ;Fetch the ASCII of the data to AL
            MOV  ASCI, AL
            MOV  AH,   4CH        ;End of the processing and return the DOS
            INT  21H
    CODE ENDS
    END STA
```

The table translation can be accomplished by following instruction sequences:

```
       LEA   BA,   TABLE
       MOV   AL,   HEX
       XLAT
       MOV   ASCI, AL
```

The XLAT instruction assumes that the base address of a byte array is in the BX register and the byte to be converted is in the AL register. The desired code value is taken from the array and put in the AL register. None of the flags are affected.

【Example 6-11】 Assuming that two words are stored in the memory locations 50H~53H, write a program to add them and store the result in the memory locations 54H,55H:

```
    DATA SEGMENT
       ORG 50H
       TEMP1 DW 5250H, 9078H
```

```
        DATA ENDS
        PROGRAM SEGMENT                         ;Define codesegment
        MAIN    PROC FAR
            ASSUME CS:PROGRAM,DS:DATA
            START:MOV    AX,    DATA            ;Starting execution addr.
                  MOV    DS,    AX              ;Main part of program
                  MOV    SI,    50H             ;Get the offset of the first data
                  MOV    DI,    52H             ;Get the offset of the second data
                  MOV    BX,    54H             ;Store the offset of the memory locations to store
                                                 the sum of 2 data to BX register
                  CLC                           ;Clear the CF
                  XOR    AX,    AX              ;Clear the AX register
                  MOV    AX,    [SI]            ;5250H→AX
                  ADC    AX,    [DI]            ;5250H + 9078H→AX
                  MOV    [BX],  AX              ;Copy the sum to DS:0054H and DS:0055H
                  MOV    AH,    4CH             ;End of the processing and return to the DOS
                  INT    21H
        MAIN ENDP                               ;End of main part of program.
        PROGRAM ENDS                            ;End of segment.
        END START                               ;End of assembly
```

[Example 6-12] Assuming two numbers (eight bytes long) are stored in the addresses begun with 50H and 58H respectively. Write a program to add them and store the result in the addresses begun with 60H.

```
        DATA SEGMENT                            ;Define data segment
            ORG  0050H
            DAT1 DQ 1234567890123456H
            DAT2 DQ 3456789012345678H
            SUM  DQ  ?
        DATA ENDS
        PROGRAM SEGMENT                         ;Define code segment
        MAIN    PROC FAR
            ASSUME CS:PROGRAM,DS:DATA
            START: MOV   AX,    DATA            ;Starting execution addr
                   MOV   DS,    AX              ;Main part of program
                   MOV   SI,    50H             ;Offset of DAT1→SI
                   MOV   DI,    58H             ;Offset of DAT2→DI
                   MOV   BX,    60H             ;Offset of sum→BX
                   MOV   CX,    4               ;Copy the count 4 to CX register
                   CLC                          ;Clear carry flag
            AA:    MOV   AX,    [SI]            ;Copy the two bytes of DAT1 to AX
                   ADC   AX,    [DI]            ;Add the two bytes of DAT1 and DAT2
                   MOV   [BX],  AX              ;Store the sum to memory locations indicated by BX
```

```
        PUSHF                    ;Push the flag register to the stack
        ADD    SI,   2           ;Increase SI by 2
        ADD    DI,   2           ;Increase DI by 2
        ADD    BX,   2           ;Increase BX by 2
        POPF                     ;Pop the flags out of the stack
        LOOP   AA                ;Decrease CX, jump AA if(CX)≠0
        MOV    AH,   4CH         ;End of the processing and return to the DOS
        INT    21H
MAIN ENDP                        ;Endp of main part of program.
PROGRAM ENDS                     ;End of segment
END START                        ;End of assembly
```

【Example 6-13】 Write a program to make delay.

```
        DAT SEGMENT              ;Define data segment
        :::
        DAT ENDS
        STA SEGMENT STACK
        DW 50 DUP(?)
        STA ENDS
        COD SEGMENT
        ASSUME CS:COD,DS:DAT,SS:STA    ;Assign segment code to CS, data to DS and stack to
                                       SS
        STAR PROC FAR            ;Define a procedure STAR
START:  PUSH   DS
        XOR    AX,   AX
        PUSH   AX
        MOV    AX,   DAT         ;Load DS with the starting address of segment data
        MOV    DS,   AX
        MOV    AX,   STA
        MOV    SS,   AX
        CALL   DELAY             ;DELAY subroutine calling, save a return address on
        RET                      the stack
        STAR ENDP                ;STAR procedure end
        DELAY PROC:              ;Define a DELAY procedure
        PUSHF                    ;Push flags to stack
        PUSH   AX                ;Push AX, CX, SI to stack
        PUSH   CX
        PUSH   SI
        MOV    CX,   0FFFFH      ;Send 0FFFFH to CX
AA:LOOP  AA                      ;(CX)-1→CX, jump to AA if (CX)≠0
        POP    SI                ;(CX)=0,pop the flags to flag register
```

```
        POP    CX                      ;Pop the values to SI, CX, AX respectively from the
        POP    AX                      stack
        POPF
        RET
    DELAY  ENDP
    COD ENDS
    END START
```

【Example 6-14】 Write a program to display the eight numbers stored in the memory:

```
DAT SEGMENT PARA 'DAT'
    A1 DB 12H,34H,56H,78H,9AH           ;Define byte variable A1 and pre-assign the data
       DB 0BCH,0DEH, 0F0H
    B1 = $ -A1                          ;Byte counts occupied by A1→B1
    B2 = B1 * 2                         ;Twice byte counts occupied by A1→B2
    A2 DB B2 DUP(?)                     ;Define byte variable A2
DAT ENDS
STAC SEGMENT PARA STACK 'STA'
    STA1 DW 100 DUP(?)
STAC ENDS
CODE SEGMENT PARA 'CODE'
ASSUME CS:CODE,DS:DAT,SS:STAC,
       ES:DAT
    STAR PROC
        PUSH   DS
        XOR    AX,    AX
        PUSH   AX
        MOV    AX,    DAT
        MOV    DS,    AX               ;Load DS with the starting address of segment DAT
        MOV    ES,    AX               ;Load ES with the starting address of segment DAT
        LEA    SI,    A1               ;Load the effective address of A1
        LEA    DI,    A2               ;Load the effective address of A2
        MOV    CX,    B1               ;Copy the count to CX register
    G1: MOV    BL,    2                ;Converts each number into an ASCII code
        MOV    AL,    [SI]
        MOV    BH,    AL
        PUSH   CX                      ;Store the count of AL to the stack
        MOV    CL,    4                ;4→AL
        ROR    AL,    CL               ;AL rotates right 4 places
        POP    CX                      ;Return the count to CX register
    G2: AND    AL,    0FH              ;Clear the high nibble of AL
        DAA
        ADD    AL,    0F0H
        ADC    AL,    40H
```

```
            MOV    [DI],  AL
            INC    DI
            MOV    AL,    BH
            DEC    BL
            JNZ    G2
            INC    SI
            LOOP   G1
            LEA    SI,    A2
            MOV    CX,    B2
            CALL   P1                      ;Procedure P1 calling
            MOV    AH,    4CH              ;End of the processing and return to the DOS
            INT    21H
    P1  PROC                               ;Define a procedure P1
            MOV    BL,    2                ;Display the characters
      G:    MOV    DL,    [SI]
            MOV    AH,    2
            INT    21H
            INC    SI
            DEC    BL
            JNZ    T
            MOV    DL,    ' '
            MOV    AH,    2
            INT    21H
            MOV    BL,    2
      T:  LOOP G
            RET
    P1  ENDP
 CODE ENDS
 END STAR
```

【Example 6-15】 Develop a procedure to complete the transform between the Uppercase letters and Lowercase letters and then display the original string and the converting results. The procedure may not change any character except the letter A-Z.

```
 TITLE EX3
 PAGE 50 70
 DAT SEGMENT PARA 'DAT'
       D0 DB 0DH,0AH
       D1 DB 'ASJDFSKASLJSLDKJSDJDSKF $'
 DAT ENDS
 STAC SEGMENT PARA STACK 'STAC'
       STA1 DW100 DUP(?)
 STAC ENDS
 CODE SEGMENT PARA 'CODE'
```

```
        ASSUME CS:CODE,DS:DAT
        ASSUME SS:STAC,ES:DAT
STAR PROC   FAR
        PUSH DS
        XOR  AX,AX
        PUSH AX
        MOV  AX,DAT
        MOV  DS,AX
        MOV  ES,AX
        CALL TT
        LEA  DI,D1
        CALL TT1
        CALL TT
        LEA  DI,D1
        CALL TT2
        CALL TT
        RET
STAR ENDP
TT1 PROC                          ;Converts all Uppercase ASCII-code letters
GO:     MOV  AL,[DI]              ;Into Lowercase ASCII-code letters
        CMP  AL,'$'
        JZ   GO2
        CMP  AL,'A'
        JB   GO1
        CMP  AL,'Z'
        JA   GO1
        ADD  BYTE PTR [DI],20H
GO1:    INC  DI
        JMP  GO
GO2:RET
TT1 ENDP
TT2 PROC                          ;Converts all Lowercase ASCII-code letters
GQ:     MOV  AL,[DI]              ;Into Uppercase ASCII-code letters
        CMP  AL,'$'
        JZ   GQ2
        CMP  AL,'a'
        JB   GQ1
        CMP  AL,'z'
        JA   GQ1
        SUB  BYTE PTR [DI],20H
GQ1:    INC  DI
        JMP  GQ
GQ2:    RET
```

```
        TT2 ENDP
        TT PROC                                              ;Display the string
            LEA   DX,   D0
            MOV   AH,   09H
            INT   21H
            RET
        TT  ENDP
        CODE  ENDS
          END  STAR
```

[Example 6-16] Suppose that an array of data contains n 16-bit numbers and is stored at memory locations beginning with A, please develop a program to sort the numbers in descending order.

Figure 6-7 is the flow chart of the example 6-16. According to the Figure 6-7. We can write the program as following:

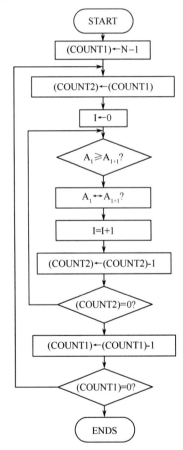

Figure 6-7   The major logic flow of Example 6-16

```
        DATAREA SEGMENT                                      ;Define data segment
            A DW N DUP ( ? )
        DATAREA ENDS
```

```
PROGRAM SEGMENT                              ;Define code segment
MAINPROC FAR                                 ;Main part of program
ASSUME      CS: PROGRAM,
            DS: DATAREA
START:                                       ;Starting execution address
            PUSH    DS                       ;Save org inal data segment
            SUB     AX, AX                   ;Push zero in AX
            PUSH    AX                       ;Save it on stack
            MOV     AX, DATAREA              ;DATAREA segment addr
            MOV     DS, AX                   ;Into DS register
;MAIN PART OF PROGRAM GOES HERE
            MOV     CX, N                    ;Set count1
            DEC     CX                       ;To n-1
LOOP1:      MOV     DI, CX                   ;Save count1 in DI
            MOV     BX, 0                    ;Clear BX
LOOP2:      MOV     AX, A[BX]                ;Load a(i) into AX and
            CMP     AX, A[BX+2]              ;Compare with a(i+1)
            JGE     COTINUE                  ;Swap if
            XCHG    AX, A[BX+2]              ;a(i) < a(i+1) and
            MOV     A[BX], AX                ;Store greater number
COTINUE:    ADD     BX, 2                    ;Increment index
            LOOP    LOOP2                    ;If not the end of a pass
                                             ;Repeat
            MOV     CX, DI                   ;Restore count1
                                             ;For the either loop
            LOOP    LOOP1                    ;If not the final pass
                                             ;Repeat
            RET                              ;Return to DOS
MAIN        ENDP                             ;End of main part of program
PROGRAM ENDS                                 ;End of code segment
            END     START                    ;End assembly
```

Tips

appendix　附录

assembler　汇编程序

Borland Turbo　Borland 公司开发的系列语言软件

chore　零碎工作

conditional-jumping/decision　有条件跳转/决策

cross-reference　交叉引用

data item　数据项

debug　调试

diagrammatically　用图表表示

EI（enable interrupt）/DI（disable interrupt）　中断允许/中断禁止

error diagnostics　错误诊断信息

executable module　可执行模块

flow control　程序流程控制

hyphen　连字号

linker　连接程序

looping construct 循环结构
looping/iteration 循环/迭代
lowercase 小写
NA (no auxiliary carry)/AC (auxiliary carry) 无辅助进位/有辅助进位
NC (no carry)/CY (carry) 无进位/有进位
NV (no overflow)/OV (overflow) 未溢出/溢出
NZ (nonzero)/ZR (zero) 非零/为零
object program 目标程序
permutation 排列
PL (plus sign)/NG (negative) 正数/负数
PO (parity odd)/PE (parity even) 奇校验/偶校验
prompt 提示符
proverbial 众所周知的
source program 源程序
sub-procedure calls 子程序调用
symbolic instruction 符号代码,符号指令
trace command 跟踪指令
UP (up direction)/DN (down direction) 增量方向/减量方向
uppercase 大写

Exercise

1. Write a program to:
Input a lowercase letter (a to z) and display the character in uppercase (A to Z).

2. Write a program to:
Input a single digit number (in the range 0 to 4) from the keyboard. Add this number to itself and display the result.

3. Develop a sequence of instructions that move the contents of data segment memory locations NUMB and NUMB + 1 into BX, DX, and SI.

4. Write a short sequence of instructions that divides the number in BL by the number in CL, and then multiplies the result by 2. The final answer must be a 16-bit number stored in the DX register.

5. Develop a procedure that converts all lowercase ASCII-code letters into uppercase ASCII-code letters. Your procedure may not change any character except the letter a-z.

6. Develop a program sequence that jumps to memory location ONE if AL = 6, TWO if AL = 7, and THREE if AL = 8.

7. Develop a procedure that reads an ASCII number from the keyboard and store it as a BCD number into memory array DATA. The number ends when anything other than a number is typed.

8. Develop a lookup table that converts hexadecimal data 00H ~ 0FH into the ASCII coded characters that represent the hexadecimal digits. Make sure to show the lookup table.

9. Develop a procedure that reads two character strings from the keyboard and compares these two strings. The procedure outputs "Match" if they are exactly identical. Otherwise it outputs "No Match".

10. Develop a procedure that accomplishes the conversion of a four-digit ASCII-coded hexadecimal number to its binary equivalent.

Chapter 7 Memory System

In a compute system, the word **memory** refers to the physical devices used to store information such as programs, sequences of instructions or data for later use. Under this definition, the memory of a computer can be divided into two categories. One category pertains to the part of the computer that holds the instructions and data that are presently being operated on, i.e., the part that the processor can access directly. The other category consists of the facilities that can store information, but the information must be transferred to the memory in the first category before it can be used by the processor. We usually refer to components in the former category as **main memory**, or simply **memory**, and components in the latter category as **mass storage**. This chapter will be concerned with main memory and will take some memory chips as examples to show how they are interfaced to the system bus.

7.1 Overview of the Memory

Figure 7-1 is the diagram of 8086/8088 system. The **memory** in the diagram is main memory (primary memory). Comparing with the secondary memory, it is fast to access but offer lower memory capacity. As we have learned in chapter 2, the main memory is comprised of groups of bits called bytes and words that are addressed as units. In 8086/8088 system, the groups of memory are 8-bit bytes.

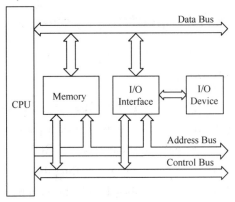

Figure 7-1　The diagram of 8086/8088 system

The term "storage" is often (but not always) used in separate computers of traditional secondary memory such as tape, magnetic disks and optical discs (CD-ROM and DVD-ROM). The term "memory" is often (but not always) associated with addressable semiconductor memory, i. e. integrated circuits consisting of silicon-based transistors, used for example as primary memory but also other purposes in computers and other digital electronic devices.

There are two main types of semiconductor memory: volatile and non-volatile. Volatile memory is computer memory that requires power to maintain the stored information. Examples of volatile memory are primary memory (typically dynamic RAM, DRAM), and fast CPU cache memory (typically static RAM, SRAM, which is fast but energy-consuming and offer lower memory capacity per area unit than DRAM). Non-volatile memory is computer memory that can retain the stored information even when not powered. Examples of non-volatile memory are flash memory and ROM/PROM/EPROM/EEPROM memory (used for firmware such as boot programs).

要点:

存储器是计算机系统的记忆设备。它用来存放计算机的程序指令、处理数据、运算结果以及各种需要计算机保存的信息,是计算机中不可缺少的一个重要组成部分。从记忆信息的角度讲,计算机中的存储器就相当于人的大脑。

存储器由一些能够表示二进制"0"和"1"的状态的物理器件组成,这些器件本身具有记忆功能,如电容、双稳态电路等。这些具有记忆功能的物理器件构成了一个个存储元(如一个电容就是一个存储元),每个存储元可以保存一位二进制信息。若干个存储元就构成了一个存储单元。通常一个存储单元由8个存储元构成,可存放8位二进制信息(即一个字节,Byte)。许多存储单元组织在一起就构成了存储器。

内部存储器(简称内存或主存)是计算机主机的组成部分之一,用来存储当前运行所需要的程序和数据,CPU可以直接访问内存并与其交换信息。相对外部存储器(简称外存)而言,内存的容量小、存取速度快。外存刚好相反,外存用于存放当前不参加运行的程序和数据,CPU不能对它直接访问,而必须通过配备专门的设备才能够对它进行读写(如磁盘驱动器等),这点是它与内存之间的一个很本质的区别。外存容量一般都很大,但存取速度相对比较慢。

半导体存储器按照工作方式的不同,可以分为只读存储器(ROM)和随机存取存储器(RAM)。

7.1.1 Non-Volatile Memory

ROM: ROM is the abbreviation of **Read-Only Memory**. In its strictest sense, ROM refers only to mask ROM which is fabricated with the desired data permanently stored in it, and thus can never be modified. There are basically four types of ROMs, the four types being distinguished by the way in which their contents are set. Setting the contents of a ROM is sometimes called programming, but should not be confused with the generation of instruction sequences discussed in the previous chapters. In one type of ROM the contents are determined by a masking operation that is performed while the chip is being manufactured. Such chips cannot be altered by the user and are referred to simply as ROMs. The contents of the second type can, if the proper equipment is available, be set by the user. They are called programmable-read-only memories (PROMs). As the masked ROMs, once a memory of this type is programmed, its contents can never be changed. The third and fourth types can not only be programmed by the user, but also can be erased and reprogrammed many times by using special equipment. They are called erasable-programmable-read-only memories (EPROMs) and electrically- erasable-read- only memories (E^2ROMs).

(1) EPROM: A shorthand of an **Erasable Programmable Read-only Memory**. It is a type of memory chip that retains its data when its power supply is switched off. In other words, it is non-volatile. It is an array of floating-gate transistors individually programmed by an electronic device that supplies higher voltages than those normally used in digital circuits. Once programmed, an EPROM can be erased by exposing it to strong ultraviolet light from a mercury-vapor light source. EPROMs are easily recognizable by the transparent fused quartz window in the top of the package, through which the silicon chip is visible, and which permits exposure to UV light during erasing.

(2) E^2PROM: It stands for **Electrically Erasable Programmable Read-Only Memory** and is a type of non-volatile memory used in computers and other electronic devices to store small amounts of data that must be saved when power is removed. It is also written as EEPROM and pronounced "e-e-prom," "double-e prom," "e-squared," or simply "e-prom". EEPROM is user-modifiable read-only memory that can be erased and reprogrammed (written to) repeatedly through the application of higher than normal electrical voltage generated externally or internally in the case of modern E^2PROMs. EPROM usually must be removed from the device for erasing and programming, whereas E^2PROMs can be programmed and erased in circuit. The E^2PROM allows multi-byte page operations. Its limited life restricts a number of times it could be reprogrammed and that is the reason E^2PROM was used for configuration information, rather than random access memory.

(3) Flash memory: Flash memory is a non-volatile computer storage chip that can be electrically erased and reprogrammed. It was developed from E^2PROM and must be erased in fairly large blocks before these can be rewritten with new data. The flash memory is technically a type of E^2PROM, the term "E^2PROM" is generally used to refer specifically to non-flash E^2PROM which is erasable in small blocks, typically bytes. Because erase cycles are slow, the large block sizes used in flash memory erasing give it a significant speed advantage over old-style EEPROM when writing large amounts of data. Flash memory now costs far less than byte-programmable EEPROM and has become the dominant memory type wherever a significant amount of non-volatile, solid state storage is needed.

要点：
只读存储器的特点是用户在使用时只能读出其中信息,不能修改或写入新的信息,断电后,其信息不会消失。根据制造工艺的不同,可分为以下几类：

(1) 掩膜式 ROM。掩膜式 ROM 存储单元中的信息由芯片制造厂在生产时一次性写入,对芯片图形通过二次光刻生产出来的,故称掩膜 ROM。其存储的内容固化在芯片内,用户可以读出,但不能改变。

(2) 可编程 ROM(PROM)。PROM 中的程序和数据是由用户自行写入的,但一经写入,就无法更改,是一次性的 ROM。

(3) 可擦除可编程 ROM(EPROM)。可由用户自行写入程序和数据,写入后的内容可用紫外线灯照射擦除,然后可以重新写入新的内容,可以多次擦除,多次使用。

(4) 电可擦除可编程 ROM(E^2PROM)。可用电信号(通常是加上一定的电压)进行清除和改写的存储器,使用方便,掉电不丢失。

7.1.2 Volatile Memory

RAM is the abbreviation of Random-Access Memory that is a form of computer data storage. It takes the form of integrated circuits that allow stored data to be accessed in any order with a worst case performance of constant time.

(1) SRAM: It stands for **Static Random-Access Memory**. It is a type of semiconductor memory where the word static indicates that, unlike dynamic RAM (DRAM), it does not need to be periodically refreshed, as SRAM uses bistable latching circuitry to store each bit. SRAM exhibits data remanence, but is still volatile in the conventional sense that data is eventually lost when the memory is not powered.

(2) DRAM: **Dynamic Random-Access Memory** is a type of random-access memory that stores each bit of data in a separate capacitor within an integrated circuit. The capacitor can be either charged or discharged. These two states are taken to represent the two values of a bit, conventionally called 0 and 1. Since capacitors leak charge, the information eventually fades unless the capacitor charge is refreshed periodically. Because of this refresh requirement, it is a dynamic memory as opposed to SRAM and other static memory. The main memory (the "RAM") in personal computers is Dynamic RAM (DRAM). DRAM is widely used in laptop, notebook and workstation computers as well as home game consoles (PlayStation3, Xbox 360 and Wii).

要点:

随机存取存储器的特点是存储器中的信息能读能写,且对存储器中任一单元的读或写操作所需要的时间基本是一样的。断电后,RAM 中的信息即消失。按其制造工艺分为以下两类:

(1) 双极型半导体 RAM。

(2) MOS 型 RAM。

MOS 型 RAM 又可分为两类:

① 静态随机存储器(SRAM)。SRAM 是利用半导体触发器的两个稳定状态表示"1"和"0"。只要电源不撤除,信息不会消失,不需要刷新电路,故称其为"静态 RAM"。其主要特点是存取时间短(几十到几百纳秒),外部电路简单,便于使用。

② 动态随机存储器(DRAM)。DRAM 是利用电容端电压的高低来表示"1"和"0",为了弥补漏电需要定时刷新,故称其为"动态 RAM"。一般微机系统中的内存采用 DRAM,配有刷新电路,每隔 1~2ms 刷新一次。DRAM 的存取速度与 SRAM 差不多。

7.1.3 Performance Index of Memory System

The principal performance indexes of a memory system are:

1. Capacity

Storage capacity refers to the number of binary digits that a memory chip can be stored. As shown in **Figure 7-2**, the capacity of a memory chip is said to be M × N if it contains M memory locations with each location consisting of N units. Similarly, a K × L memory module is one that consists of K locations, each having L bits.

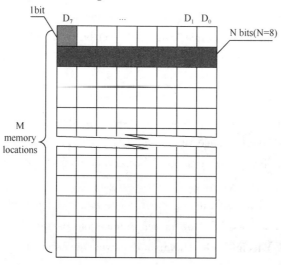

Figure 7-2 The diagram of a memory chip

[Example 7-1] Please calculate the storage capacity of the memory chip 6116 if it has 11 address lines and each memory location has 8 bits.

The M of this memory chip is 2^{11}, while each cell has 8 bits ($N=8$).

Then the capacity of the 6116 chip $= 2^{11} \times 8$
$$= 2 \times 1024 \times 8$$
$$= 2K \times 8$$
$$= 2KB$$

[Example 7-2] Please design a 64 KB memory system by using chip 6116(2K × 8) and chip 4416(16K × 4) respectively.

For the chip 6116:
$(64K \times 8) \div (2K \times 8) = 32$
So 32 chips are needed to extend 64KB.

For the chip 4416:
$(64K \times 8) \div (16K \times 4) = (64K \div 16K) \times (8 \div 4) = 8$
So 8 chips are needed to extend 64KB.

The Chip storage capacity is different with the storage capacity of a microcomputer. Most computer systems include both ROM and RAM modules and the memory in which normally consists of many memory chips. If nonvolatile RAM is needed, then either core memory must be used or backup power must be available so that the volatile semiconductor memory can be maintained in a powered-up state. ROM is generally used whenever possible because it is less expensive, nonvolatile, more reliable, and impervious to noise, and its simple construction lends itself to high cell density.

2. Speed

The speed of a memory is gauged by its **access time**, which is defined as the time delay between the receipt of a stable address input and the data output. This time depends on many factors and is even related to the bit capacity of the memory devices. High-speed transistors tend to require more chip space so that fewer cells can be included on each chip. Also, high-speed devices, which are normally made using a bipolar technology, are more expensive and consume more power.

3. Power Consumption

Power consumption can be very important in systems that must sometimes be powered by batteries or solar cells (e. g., space vehicles). Using low power consumption of memory chip not only can reduce the cost of power supply, but also can improve the reliability of the memory system.

The technology used in constructing the IC determines the power required for each cell. The technology most often used in low-power applications is the complementary metal-oxide semiconductor (CMOS) technology. The main disadvantage associated with CMOS is that a fair amount of chip space is needed for each cell; thus the capacity of each device is reduced. Unfortunately, power consumption and speed tend to be proportional and it is difficult and expensive to achieve both low power and high speed. The technology that has been found to be a good compromise of speed, power consumption, capacity per device, and cost is the high-density MOS (HMOS) technology.

4. Reliability

The reliability of the memory is measured by MTBF (Mean Time Between Failures). For a memory system, the longer the MBTF means the higher reliability it owns. Because the reliability of an IC is high once it has been thoroughly tested, the reliability of a module is strongly related to the number of solder connections and board complexity. Therefore, reliability increases as the total pin count decreases, thus it supplys an additional incentive to minimize the number of memory and supporting devices in module.

要点：

存储器芯片的主要技术指标如下：

（1）存储容量。存储容量是指一块存储芯片上所能存储的二进制位数，用"存储单元个数×每存储单元的位数"来表示。

（2）存储速度。存储器的存取速度是影响计算机运算速度的主要因素，用两个参数来衡量：

① 取时间 TA（Access Time）。定义为启动一次存储器操作（读或写），到完成该操作

所经历的时间。

② 存储周期 TMC(Memory Cycle)。定义为启动两次读(或写)存储器操作之间所需的最小时间间隔。

(3) 功耗。

(4) 可靠性。

7.2　Memory Devices

7.2.1　SRAM 6264

The 6264 is a high-performance CMOS static RAM integrated circuit. It has a capacity of 64 Kbit (8 KB). It is produced by a wide variety of different vendors, including Hitachi, Hynix, and Cypress Semiconductor. **Figure 7-3** is the pins assignments of a 6264 chip.

As shown in **Figure 7-3**, it has 13 address lines A_0-A_{12} and 8 bidirectional data lines D_0-D_7. Easy memory expansion is provided by an active LOW chip enable ($\overline{CS_1}$), an active HIGH chip enable (CS_2), and active LOW output enable ($\overline{OE}$) and three-state drivers.

Figure 7-4 is writing operation timing diagram of 6264. An active LOW write enable signal ($\overline{WE}$) controls the writing/reading operation of the memory. When $\overline{CS_1}$ and $\overline{WE}$ inputs are both LOW and CS_2 is HIGH, data on the eight data input/output pins (I/O_0 through I/O_7) is written into the memory location addressed by the address present on the address pins (A_0 through A_{12}).

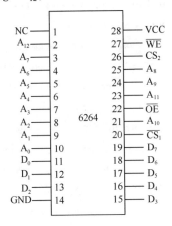
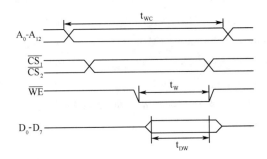

Figure 7-3　Pins assignments of 6264 chip　　Figure 7-4　Writing operation timing diagram of 6264

Reading the device is accomplished by selecting the device and enabling the outputs, $\overline{CS_1}$ and $\overline{OE}$ active LOW, CS_2 active HIGH, while $\overline{WE}$ remains inactive or HIGH. Under these conditions, the contents of the location indicated by the address pins A_0-A_{12} are present on the eight data input/output pins D_0-D_7.

The input/output pins remain in a high-impedance state unless the chip is selected. **Table 7-1** is the writing/reading operation truth table of 6264. And **Figure 7-5** is the connection diagram of 6264 with a CPU.

Table 7-1 The writing/reading operation of 6264

$\overline{WE}$	$\overline{CS_1}$	CS_2	$\overline{OE}$	D_0-D_7
0	0	1	×	Writing Operation
1	0	1	0	Reading Operation
×	0 or 1	0 or 1	×	High-impedance State

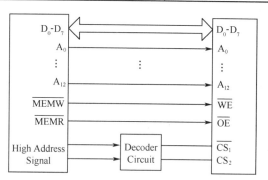

Figure 7-5 Connection diagram of 6264 with a microprocessor

In **Figure 7-5**, there is an address decoder to attach the 6264 chip to the microprocessor. With an address decoder, two or more memory devices can be connected to a microprocessor. Decoding the address sent from the microprocessor makes the memory function at a unique section or partition of the memory map. The simple NAND gate decoder and the integrated circuit decoders such as 3-to-8 line decoder (74LS138), the dual 2-to-4 line decoder (74LS139) and PLD programmable decoder can be found in some microprocessor systems.

When a 8K × 8 SDRAM 6264 is used, address lines A_0-A_{12} of the 8086 are connected to address inputs A_0-A_{12} of the 6264. The remaining seven address pins A_{13}-A_{19} are connected to the inputs of a NAND gate decoder. The decoder selects this 6264 from one of many 8K-byte sections of the entire 1M-byte address range of the 8086/8088 microprocessor. **Figure 7-6** is the example of connection diagram of 6264 to a Microprocessor with the NAND gate decoders. There is quite different between (a) and (b). For (a), A_{13}-A_{19} are all used to decode and the 6264 has a unique section **F0000H-F1FFFH** of the memory map, but for (b), the 6264 could be one of two sections of the memory map such as **F0000H-F1FFFH** or **B0000H-B1FFFH** for A_{18} is not used.

Figure 7-7 is the example of connection diagram of 6264 to a Microprocessor with the 74LS138 decoder.

As we learned before, the truth table of 74LS138 shows that only one of the eight outputs ever goes low at any time. For any of the decoder's outputs to go low, the three enable inputs ($\overline{G_{2A}}$, $\overline{G_{2B}}$, and G_1) must all be active. To be active, the $\overline{G_{2A}}$ and $\overline{G_{2B}}$ inputs must both be low (logic 0), and G_1 must be high (logic 1). Once the 74LS138 is enabled, the address inputs (C, B, and A) select which output pin of 74LS138 goes low. Notice that the outputs of the decoder can be connected to eight different memory devices. Here the decoder selects one 8K-

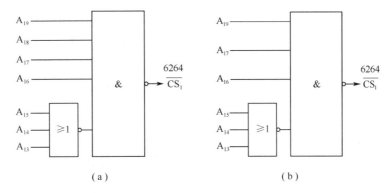

Figure 7-6　Connection diagram of 6264 to a microprocessor with the NAND gate decoders

byte memory block. This figure also illustrates the address range of the memory device and the common connections to the memory device.

In this circuit, a 3-input NAND gate is connected to address bits A_{17}-A_{15}. When all three address inputs are high, the output of this NAND gate goes low and enables $\overline{G_{2B}}$ of the 74LS138. And the $\overline{G_{2A}}$ is activated when the pin A_{19} is zero. So, the SRAM 6264 in this figure has a unique section **38000H ~ 39FFFH** of the memory map.

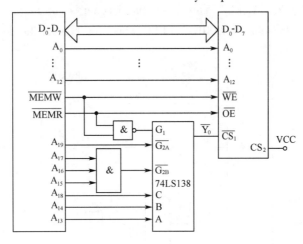

Figure 7-7　Connection diagram of 6264 to a microprocessor with the 74LS738 decoder

【**Example 7-3**】Please determine the address range covered by the SRAM 6264 in the memory map in **Figure 7-8**.

A simple logic circuit is used to decode the memory address. The output of the NAND decoder is a logic 0; the active low is connected to the $\overline{CS_1}$ pin that selects (**enable**) the 6264. The $\overline{OE}$ pin is activated by the 8086/8088 $\overline{RD}$. **Example 7-3** illustrates that the address range of this 6264 is determined by the externally decoded address bits (A_{19}-A_{13}) and the address lines of the 6264 (A_{12}-A_0). As shown in **Figure 7-8**, the address bits A_{19}-A_0 are listed as following:

$A_{19} A_{18} A_{17} A_{16} A_{15} A_{14} A_{13} A_{12}$　　A_{11}-A_0

0　0　1　1　1　1　1　0/1　　000H-FFFH

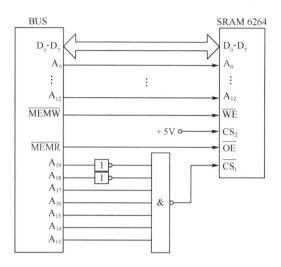

Figure 7-8 Connection diagram of 6264 to a microprocessor with the NAND gate decoder

The address bits A_{12}-A_0 are first written as 0s to locate the lowest address and then as 1s to find the highest address. So this 6264 is located at memory locations 3E000H-3FFFFH in hexadecimal. The location of this 8K-byte section of the memory will be changed with the connection of its high address bits A_{19}-A_{13}. Of course, the decoder is not unique. You can locate the 6264 at same memory locations 3E000H-3FFFFH when you change the decoder as that shown in **Figure 7-9**.

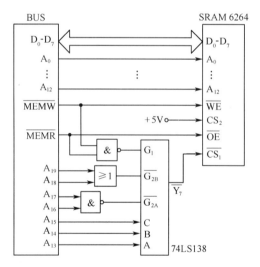

Figure 7-9 Connection diagram of 6264 to a microprocessor with the 74LS138 decoder

Notice that in **Example 7-3**, all of the address lines of 8086/8088 are used to connect the circuit. This memory chip occupies a unique section of the memory map. If only part of the address lines A_{19}-A_{13} are used to decode, the situation will be quite different. The address range occupied by this 6264 chip has multiple choices. **Example 7-4** illustrates the address range of a SRAM 6264 while only parts of the A_{19}-A_{13} are used to decode.

【**Example 7-4**】 Please determine the address range covered by the SRAM 6264 in the memory map in **Figure 7-10**.

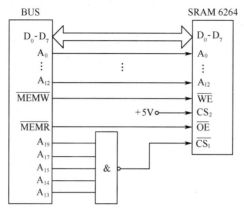

Figure 7-10 Connection diagram of 6264 to a microprocessor

As shown in **Figure 7-10**, the high address bits A_{18} and A_{16} are not used in the decode circuit. So there are 4 different memory locations for this 6264.

$A_{19}A_{18}A_{17}A_{16}A_{15}A_{14}A_{13}A_{12}$ $A_{11}\text{-}A_0$
1 X 1 X 1 1 1 0/1   000H-FFFH

A_{18}	A_{16}	Address Range
1	1	FE000H-FFFFFH
1	0	EE000H-EFFFFH
0	1	BE000H-BFFFFH
0	0	AE000H-AFFFFH

要点：

以典型的 SRAM 芯片 6264 为例，说明 SRAM 的外部特性及工作过程。

（1）6264 引线及功能。

① $A_0 \sim A_{12}$——13 根地址线。

② $D_0 \sim D_7$——8 根双向数据线。

③ $\overline{CS_1}$、CS_2——片选信号线。当$\overline{CS_1}$为低电平、CS_2为高电平时，该芯片被选中，CPU 才可以对它进行读写。

④ $\overline{OE}$——输出允许信号。只有当$\overline{OE}$为低电平时，CPU 才可以从芯片中读取数据。

⑤ $\overline{WE}$——写允许信号。当$\overline{WE}$为低电平时，允许数据写入芯片；当$\overline{WE}=1$，$\overline{OE}=0$ 时，允许数据从该芯片读出。

（2）6264 写入数据的过程。

将写入单元的地址送到 $A_0 \sim A_{12}$ 上：

① 将要写入的数据送到数据线上。

② 使片选信号($\overline{CS_1}$、CS_2)同时有效。

③ 在$\overline{WE}$端加上有效的低电平，$\overline{OE}$端可以任意。

④ 写入过程的时序如图 7-4 所示。

⑤ 从芯片中读出数据的过程与写操作类似：将要读出单元的地址送到 $A_0 \sim A_{12}$ 上；使片选信号（$\overline{CS1}$、CS_2）同时有效。与写操作不同的是，此时，要使读允许信号 $\overline{WE} = 1$，$\overline{OE} = 0$。这样，选中单元的内容就可以从 6264 的数据线读出。

（3）SRAM 芯片如何实现与系统的连接呢？

将一个存储器芯片接到总线上，除部分控制信号及数据信号线的连接外，主要是如何保证该芯片在整个内存中占据的地址范围能够满足用户的要求。芯片的选片信号是由高位地址信号和控制信号的译码产生的，事实上，也就是高位地址信号决定了这个芯片在整个内存中占据的地址范围。

下面介绍芯片如何利用地址译码方式实现译码。

① 全地址译码。所谓全地址译码，就是构成存储器时要使用全部 20 位地址总线信号，即用全部的高位地址信号作为译码器的输入，地位地址信号接存储芯片的地址输入线，从而使得存储器芯片的每一个单元都占据一个唯一的内存地址。

② 部分地址译码。顾名思义，部分地址译码就是仅把地址总线的一部分地址信号线与存储器连接，通常是用部分高位地址信号（而不是全部）作为片选译码信号，使得被选中得存储器芯片占有几组不同的地址范围。

7.2.2 SRAM 6116

The 6116 Static RAM is 16,384 bit memories organized as 2,048 words by 8 bits and operates on a single +5V supply. Unlike 2016's and equivalents are generally NMOS or MOS process parts, the 6116 and compatibles are built from CMOS technology. All provide three state outputs, TTL compatible, and allow for direct interfacing with common system bus structures. **Figure 7-11** is the pins assignments of a 6264 chip.

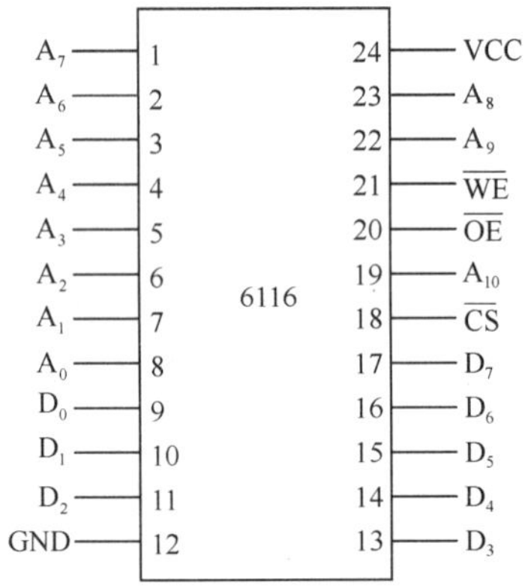

As shown in **Figure 7-11**, SRAM 6116 has 11 address lines A_0-A_{10} and 8 bidirectional data lines D_0-D_7. Easy memory expansion is provided by an active LOW chip enable ($\overline{CS}$) and active LOW output enable ($\overline{OE}$). An active LOW write enable signal ($\overline{WE}$) controls the writing/reading operation of the memory.

Figure 7-11 Pins assignments of 6116 chip

The input/output pins remain in a high-impedance state unless the chip is selected. **Table 7-2** is the writing/reading operation truth table of 6116. And **Figure 7-12** is the connection diagram of 6116 with a CPU.

Table 7-2 The writing/reading operation of 6116

$\overline{CS}$	$\overline{OE}$	$\overline{WE}$	D_0-D_7
0	×	0	Writing Operation
0	0	1	Reading Operation
1	×	×	High-impedance State

【Example 7-5】 Please design a 4KB memory space which locates at 78000H-78FFFH in the memory map by using SRAM 6116.

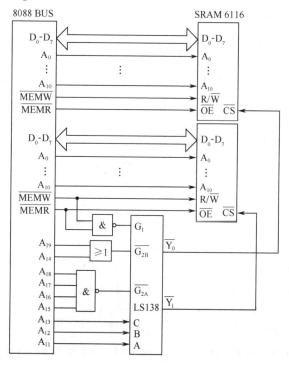

Figure 7-12　Connection diagram of 6116 to a microprocessor

SRAM 6116 is a 2KB memory device. To construct 4KB memory space, two pieces of SRAM 6116 chip are needed. In order to attach these two memory devices to the microprocessor 8088, the 74LS138 is adopted as a decoder. The 74LS138 decoder is used to select 2 SRAM 6116 memory components. The address 78000H ~ 78FFFH can be represented by binary numbers shown as following.

A_{19} A_{18} A_{17} A_{16} A_{15} A_{14} A_{13} A_{12} A_{11} A_{10} A_9 A_8 A_7 A_6 A_5 A_4 A_3 A_2 A_1 A_0
0　1　1　1　1　0　0　0　0　0　0　0　0　0　0　0　0　0　0　0
⋮　　　　　　　⋮　　　　　　　　　⋮
0　1　1　1　1　0　0　0　1　1　1　1　1　1　1　1　1　1　1
0　1　1　1　1　0　0　0　1　0　0　0　0　0　0　0　0　0　0　0
⋮　　　　　　　⋮　　　　　　　　　⋮
0　1　1　1　1　0　0　0　1　1　1　1　1　1　1　1　1　1　1

For $\overline{Y_0}$ or $\overline{Y_1}$ outputs low, the three enable inputs ($\overline{G_{2A}}$, $\overline{G_{2B}}$, and G_1) must all be active. Once the 74LS138 is enabled and the address input (C, B, and A) is 000 or 001, the $\overline{Y_0}$ or $\overline{Y_1}$ goes low respectively.

The address bits A_{10}-A_0 are first written as 0s to locate the lowest address and then as 1s to find the highest address. So the first 6116 in **Figure 7-12** is located at memory locations 78000H-787FFH in hexadecimal and the second one is covered at 78800H-78FFFH.

Figure 7-12 is the connection diagram of 6116 to a Microprocessor with the 74LS138 decoder. It illustrates the address range of the memory device and the common connections between the microprocessor and the memory device.

7.2.3 DRAM 2164A

The 2164A Dynamic RAM is 64K bit memories organized as 64K by 1 bit. **Figure 7-13** is the pins assignments of a 2164A chip

As shown in **Figure 7-13**, it has 8 address lines A_0-A_7 and 2 data lines D_{IN} and D_{OUT}. An active LOW write enable signal ($\overline{WE}$) controls the writing/reading operation of the memory. The 8 address lines A_0-A_7 are multiplex used. This means that the row address and the column address must share the same pins and be received one after the other. The row address is strobed by a negative-going pulse on the $\overline{RAS}$ (Row Address Strobe) pin and with the pin $\overline{RAS}$ (Row Address Strobe) still low, the column address is strobed by a negative-going pulse on $\overline{CAS}$ (Column Address Strobe) pin. The row and column address specify one bit of the 64K.

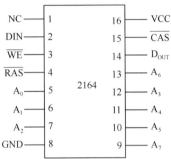

Figure 7-13 Pins assignments of 2164A chip

The timing diagrams for the read and write cycles are shown in **Figure 7-14.** For a read cycle, $\overline{WE}$ must be inactive before the $\overline{CAS}$ pulse is applied and remain inactive until the $\overline{CAS}$ pulse is over. After the column address is strobed, $\overline{RAS}$ is raised with $\overline{CAS}$ low, the data bit is

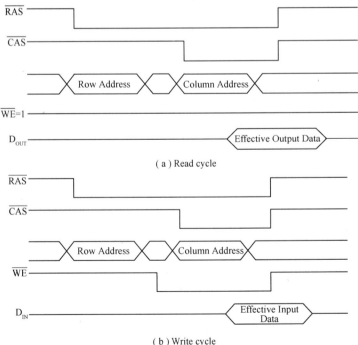

Figure 7-14 Timing diagrams of 6116

made available on DOUT. For a write cycle the DIN signal should be applied by the time $\overline{CAS}$ goes low, but after the $\overline{WE}$ pin goes low. The write is performed through the DIN pin while $\overline{RAS}$, $\overline{CAS}$, and $\overline{WE}$ are effective. The DOUT pin is held at its high-impedance state throughout the write cycle.

要点：

2164A 是一块 64K×1bit 的 DRAM 芯片，图 7-13 所示为 2164A 的引脚图。

（1）$A_0 \sim A_7$：地址输入线。2164A 芯片的地址引线是复用的。虽然 2164 的容量是 64K 个单元，但它只有 8 根线，在存取 DRAM 芯片的某单元时，其操作的过程是将存取的地址分两次输入到芯片中，每一次都由同一组地址线输入。两次送到芯片上的地址分别称为行地址和列地址，它们被锁存到芯片内部的行地址锁存器和列地址锁存器中。

（2）D_{IN} 和 D_{OUT}：芯片的数据输入、输出线。

（3）$\overline{RAS}$：行地址锁存信号。该信号将行地址锁存在芯片内部的行地址锁存器中。

（4）$\overline{CAS}$：列地址锁存信号。该信号将列地址锁存在芯片内部的行地址锁存器中。

（5）$\overline{WE}$：写允许信号。当它为低电平时，允许将数据写入。反之，当 $\overline{WE}$ 为高电平时，可以从芯片中读出数据。

7.2.4 EPROM 2764A

The EPROM 2764A is an 8K×8 erasable-programmable-read-only memory device. Generally speaking, the contents of an EPROM are determined by charge distribution. EPROMs are programmed by charge injection, and once programmed the charge distribution is maintained until it is distributed by some external energy source such as an ultraviolet light. An EPROM is programmed by applying an address to the address pins and high or low voltages to all of the data output pins, and then applying the proper voltages and pulses to the supply and control pins.

Similar with SRAM 6264, EPROM 2764A also has 13 address lines A_0-A_{12} and 8 data lines D_0-D_7. Easy memory expansion is provided by an active LOW chip enable ($\overline{CE}$). The active LOW output enable ($\overline{OE}$) controls the reading operation of the memory. The 21V is applied to the V_{cc} pin while the $\overline{CE}$ pin is held low. The address of the byte to be programmed is put on pins A_{12}-A_0 and the data byte is simultaneously applied to D_7-D_0. The data byte is then written into the address byte by pulsing $\overline{PGM}$ with +5V. The contents of each byte should be verified after it has been written. Its pin assignment is shown in **Figure 7-15**.

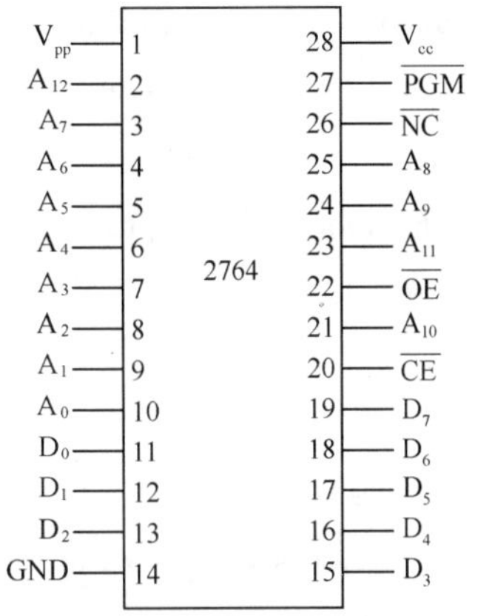

Figure 7-15 Pins assignments of 2764A

Figure 7-16 is reading operation timing diagram of 2764. An active LOW output enable signal ($\overline{OE}$) controls the reading operation of the memory. When $\overline{CE}$ and $\overline{OE}$ inputs are both LOW, data on the eight data input/output pins (D_0 through D_7) is read from the memory location addressed by the address present on the address pins (A_0 through A_{12}).

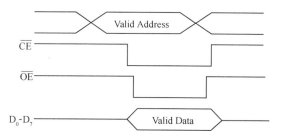

Figure 7-16　Reading cycle timing diagrams of 2764

【**Example 7-6**】 Please determine the address range covered by 2764 device in the memory map in **Figure 7-17**.

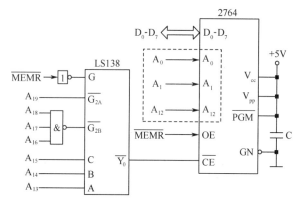

Figure 7-17　Connection diagram of 2764 to a microprocessor 8088

Figure 7-17 is the connection diagram of 2764 to a Microprocessor 8088 with the 74LS138 decoder. It illustrates the address range of the memory device and the common connections between the microprocessor and the memory device.

The 74LS138 decoder is used to select one 2764 memory component. For $\overline{Y_0}$ outputs low, the three enable inputs ($\overline{G_{2A}}, \overline{G_{2B}}$, and G_1) must all be active. Once the 74LS138 is enabled and the address input (C, B, and A) is 000, the $\overline{Y_0}$ goes low. As shown in **Figure 7-17**, the address bits A_{19}-A_0 are listed as following:

A_{19}　A_{18}　A_{17}　A_{16}　A_{15}　A_{14}　A_{13}　A_{12}-A_0
　0　　 1　　 1　　 1　　 C　　 B　　 A

The address bits A_{12}-A_0 are first written as 0s to locate the lowest address and then as 1s to find the highest address. So the 2764 in **Figure 7-17** is located at memory locations 70000H-71FFFH in hexadecimal.

要点：

（1）EPROM276 是一块 8K×8bit 的芯片,其引脚与 SRAM 6264 完全兼容。

① $A_0 \sim A_{12}$:13 根地址输入线。

② $D_0 \sim D_7$:8 根双向数据线。

③ $\overline{CE}$:片选信号,低电平有效。

④ $\overline{OE}$：输出允许信号。低电平有效，当$\overline{OE}=0$时，芯片中的数据可由$D_0 \sim D_7$端输出。
⑤ $\overline{PGM}$：编程脉冲输入端。对编程时，在该端加上编程脉冲。读操作时，$\overline{PGM}=1$。
⑥ V_{pp}：编程电压输入端。编程时该端需加高电压。

（2）2764在数据读出时的工作过程。

先把要读出的存储单元地址送到A0～A12地址线上，然后使$\overline{CE}=0$、$\overline{OE}=0$，就可以在芯片的$D_0 \sim D_7$上读出需要的数据。读出过程的时序如图7-16所示。

7.3 Memory Module Design

7.3.1 Memory Pin Connections

Pin connections to all memory devices are the address inputs, data outputs or inputs/outputs, some type of selection input, and at least one control input used to select a read or write operation. See **Figure 7-18** for ROM and RAM generic-memory devices.

1. Address Connections

All memory devices have address inputs that select a memory location within the memory device. Address inputs are almost always labeled from A_0, the least-significant address input, to A_n, where subscript n can be any value but is always labeled as one less than the total number of address pins. For example, a memory device with 10 address pins has its address pins labeled from A_0 to A_9. The number of address pins found on a memory device is determined by the number of memory locations found within it.

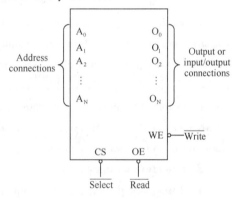

Figure 7-18 A pseudo-memory component illustrating the address, data, and control connections

Today, the more common memory devices have between 1K (1024) to 64M (67,108,864) memory locations, with 256M memory location devices on the horizon. A 1K memory device has 10 address pins (A_0-A_9); therefore, 10 address inputs are required to select any of its 1024 memory locations. It takes a 10-bit binary number (1024 different combinations) to select any single location on a 1024-location device. If a memory device has 11 address connections (A_0-A_{10}), it has 2048 (2K) internal memory locations. The number of memory locations can thus be extrapolated from the number of address pins. For example, a 4K memory device has 12 address connections, an 8K device has 13, and so forth. A device that contains 1M locations requires a 20-bit address (A_0-A_{19}).

A 400H represents a 1K-byte section of the memory system. If a memory device is decoded to begin at memory address 10000H and it is a 1K device, its last location is at address 103FFH-one location less than 400H. Another important hexadecimal number to remember is a 1000H, because 1000H is 4K. A memory device that contains a starting address of 14000H that is 4K bytes long, ends at location 14FFFH—one location less than 1000H. A third number is 64K, or 10000H. A memory that starts at location 30000H and ends at location 3FFFFH is a 64K byte memory. Finally, because 1M of memory is common, a 1M memory contains 100000H memory locations.

2. Data Connections

All memory, devices have a set of data outputs or input/outputs. The device illustrated in **Figure 7-18** has a common set of input/output (I/O) connections. Today, many memory devices have bi-directional common I/O pins.

The data connections are the points at which data are entered for storage or extracted for reading. Data pins on memory devices are labeled D_0 through D_7 for an 8-bit-wide memory device. In this sample memory device, there are eight I/O connections, which means that the memory device stores eight bits of data in each of its memory locations. An 8-bit-wide memory device is often called a byte-wide memory. Although most devices are currently eight bits wide, some devices are 16 bits, four bits, or just one bit wide.

Generally speaking, a memory catalog listings of memory devices often refer to memory locations times bits per location. For example, a memory device with 1K memory locations and eight bits in each location is often listed as a 1K × 8 by the manufacturer. A 16K × 1 is a memory device containing 16K × 1-bit memory locations. Memory devices are often classified according to total bit capacity. For example, a 1K × 8-bit memory device is sometimes listed as an 8K memory device, or a 64K × 4 memory is listed as a 256K device. These variations occur from one manufacturer to another.

3. Selection Connections

Each memory device has an input-sometimes more than one-that selects or enables the memory device. This type of input is most often called a **chip select** ($\overline{CS}$), **chip enable** ($\overline{CE}$), or simply **select** ($\overline{S}$) input. RAM memory generally has at least one $\overline{CS}$ or $\overline{S}$ input, and ROM has at least one $\overline{CE}$. If the $\overline{CE}$, $\overline{CS}$ or $\overline{S}$ input is active (a logic 0, in this case, because of the over-bar), the memory device performs a read or write operation; if it is inactive (a logic 1, in this case), the memory device cannot do a read or a write because it is turned off or disabled. If more than one CS connection is present, all must be activated to read or write data.

4. Control Connections

All memory devices have some forms of control input or inputs. A ROM usually has only one control input, while a RAM often has one or two control inputs.

The control input most often found on a ROM is the **output enable** ($\overline{OE}$) or **gate** ($\overline{G}$) connection, which allows data to flow out of the output data pins of the ROM. If $\overline{OE}$ and the selection input ($\overline{CS}$) are both active, the output is enabled; if $\overline{OE}$ is inactive, the output is disabled at its high-impedance state. The $\overline{OE}$ connection enables and disables a set of three-state buffers 10-cated within the memory device and must be active to read data.

A RAM memory device has either one or two control inputs. If there is only one control input, it is often called $R/\overline{W}$. This pin selects a read operation or a write operation only if the device is selected by the Selection input ($\overline{CS}$). If the RAM has two control inputs, they are usually labeled $\overline{WE}$(or $\overline{W}$), and $\overline{OE}$(or $\overline{G}$). Here, $\overline{WE}$(write enable) must be active to perform a memory write, and $\overline{OE}$ must be active to perform a memory read operation. When these two controls ($\overline{WE}$ and $\overline{OE}$) are presented, they must never both be active at the same time. If both control inputs are inactive (logic 1s), data are neither written nor read, and the data connections are at their high-impedance state.

7.3.2 Memory Module Design

The general design of a memory module is shown in **Figure 7-19**.

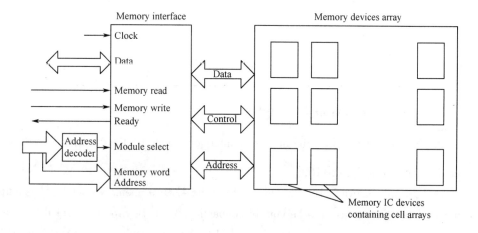

Figure 7-19   The general design of a memory module

It consists primarily of an interface and an array of memory devices, each of which contains an array of memory cells with each cell being the circuitry needed to store 1 bit. The cells in a device may be accessed separately or in groups, but either way the following relations must hold:

No. of the devices in a row = No. of the bits in a words / No. of cells in a group

No. of the devices in a column = No. of the words in a module / No. of groups in a device

A memory device is said to be an $M \times N$ device if it contains M groups with each group consisting of N cells. Similarly, a $K \times L$ module is one that consists of K words, each having L bits. **Table 7-3** gives some memory module designs examples which utilize typical memory devices.

Table 7-3   Representative memory device arrays

Memory size	Type of RAM	No. of chips in each column	No. of chips in each row	No. of chips in the module
4K × 8	1K × 1	4	8	32
	4K × 1	1	8	8
	256 × 4	16	2	32
	1K × 4	4	2	8

(续)

Memory size	Type of RAM	No. of chips in each column	No. of chips in each row	No. of chips in the module
4K × 16	1K × 1	4	16	64
	4K × 1	1	16	16
	256 × 4	16	4	64
	1K × 4	4	4	16
16K × 8	1K × 4	16	2	32
	4K × 1	4	8	32
	8K × 1	2	8	16
	16K × 1	1	8	8
64K × 8	16K × 1	4	8	32
	64K × 1	1	8	8

7.3.3 Memory Expansion Examples

【**Example 7-7**】 Please draw the necessary interfacing logic for the 4K-byte SRAM module constructed of 2K × 8 devices 6116.

Figure 7-20 utilizes two 6116 chips to construct the 4K-byte memory module. The address connections, data connections and the control connections of these two chips are the same except their selection connections. The decoder 74LS138 is used to select one of these two chips. Its address inputs C, B and A connect to the microprocessor address pins A_{13}-A_{11}. And for the first 6116 chip (No. $1^\#$), its address range is 00 0000 0000 0000-00 0111 1111 1111 (000H-07FFH), while the second one (No. $2^\#$) covers 00 1000 0000 0000-00 1111 1111 1111 (0800H-0FFFH).

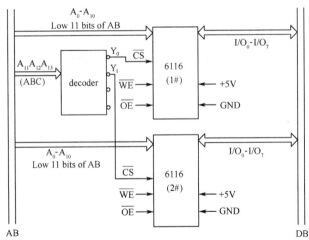

Figure 7-20 The diagram that uses two 6116 chips for a 4K-byte section of memory in an microprocessor 8088 system

【Example 7-8】Please draw the necessary interfacing logic for the 1K-byte memory module constructed of $1K \times 4$ devices 2114.

The pin configuration of 2114 is shown in **Figure 7-21**. It has ten address inputs and four data input/output connections. So it could access 2^{10} memory locations and each has 4 bits long.

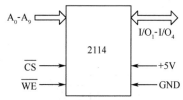

Figure 7-21 The pin configuration of the 2116

Figure 7-22 utilizes two 2114 chips to construct the 1K-byte memory module. The address connections, selection connections and the control connections of these two chips are the same except their data connections. For the first 2114 chip (No. $1^\#$), its data lines are connected to the upper 4 bits (D_4-D_7) of the 8088 data pins, while these of the second one (No. $2^\#$) are linked to lower 4 bits (D_0-D_3).

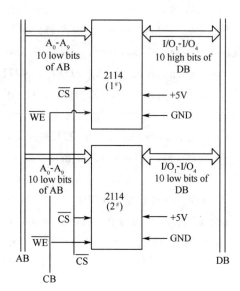

Figure 7-22 The diagram that uses two 2114 chips for a 1K-byte section of memory in an microprocessor 8088 system

【Example 7-9】**Figure 7-23** uses two 6264 and one 27128 memory devices for a memory section in an 8088 microprocessor-based system. Please indicate the address range for each memory device.

As shown in **Figure 7-23**, a 6264 has thirteen address inputs and eight data input/output connections and a 27128 has fourteen address inputs and eight data input/output connections. Their data connections, address connections, selection connections and the control connections of these chips are almost the same except that the A_{13} will take part in decoding for two 6264.

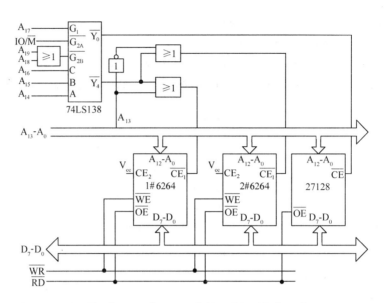

Figure 7-23 The diagram that uses 6264 and 27128 chips for a memory section in a microprocessor 8088 system

There is no $\overline{WE}$ signal since the 27128 is a ROM. The decoder 74LS138 and a simple NAND gate decoder are used to select one of these three chips. The address inputs C, B and A of 74LS138 connect to the microprocessor address pins A_{16}-A_{14}. To ensure the 74LS138 is active, the $A_{19}A_{18}=00$ and $A_{17}=1$, and when the $A_{16}A_{15}A_{14}=000$, the 27128 is selected. That is to say, the address range of this 27128 is 0010 0000 0000 0000 0000-0010 0011 1111 1111 1111 (20000H-23FFFH).

For the 6264 chips, the $A_{19}A_{18}A_{17}$ are same as the 27128 device. The A_{13} and $\overline{Y_4}$ are only inputs for the NAND decoder to determine which one of these two 6264 chips is selected. For the first 6264 chip (No. 1#), $A_{13}=0$ and $\overline{Y_4}=0$ ($A_{16}A_{15}A_{14}=100$) make it active, so its address rang is from 0011 0000 0000 0000 0000 to 0011 0001 1111 1111 1111 (30000H to 31FFFH). For the second one (No. 2#), it will be active when $A_{13}=1$ and $\overline{Y_4}=0$ and thus covers the memory locations from 0011 0010 0000 0000 0000 to 0011 0011 1111 1111 1111 (32000H to 33FFFH).

要点：

存储器扩展技术：

任何存储芯片的存储容量都是有限的。要构成一定容量的内存,往往单个芯片不能满足字长或存储单元个数的要求。这时,就需要多个存储芯片进行组合,以满足对存储容量的需求,这种组合称为存储器的扩展。扩展时要解决的问题包括位扩展、字扩展和字位扩展。

(1) 位扩展。位扩展保持总的地址单元数(存储单元个数)不变,但每个地址单元中的位数增加。位扩展的电路连接方法是:将每个存储芯片的地址线和控制线(包括选片信号线、读/写信号线等)全部连接在一起,而将它们的数据线分别引出连接至数据总线的不同位上。

(2)字扩展。字扩展是对存储器容量的扩展。此时,存储芯片上每个存储单元的字长已满足要求,而只是存储单元的个数不够,需要增加的是存储单元的数量,这就是字扩展。字扩展的电路连接方法是:将每个存储芯片的地址信号、数据信号和读/写信号等控制线按信号名称全部并联在一起,只将选片端分别引出到地址译码器的不同输出端,即用片选信号来区别各个芯片的地址。

(3)字位扩展。在构成一个实际的存储器时,往往需要同时进行位扩展和字扩展才能满足存储容量的需求。扩展时需要的芯片数量可以这样计算:要构成一个容量 M×N 位的存储器,若要使用 1×k 位的芯片,则构成这个存储器需要 (M/1)×(N/k) 个这样的存储器芯片。

Tips

pertain 与……相关,属于,适用(于)
comprise 包括,由……组成
magnetic disk 磁盘
volatile memory 非永久性存储器,易失存储器
non-volatile memory 永久性存储器,非易失存储器
flash memory 闪存
retain 保持
abbreviation 缩写词
erase 擦掉,抹去,清除
semiconductor 半导体
gauge 测量,测定
bipolar 双极的,有两极的
proportional 与……成比例的
solder 焊,焊接
incentive 动机,诱因,刺激
expansion 扩展
three-state driver 三态驱动器
inactive 非激活的,闲置的
decoder 解码器
partition 划分
equivalent 相等的
NMOS (N-channel Metal Oxide Semiconductor) N 沟道金属氧化物半导体

MOS (Metal Oxide Semiconductor) 金属氧化物半导体
CMOS (Complementary Metal Oxide Semiconductor) 互补金属氧化物半导体
compatible 兼容的
high-impedance 高阻态,高阻抗
multiplex 多元的
strobe 选通信号,选通
negative-going edge 下降沿,负沿
charge injection 电荷注入
charge distribution 电荷分布
ultraviolet 紫外的
simultaneously 同时发生的,同时存在的
subscript 下标,脚注
extrapolate 推断,推知
variation 变化,变动
manufacturer 制造商
data connection 数据线连接方式
address connection 地址线连接方式
selection connection 选通线连接方式
control connection 控制线连接方式

Exercise

1. Complete the following table.

Memory Size	Type of Device	Device Array	
		Number of Rows	Number of Columns
4K × 8	2K × 4		
64K × 8	32K × 1		
8K × 16	4K × 4		
1M × 16	64K × 1		

2. For a 32K-byte memory with a single error detection (parity) bit for each byte, determine the configuration of the chip array in terms of the numbers of memory chips in each row, in each column, and in the entire module using 8K × 1 RAMs. Repeat for 16K × 1 RAMs.

3. List the number of data items stored in each of the following memory devices and the number of bits in each datum:

(a) 2k × 4

(b) 1k × 1

(c) 4k × 8

(d) 16k × 1

(e) 64k × 4

4. What is the purpose of the $\overline{CS}$ or $\overline{CE}$ pin on a memory component?

5. What is the purpose of the $\overline{OE}$ pin on a memory device? What is the purpose of the $\overline{WE}$ pin on a RAM?

6. What is the purpose of the $\overline{CAS}$ and $\overline{RAS}$ inputs of a DRAM?

7. Why are memory address decoders important?

8. How many bytes of storage are there between the address 20000H and 8BFFFH? How many memory devices 6264 are needed to construct the same size memory section?

9. Suppose the connection diagram between the 74LS138 decoder and the CPU memory devices is shown in **Figure 7-24**, please write down the memory address range for each memory device which connects to $\overline{Y_0}$, $\overline{Y_3}$, $\overline{Y_5}$ and $\overline{Y_7}$ respectively.

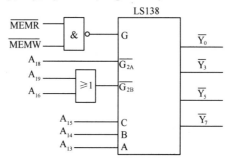

Figure 7-24 The connection of decoder

10. Design a decoder circuit and connect to 2 memory chips 6264 to select the memory for address range 61000H-61FFFH.

Chapter 8 I/O Interfaces

The I/O and memory interfaces are the counterparts to the bus control logic. What go between the bus control logic and the interfaces are simply the conductors in the bus; therefore, the interfaces must be designed to accept and send signals that are compatible with the bus control logic and its timing. I/O Interface is a logic unit to complete data exchange, synchronization and coordination, speed matching and data format conversion between host computer and external equipment. It is an important part of a computer system.

As we know, there are various kinds of external equipment, such as mechanical, electronic and other forms. The information transferred between CPU and I/O device includes data, control signal and status information. It can be digital, analog, switch or pulse value. Moreover, there is a large difference in the speed of data transmission for different external devices, for example, input speed of keyboard is for a second but for a magnetic disk, is up to 1MB/s.

An I/O interface must be able to:

(1) Interpret the address and memory-I/O select signals to determinate whether or not it is being referenced and, if so, determine which of its registers is being accessed.

(2) Determine whether an input or output is being conducted and accept output data or control information from the bus or place input data or status information on the bus.

(3) Input data from or output data to the associated I/O device and convert the data from parallel to the format acceptable to the I/O device, or vice versa.

(4) Send a ready signal when data have been accepted from or placed on the data bus, thus informing the processor that a transfer has been completed.

(5) Send interrupt requests and, if there is no interrupt priority management in the bus control logic, receive interrupt acknowledgments and send an interrupt type.

(6) Receive a reset signal and reinitialize itself and perhaps, its associated device.

Figure 8-1 contains a block diagram of a typical I/O interface. The function of an I/O interface is essential to translate the signals between the system bus and the I/O device and provide the buffers needed to satisfy the two set of timing constraints.

As shown in **Figure 8-1**, an interface can be divided into two parts, a part that interfaces to the I/O device and a part that interfaces to the system bus. Although little can be said about the I/O device side of the interface without knowing a lot about the device, the bus sides of all devices in a given system are very similar because they connect to the same bus. To support the main interface logic there must be data bus drivers and receivers, logic for translating the interface control signals into the proper handshaking signals, and logic for decoding the addresses that appear on the bus. In an 8086/8088 system, 8286 transceivers could drive the data bus, just as they are used to drive the bus at its bus control logic end. However, the main interface devices may have built-in drivers and receivers that are sufficient for small, single-board systems.

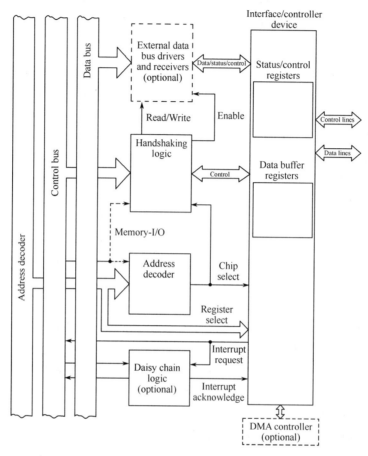

Figure 8-1　Typical diagram of an I/O interface

要点：

输入/输出(Input/Output)设备的作用:输入/输出设备是完成计算机与外界的数据交换,主机和外部设备两者之间的同步与协调,速度的匹配和数据格式转换的逻辑部件。它是计算机系统的重要组成部分。

外部设备的种类繁多,可以是机械式的、电子式的以及其他形式的;输入输出的信息也不同,可以是数字量、模拟量或开关量;信息传输的速度也有很大区别,如手动键盘输入速度为秒级,而磁盘输入可达1MB/s。

8.1　I/O Instructions

IN and OUT are the instructions that transfer information from or to an I/O device. **Table 8-1** lists all instruction formats for 8086/8088 microprocessor.

Table 8-1　IN and OUT instruction

Instruction	Data Width	Function
IN　　AL,　　port 8	8	A byte is input from port port 8 into AL
IN　　AX,　　port 8	16	A word is input from port port 8 into AX
IN　　AL,　　DX	8	A byte is input from the port addressed by DX into AL
IN　　AX,　　DX	16	A word is input from the port addressed by DX into AX
OUT　port 8,　AL	8	A byte is output from AL to port 8
OUT　port 8,　AX	16	A word is output from AX to port 8
OUT　DX,　　AL	8	A byte is output from AL to the port addressed by DX
OUT　DX,　　AX	16	A word is output from AX to the port addressed by DX

Notice that:

(1) Both IN and OUT instructions transfer data between an I/O device and the microprocessor's accumulator (AL, or AX).

(2) The I/O address can be stored in register DX as a 16-bit I/O address or in the byte (port 8) immediately following the opcode as an 8-bit I/O address. The 8-bit form (port 8) appeared in the instruction is called a fixed address. The 16-bit I/O address in DX is called a variable address.

Whenever data are transferred by using the IN or OUT instruction, the I/O address, often called a port number (or simply port), appears on the address bus. The external I/O interface decodes the port number in the same manner that it decodes a memory address. The 8-bit fixed port number (port 8) appears on address bus connections A_7-A_0 with bits A_{15}-A_8 equal to 00000000_2. The address connections above A_{15} are undefined for an I/O instruction. The 16-bit variable port number (DX) appears on address connection A_{15}-A_0. This means that the first 256 I/O port addresses (00H-FFH) are accessed by both of the fixed and variable I/O instructions, but any I/O address from 0100H-FFFFH is only accessed by the variable I/O address. Interfaces can be categorized according to the way in which they communicate with their I/O devices. Some I/O devices send and receive their information serially, while some work in parallel.

要点:

8088/8086 指令系统提供的 I/O 指令有两类:一类是直接寻址的 I/O 指令,即在指令中直接给出 I/O 接口的端口地址;另一类是间接寻址的 I/O 指令,即用 DX 寄存器来指出 I/O 接口的端口地址。采用直接寻址的端口地址范围是 00~FFH,共 256 个端口地址;采用间接寻址的端口地址范围是 0000H~FFFFH,共 64K 个端口地址。

什么是端口?

在一个外设接口电路中用数据寄存器存放数据,用状态寄存器来存放状态信息,用控制命令寄存器存放控制信息。这些寄存器称为 I/O 端口,每一个端口有一个端口地址,CPU 寻址必须对端口进行而不能笼统的对某个外设。另外,端口寄存器往往是 8 位的,通常一个外设的数据端口也是 8 位的,而状态与控制端口往往只用其中的一位或两位,故不同的外设的状态和控制信息可以共用一个端口。

8.2　I/O Interfacing Methods

In 8086/8088, there are two different methods of interfacing I/O to the microprocessor: isolated I/O and memory-mapped I/O. Sometimes we called the isolated I/O as direct or I/O-mapped I/O. **Figure 8-2** shows the methods of interfacing I/O to microprocessor.

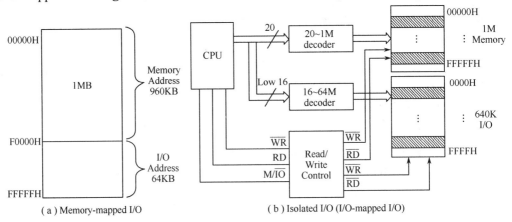

Figure 8-2　I/O interfacing methods

As shown in **Figure 8-2(a)**, a memory-mapped I/O device is treated as a port of memory locations in the memory map. It occupies 64KB memory space and therefore reduces the amount of memory available to applications. But for isolated memory (I/O-mapped I/O), as shown in **Figure 8-2(b)**, the I/O locations are isolated from the memory system in a separate I/O address space. That is to say, both of them are addressing from 0H. In isolated memory, we could encounter a new problem. Let's see the following example:

【Example 8-1】Suppose the address of a unit is 20H, in an isolated memory system, how to differentiate it is a location in I/O space or in memory space?

To distinguish between the memory operation and I/O operation, the following instructions can be used.

(1) MOV AL, [20H]
(2) IN　 AL, [20H]

The first instruction address [20H] in the memory locations, while the second address [20H] in I/O ports. That is to say, for isolated I/O, the data transferred between I/O and the microprocessor must be accessed by IN and OUT instructions. But for memory-mapped I/O, it's quite different. It does not use the IN and OUT instructions. It uses the same instructions as that transferring data between the memory and the microprocessor.

要点:

I/O 端口的编址方式有两种:存储器映射方式(Memory-Mapped)和 I/O 映射(I/O Mapped)方式。

存储器映射方式把一个 I/O 端口看成存储器统一编址，I/O 端口空间是存储空间的一部分，因此所有访问存储器单元的指令都可以用来访问端口而无需设置专门的 I/O 类指令，如 M6800、R6502 系列微机都采用这种方式。

I/O 映射中，I/O 端口构成单独的 I/O 空间，不占用存储器地址，CPU 设置专门的 I/O 指令来访问端口，如在 8086/8088 中的 IN 和 OUT 指令。存储器地址和 I/O 端口地址可重叠，由于二者访问指令不同，因此不会产生混淆。Z80、8086/8088 等系列微机都采用这种方式。

8.3 Serial Interface and Serial Communication

8.3.1 Serial Interface

Many I/O devices transfer information to or from a computer serially, i.e., one bit at a time over a single conductor pair or communication channel, with each bit occupying an interval of time having a specified length. There are several serial interface devices and typically they are constructed as shown in **Figure 8-3**. The status register would contain some information concerning the state of the current transmission, and the control register is for holding the information that determines the operating mode of the interface. The data-in buffer is paired with a serial input/parallel output shift register. During an input operation, the bits are brought into the shift register one at a time and, after a character has been received, the information is transferred to the data-in buffer register, where it waits to be taken by the CPU. (Although a single datum is not necessarily an alphanumerically coded character, it normally is and, therefore, it will be referred to as a character.) Similarly, the data-out buffer is associated with a

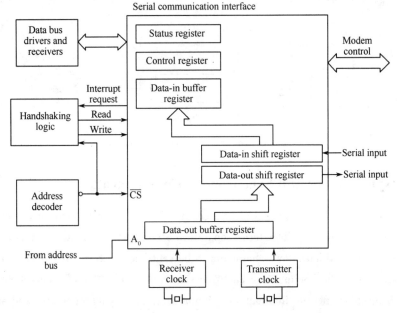

Figure 8-3 Serial interface

parallel input/serial output shift register. An output is performed by sending data to the data-out buffer, transferring it to the shift register, and then shifting it onto the serial output line.

The CPU could address the four port registers. Because both of the status register and Data-in buffer register can only be read from while the control register and Data-out buffer register can only be written into, an active signal on the read or write line would indicate either the read only register or write only register. Moreover, A_0 could be used to distinguish between these two read-only registers or write-only registers.

8.3.2 Basic Serial Transmission Lines

The interface shown in **Figure 8-3** has separate lines for sending and receiving information. **Figure 8-4** shows us there are two transmission modes for serial communications. When different lines are used for the two signal directions the communication system is said to be full duplex, as shown in **Figure 8-4(a)**, such a system can transmit and receive at the same time. The alternative, called half duplex, is to use the same line for both inputting and outputting. As shown in **Figure 8-4(b)**, in this system, it can transmit bi-directional, but not the same time. If the computer has been receiving characters and then wishes to send characters, or vice versa, the communication link must be turned around, a process that requires time. Although full duplex avoids turning around and provides echoing, it does require an extra line. The two modes of communication are illustrated in **Figure 8-4**.

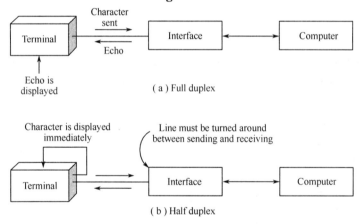

Figure 8-4 Basic transmission modes

8.3.3 Asynchronous and Synchronous Communication

There are **two basic types** of serial data communication. One is **asynchronous** serial communication in which special bit patterns separate the characters and the other is **synchronous** serial communication which allows characters to be sent back to back but must include special "sync" characters at the beginning of each message and special "idle" characters in the data stream to fill up time when no information is being sent. An asynchronous transmission may include dead time of arbitrary lengths between the characters, while in a synchronous transmission the characters must be precisely spaced even though some of the characters may contain no information. Although both types may waste time spending useless bits, the maximum information rate of a synchronous line is higher than that of an asynchronous line with the same bit rate because the asynchronous transmission must include extra bits with each character. On the other hand, the clocks at the opposing ends of an asynchronous transmission line do not need to have exactly the same frequency (as long as they are within permission limits) because the special patterns allow for re-synchronization at the beginning of each character. For a synchronous transmission the activity must be coordinated by a single clock since it is the clock that determines the position of each bit. This means that the clock timing must be transmitted as well as the data.

1. Asynchronous Communication

Figure 8-5 illustrates the format of an asynchronous character. A character contains several extra bits as well as the coded information. When the transition signal falls to state "0" or space from state "1", it marks the beginning of the character. So when the transition begins, the first bit is always 0 and is called the start bit. It is followed by from 5 to 8 information bits. The last bits are stop bits. There may be 1, $1\frac{1}{2}$, or 2 stop bits.

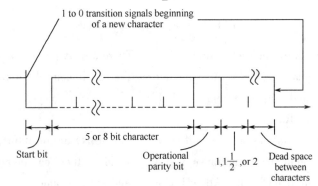

Figure 8-5 Format of a standard asynchronous character

2. Synchronous Communications

For synchronous communications, there are two character formats, the character-oriented and bit-oriented.

The character-oriented transmissions must begin with a series of sync characters. The data formats of character-oriented are illustrated in **Figure 8-6**.

The single and double syn transmissions are internal synchronizations, as shown in **Figure 8-6(a) and Figure 8-6(b)**. The difference between them is how many "sync" characters, they add before the data transmission. The receiver will examine each bit as the data arrives and it will assume a transmission has begun only when a sequence of bit exactly matches the bits of sync character. For the external synchronization, as shown in **Figure 8-6(c)**, the "sync" character is sent by a specified control line to realize the synchronization of transmitter and receiver. No matter what the synchronization mode is, they all use two bytes "CRC" (Cyclic Redundancy Code) character to end a frame information.

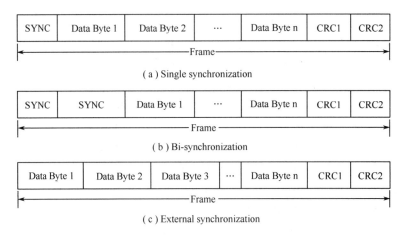

Figure 8-6　The data format of character-oriented

The data format of bit-oriented based on SDLC is given in **Figure 8-7**.

Figure 8-7　The format of bit-oriented

The information is transmitted in unit of frame. There are six parts in a frame, which consists of a byte start mark "7EH", an 8-bit address, an 8-bit control, n bits information, 2 "CRC" characters, and a byte end mark "7EH".

3. Transmission Rates

Transmission rates are measured in **bits per second** (**bps**) and in **baud**, which means the number of discrete conditions being transmitted per second. If there can be only one of two possible conditions at any point in time, then bps and baud are the same.

The standard baud rates that are most often used are 110, 300, 600, 1200, 1800, 2400, 4800, 9600, and 19200. Most CRT terminals are capable of handling any one of these rates up to 9600 baud, while printing terminals are limited by the speed of the print mechanism.

As an example, consider an asynchronous transmission in which each character contains a start bit, 7 information bits, a parity bit, and 1 stop bit. If the baud rate of the line were 1200, then the maximum number of characters per second that could be transmitted would be 1200/10 = 120. The maximum rate could be attained only when there is no dead time between characters. As a comparison, a synchronous line operating at 1200 baud and with no parity could transmit four sync characters and a 100-character massage in

$$7(100+4)/1200 = 0.6067 \text{ second}$$

This would mean that up to 100/0.6067 = 165 useful characters per second could be transmitted.

要点：

（1）计算机通信有两种方式:并行通信和串行通信。并行通信中,数据各位通过各自的信号线同时传送。串行通信中,数据一位一位顺序通过同一信号线进行传输。串行通信线路简单,适用于远距离传输,缺点是传输速度慢。

(2) 串行通信数据的收发可采用异步和同步两种基本的工作方式。

① 异步通信方式。异步通信所采用的数据格式是由一组不定"位数"数组组成。第1位称起始位,宽度为1位,低电平;接着传送一个字节(8位)的数据,以高电平为"1",低电平为"0";最后是停止位,宽度可以是1位、1.5位或2位,在两个数据之间可以有空闲位。

② 同步通信方式。同步通信控制规程可分为两类,面向字符型(Character-Oriented)和面向比特型(Bit- Oriented)。面向字符型的同步通信数据格式有单同步、双同步和外同步三种形式。单同步是指在传送数据块之前先传送一个同步字符SYNC,接收端检测到该同步字符后开始接收数据;双同步则先传送两个字符SYNC1和SYNC;外同步格式中数据之前不包含同步字符,而使用一条专用控制线来传送同步字符,以实现收发双方的同步操作。任何一帧信息都以两个字节的循环控制码CRC为结束。

面向比特型的同步通信数据格式:根据同步数据链路控制规程SDLC,面向比特型的数据以帧为单位传输,每帧由6个部分组成:第1部分是开始标志7EH;第2部分是一个字节的地址场;第3部分是一个字节的控制场;第4部分是信息场,即需要传送的数据是bit的集合;第5部分是两个字节的帧校验场,及循环冗余校验码CRC;第6部分是结束标志7EH。

(3) 波特率。每秒传输的二进制数据的位数(bit数)称为传送速率,即波特率。国际上规定了标准波特率系列,它们为150、300、600、1200、2400、4800、9600、19200等。计算机之间的异步通信率一经确定后,一般不应变动,但通信的数据是可变动的,也就是数据组之间的空闲位是可变的。

8.4 8251A Programmable Communication Interface

The Intel's 8251A programmable communication interface is a serial interface device. It's capable of being programmed for asynchronous or synchronous communication.

8.4.1 The Architecture of the 8251A

The architecture of the 8251A is shown in **Figure 8-8**. It consists of transmitter, receiver, data bus buffer, Read/ Write control logic circuit and modem circuit. **Figure 8-9** is the pins assignment of 8251A.

1. The Transmitter of 8251A

The transmitter of 8251A contains data-out buffer register, transmitter shift register and transmitter control circuit. The data bits put in the data-out buffer register by the CPU are transferred to the transmitter shift register and then are shifted out through the T_XD pin. If 8251A is programmed in synchronous communication, the data bit will be sent out along with the necessary synchronization bits. When the 8251A enters a synchronous mode, the data bits are output serially through the T_XD pin after the start bit and stop bit adding.

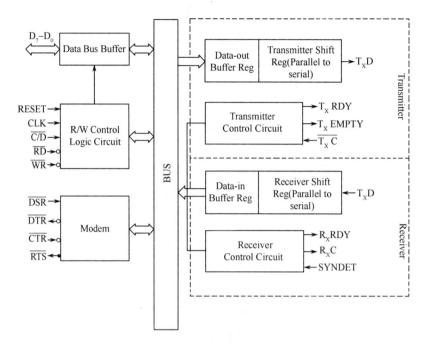

Figure 8-8 Architecture of the 8251A

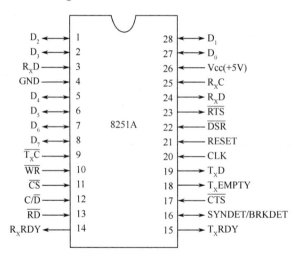

Figure 8-9 Pins assignment of the 8251A

The pins related with transmitter are listed as following:

(1) $R_X RDY$: Data transmission line. It outputs data serially.

(2) $T_X RDY$: The ready signal of transmitter. It shows the data-out buffer register is empty.

(3) $T_X EMPTY$: The idle signal of transmitter. It indicates that the transmitter shift register is empty. When $T_X EMPTY$ is set 1, the CPU can write the data into the data-out buffer register.

As shown in **Table 8-2**, the $T_X RDY$ and $T_X EMPTY$ work together to determin the states of

8251 transmitter.

Table 8-2 States of 8251 transmitter

T_XRDY	T_XEMPTY	Status of Transmitter
0	0	Data-out buffer register. is full, Transmitter shift register. is full
1	0	Data-out buffer register. is empty, Transmitter shift register. is full
1	1	Data-out buffer register. is empty, Transmitter shift register. is empty
0	1	Impossible

(4) $\overline{T_X C}$: The clock signal of transmitter. It is inputted from outside. In synchronous mode, the clock frequency of $T_X C$ should be equal to the baud rate of transmitter, but in asynchronous mode, it varies with the baud rate factor which defined by program.

2. The Receiver of 8251A

The receiver of 8251A contains three parts: data-in buffer register, receiver shift register and receiver control circuit.

The serial bit stream outside arriving on the RxD pin bit by bit is shifted into the receiver shift register and then the data bits are transferred to the data-in buffer register, where they can be input by the CPU parallel. If the 8251A is in a synchronous mode, the synchronization character should be checked. The receiver begins to receive data serially only when the synchronization has been achieved. If it is in an asynchronous mode, the start bit and stop bit should be recognized and deleted at first. Then the R_XRDY line will output a high level signal which indicates the data has been already to transmit to CPU. The data receiving rate of 8251A is determined by the clock frequency inputted by $\overline{R_X C}$ pin.

The pins related with the receiver are listed as following:

(1) $R_X D$: Data receiving line. It inputs serial data.

(2) R_XRDY: The ready signal of receiver. It implies that a character has been received at the data-in buffer register and is ready to be input to CPU. If the interrupt mode adopting, the R_XRDY pin is used to send interrupt request signal to CPU. It will be in an invalid state when CPU takes away the character.

(3) STNDET/BRKDET: The detection signal of a dual function, high level effectively. For synchronous mode, the pin (SYNDET) is used as a synchronous detection. When the 8251A works in an internal synchronous mode and the $R_X D$ has got a single or double synchronous character, the pin SYNDET outputs a high level to indicate that the synchronization is accomplished and the data follow-up is valid, If the search for sync characters is conducted by an external device, then SYNDET can be used to input a signal, indicating that a match has been found by the external device. The receiver begins receiving data when the SYNDET is valid. At asynchronous mode, the pin BRKDET can only be an output. It is used for break detecting. For example, the BRKDET goes high to designate that data is broken currently whenever a character consisting of eight 0s is received consecutively.

(4) $\overline{R_X C}$: The clock signal of receiver. It is an external input signal. It determines the da-

ta receiving rate of 8251A. The clock frequency of the receiver is equal to the data receiving rate at the synchronization mode. At the asynchronous mode, it changes with the baud rate factor denoted by program. It is very similar to the transmitter clock $T_X C$.

3. Data Bus Buffer

The data bus buffer is an information exchange channel between CPU and 8251A. It consists of three 8-bits buffer registers, two of which are used to store the data and state of 8251A. When the IN instruction is executed, the CPU will read the data and state from these two registers. Another buffer register is applied to save the data or control signal written by CPU. The OUT instruction can realize the write operation to this register.

4. Read/Write Control Circuit

The Read/Write control circuit is used to receive the control signals which determine what state 8251A is in and then send these signals to related components of 8251A. It is the controller of 8251A. The control pins are as following:

(1) RESET: Reset signal. It is an input signal to 8251A. When it is active high, all registers in 8251A are reset, and the receiving and transmitting lines are forced to stay in idle state.

(2) CLK: Clock signal. It is an input signal to 8251A. It is used to generate internal timing signal of 8251A. The frequency of CLK signal must be 30 times greater than that of transmitter ($T_X C$) and receiver ($R_X C$) in the synchronization mode and 4.5 times in asynchronous mode. The CLK frequency in 8251A should be in the range from 0.74 to 3.1MHz.

(3) $\overline{CS}$: Chip select signal from CPU. The $\overline{CS}$ is effective when it is active low. It indicates that the 8251A chip is strobed.

(4) $\overline{RD}/\overline{WR}$: Read or write control signal from CPU. It is effective when it is active low.

(5) C/$\overline{D}$: Control or data signal. When C/$\overline{D}$ pin is high, it designates that the information transferred via the data bus currently is a control or state signal. Otherwise it is the data.

The 8251A is associated with only two port addresses. The C/$\overline{D}$ pin is connected to the address line A_0 and A_0 differentiates the two port addresses. The 8251A internally interprets the C/$\overline{D}$, $\overline{RD}$ and $\overline{WR}$ signals as **Table 8-3**.

Table 8-3 8251A transfer description

C/$\overline{D}$ = (A_0)	$\overline{RD}$	$\overline{WR}$	Transfer Description
0	0	1	Data input from the data-in buffer
0	1	0	Data output to the data-out buffer
1	0	1	Status register is put on data bus
1	1	0	Data bus is put in mode, control or sync character register

In **Table 8-3** 1 means that the pin is high and 0 means that it is low. The any other combinations will lead the three-state D_7-D_0 pins to go into their high-impedance states.

5. MODEM (Modulation and Demodulation) Control Circuit

When an 8251A is used to realize the long-distance serial communication, the MODEM is needed to complete the conversion between digital signals and analog signals. The interface pins between terminals and 8251A are listed as following:

(1) $\overline{\text{DTR}}$(Data Terminal Ready): The ready signal of data terminal output to the modem. When it is active low, it denotes the CPU is ready to receive data. It can be defined by program and the pin $\overline{\text{DTR}}$ is effective when the bit DTR in control register is set to 1.

(2) $\overline{\text{DSR}}$(Data Set Ready): The ready signal of data device input from the modem. When it is active low, it indicates the modem or the external device is ready to transmit data. It is actually an answer signal to $\overline{\text{DTR}}$. The CPU can read status register of 8251A to detect DSR bit by the IN instruction. When the bit DSR is 1, it designates the pin $\overline{\text{DSR}}$ is effective.

(3) $\overline{\text{RTS}}$(Request-to-send): Request to send signal output to the modem. The pin $\overline{\text{RTS}}$ is active low designates that CPU is ready to transmit data. It is effective when the RTS bit of control register is 1.

(4) $\overline{\text{CTS}}$(Clear-to-send): Clear to send signal inputting from the modem. When $\overline{\text{CTS}}$ pin is effective, it denotes the modem is ready to receive data. The transmitter can serially send the data only when T_XEN bit of the control register is 1 and the pin $\overline{\text{CTS}}$ is valid. $\overline{\text{CTS}}$ is the acknowledge signal of $\overline{\text{RTS}}$.

要点：

(1) 8251A 是一个通用串行输入/输出接口，可用来将 86 系列 CPU 以同步或异步方式与外部设备进行串行通信。它能将并行输入的 8 位数据变换成逐位输出的串行信号，也能将串行输入数据变换成并行数据，一次传输给处理机。

(2) 8251A 由发送器、接收器、数据总线缓冲寄存器、读/写控制电路及调制/解调控制电路五部分组成。

① 发送器。8251A 的发送器包括发送缓冲器、发送移位器以及发送控制电路三部分。与发送器有关的引脚信号如下：

T_XD——数据发送线，输出串行数据。

T_XRDY——发送器已准备好信号。

T_XEMPTY——发送器空闲信号。8251A 发送器状态见正文表格。

T_XC——发送器时钟，由外部输入。

② 接收器。8251A 的接收器包括接收缓冲器、接收移位寄存器以及接收控制电路三部分。

外部通信数据从 R_XT 端，逐位进入接收位移寄存器中。如果是同步方式，则要检测同步字符，确认已经达到同步，接收器才可开始串行接收数据，待一组数据接收完毕，便把位移寄存器中的数据并行置入接收缓冲寄存器中；如果是异步方式，则应识别并删除起始位和停止位。这时，R_XDRY 线输出高电平，表示接收器已经准备好数据，等待向 CPU 输出。与接收器有关的引脚信号如下：

R_XD——数据接收线，输入串行数据。

R_XRDY——接收器已准备好信号。表示接收缓冲寄存器中以接收到一个数据符号，等待向 CPU 输入。

SYNDET/BRKDET——双功能的检测信号，高电平有效。

对于同步方式，SYNDET 是同步检测端。若采用内同步，当 R_xD 端上收到一个或两个同步字符时，SYNDET 输出高电平，表示已达到同步，后续接收到的就是有效数据，若采用外同步，外同步字符从 SYNDET 端输入，当 SYNDET 输入有效，表示已达到同步，接收器可开始接收有效数据。

对于异步方式，BRKDET 用于检测线路是处于工作状态还是断缺状态。当 R_xD 端上连续收到 8 个"0"信号，则 BRKDET 变成高电平，表示当前处于数据断缺状态。

R_xC——接收器时钟，由外部输入。这时钟频率决定 8251A 接收数据的速率。

③ 数据总线缓冲器。数据总线缓冲器是 CPU 和 8251A 之间信息交换的通道。它包含 3 个 8 位缓冲寄存器，其中两个用来存放 CPU 向 8251A 读取的数据及状态，当 CPU 执行 IN 指令时，便从这两个寄存器读取数据字及状态字。另一个缓冲器用来存放 CPU 向 8251A 写入的数据或控制字。当 CPU 执行 OUT 指令时，可向这个寄存器写入，由于两者共用同一个缓冲寄存器，这就要求 CPU 在向 8251A 写入控制字时，该寄存器中无将要发送的数据。为此，该接口电路必须要有一定的措施来防止混淆。

④ 读/写控制电路。读/写控制电路用来接收一系列的控制信号，由它们可确定 8251A 处于什么状态，并向 8251A 内部各功能部件发出相关的控制信号，因此它实际上是 8251A 的内部控制器。

读/写控制电路接收的控制信号如下：

RESET——复位信号。向 8251A 输入，高电平有效，此时，迫使 8251A 中各寄存器处于复位状态。

CLK——主时钟。向 8251A 输入。CLK 信号用来产生 8251A 内部的定时信号。对于同步方式，CLK 必须大于发送时钟和接收时钟频率的 30 倍。对于异步方式，CLK 必须大于发送和接收时钟的 4.5 倍。

$\overline{CS}$——选片信号。由 CPU 输入，低电平有效。

$\overline{RD}/\overline{WR}$——读和写控制信号。由 CPU 输入，低电平有效。

$C/\overline{D}$——控制/数据信号。$C/\overline{D}=1$，表示当前通过数据总经传送的是控制字或状态信息；$C/\overline{D}=0$，表示当前通过数据总线传送的是数据；均可由一位地址码来选择。

⑤ 调制解调控制电路。当使用 8251A 实现远距离串行通信时，8251A 的数据输出端要经过调制器将数字信号转换模拟信号，数据接收端收到的是经过解调器转来的数字信号，因此 8251A 要与调制解调器直接相连，它们之间的接口信号如下：

$\overline{DTR}$——数据终端准备好信号，向调制解调器输出，低电平有效。表示 CPU 已准备好接收数据。

$\overline{DSR}$——数据装置准备好信号，由调制解调器输入，低电平有效。表示调制/解调器或外设已准备好发送数据。

$\overline{RTS}$——请求发送信号。向调制解调器输出，低电平有效。$\overline{RTS}$有效，表示 CPU 已经准备好发送数据。

$\overline{CTS}$——清除发送信号。由调制解调器输入，低电平有效。$\overline{CTS}$有效，表示调制解调器已作为接收数据准备，只要控制字中的 TXEN=1，$\overline{CTS}$有效时，发送器才可串行发送数据。

8.4.2 The Application of the 8251A

The initialization operation of 8251A programmable communication interface should be done before it works. The 8251A interface is diagrammed in **Figure 8-10**. It is capable of being programmed for asynchronous or synchronous communication.

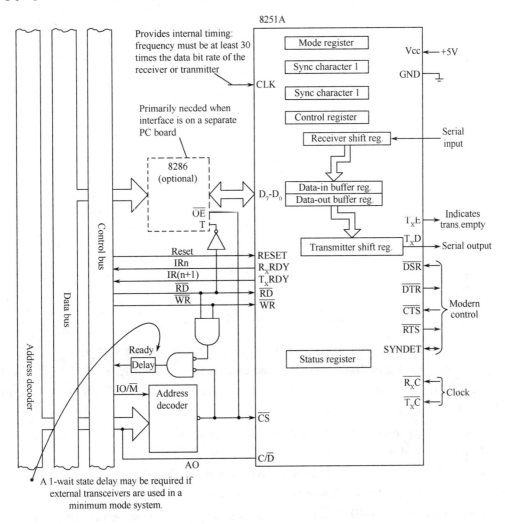

Figure 8-10 8251A serial communication interface

1. The Mode Register

The mode register is initialized by the executing program to determine whether the 8251A is in asynchronous mode or synchronous mode and the format of the characters being received and transmitted. **Figure 8-11** is the format of the mode register of 8251A.

As shown in **Figure 8-11**, there is an 8-bit byte to define the work mode of the 8251A.

(1) D_1D_0: Indicates the 8251A works in synchronous or asynchronous mode. In the synchronous mode the baud rates of the transmitter and receiver, which are the shift rates of the shift registers, are the same as the frequencies of the signals applied to $\overline{T_XC}$ and $\overline{R_XC}$, respectively, but in the asynchronous mode the three remaining possible combinations for the D_1D_0 in the mode register dictate the baud rate factor. The relationship between the frequencies of the

267

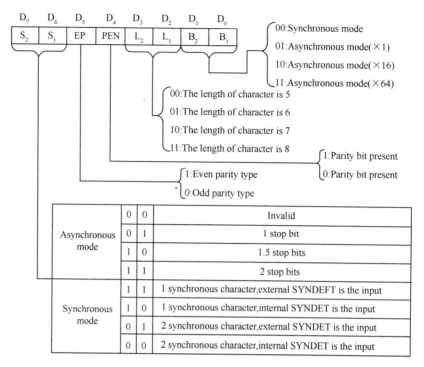

Figure 8-11 Format of the mode register

$\overline{T_XC}$ and $\overline{R_XC}$ clock input and the baud rates of the transmitter and receiver is:

Clock frequency = Baud rate factor × Baud rate

(2) D_3D_2: No matter in the asynchronous or synchronous mode, D_3 and D_2 indicate the number of data bits in each character.

(3) D_4: Indicates whether or not there is to be parity bit.

(4) D_5: Specifies the type of parity (odd or even).

(5) D_7D_6: The meaning of D_7D_6 is different with the communication mode. For the asynchronous mode these two bits indicate the number of stop bits, but for the synchronous mode D_6 determines whether the SYNDET pin is to be used as an input or as an output and D_7 indicates the number of sync character(s).

2. The Control Register

The control register, which is also set by the program, controls the operation of the interface. The format for the control register is given in **Figure 8-12**.

(1) D_0: Transmit enable flag. It indicates whether the transmitter is ready to transmit. It should be 1 before data can be output.

(2) D_1: Data terminal ready flag. It specifies whether the data terminal is ready or not. It is set to 1 when the data terminal is ready and forces the $\overline{DTR}$ pin to 0.

(3) D_2: Receive enable flag. It must be 1 before data can be received.

(4) D_3: Send break character flag. It equals to 1 and forces T_XD to 0, thus causing break characters to be transmitted.

(5) D_4: Error clearing flag. Setting it to 1 causes all the error bits in the status register to

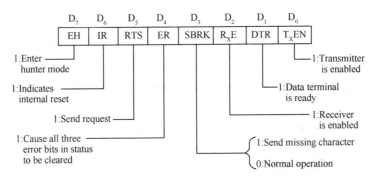

Figure 8-12 Format of the control register

be cleared (the bits that are set when framing, overrun, and parity errors occur).

(6) D_5: Request to send flag. It is used for sending a Request to Send signal to a modem. A 1 putting in D_5 will force the $\overline{RTS}$ pin output to be active. It implies the CPU is ready to send out data.

(7) D_6: Internal reset flag. An one here reinitializes the 8251A and indicates internal reset.

(8) D_7: Enter hunt mode flag. If it is 1, it designates to enter the hunt mode. D_7 is used only with the synchronous mode. When set, it causes the 8251A to begin a bit-by-bit search for a synchronous character or sync characters.

3. The Status Register

The status register offers some information of 8251A available to the executing program. The format of the status register is given in **Figure 8-13**.

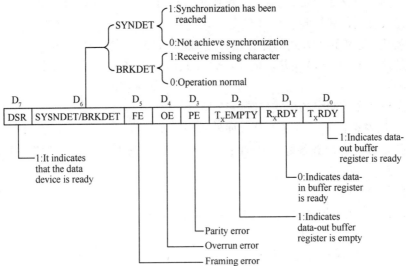

Figure 8-13 Format of the status register

(1) D_0: $T_X RDY$ indicates that the data-out buffer is empty. Unlike the $T_X RDY$ pin, this bit is not affected by the $\overline{CTS}$ input pin or the $T_X EN$ control bit.

(2) D_1: $R_X RDY$ reflects the signal on the $R_X RDY$ pin. It indicates that a character has

been received and is ready to be input to the processor.

(3) D_2: T_XEMPTY indicates that the transmitter shift register is waiting to be sent a character from the data-out buffer register. During synchronous transmissions, while this bit is set, the transmitter will take its data from the sync character registers until data are put in the data-out buffer register.

(4) D_3: PE(Parity error) indicates parity error when it is set to 1.

(5) D_4: OE(Overrun error) specifies overrun error when it is set to 1.

(6) D_5: FE(Frame error) denotes framing error when it is set to 1.

(7) D_6: SYNDET/BRKDET reflects the signal on SYNDET/BRKDET pin. It designates whether the synchronization is accomplished in synchronous mode or executes break detecting in asynchronous mode.

(8) D_7: DSR(Data Set Ready). When DSR is 1, it reflects the ready state of the modem and the $\overline{DSR}$ pin is effective.

4. The Sync Character Registers

The sync character registers are used for storing the sync characters needed for synchronous communication.

要点：

(1) 方式选择控制字的格式如图8-11所示。

B_2B_1用来定义8251A的工作方式是同步还是异步，若为异步方式还可由B_2B_1的取值来确定传送速率。

L_2L_1位用来定义数据字符的长度。

PEN位用来定义是否带奇偶校验，称作校验允许位。在PEN=1情况下，由EP位定义是采用奇校验还是偶校验。

S_2S_1位用来定义异步方式的停止位长度(1位，1.5位或2位)。对于同步方式，S_1位用来定义是外同步(S_1=1)还是内同步(S_1=0)，S_2位用来定义是单同步(S_2=1)还是双同步(S_2=0)。

(2) 操作命令控制字的格式如图8-12所示。

T_XEN=1，发送器才能通过T_XD线向外部串行发送数据。

DTR位是数据终端准备好位。DTR=1，表示CPU已准备好接收数据，这时，$\overline{DTR}$引线端输出有效。

R_XE位是允许接收位。

SBRK位是发送断缺字符位。SBRK=1，通过TXD线一直发送"0"信号。正常通信过程中，SBRK位应保持为"0"。

ER位是清除错误标志位。8251A设置3个出错标志，分别是奇偶校验标志PE，越界错误标志OE和帧校验错标志FE。ER=1时，三者都清零。

RTS位是请求发送信号。RTS=1，迫使8251A输出$\overline{RTS}$有效，表示CPU已作好发送数据准备，请求向调制/解调器或外部设备发送数据。

IR位是内部复位信号。IR=1，迫使8251A回到接收方式选择控制字的状态。

EH位为跟踪方式位，只对同步方式有效，EH=1，表示开始搜索同步字符，因此对于同步方式,一旦允许接收,必须使 EH = 1,并且使 ER = 1,清除全部错误标志,才能开始搜索同步字符。此后写入 8251A 的操作命令字。只有外部复位命令 RESET = 1 或内部复位命令 IR = 1 才能使 8251A 回到接收方式选择命令字状态。

(3) 状态控制字的格式如图 8-12 所示。

$T_xRDY = 1$ 表示发送器准备好。

$R_xRDY = 0$ 表示接收器准备好。

$T_xEMPTY = 1$ 表示发送器为空。

PE 表示是否存在奇偶错的标志位。

DE 表示是否存在溢出错的标志位。

FE 表示是否存在帧校验错的标志位。

SYNDET/BRKDET 引脚,与同名引线端的状态完全相同,供 CPU 查询。

DST = 1 表示数据装置准备好。

CPU 可在 8251A 工作过程中利用 IN 指令读取当前 8251A 的状态控制字。

8.4.3 The Initialization of the 8251A

The initialization of 8251A usually follows reset operation. You should develop some instruction sequences to initialize the mode register at first. If the 8251A works in asynchronous output mode, defining the control register of 8251A is the second step before the data transmission. Otherwise, the one or more ASCII sync characters will be searched and output before the command written to the control. A flowchart of the initialization of 8251A for output is given in **Figure 8-14**.

From **Figure 8-14**, no matter which mode the 8251A works in, the first output is the mode register after 8251A reset operation. If the 8251A works in asynchronous mode, the following output is the control register which defines the operation of the interface. If D_6 in the control register is set, it will cause the 8251A to be reinitialized and the reset sequence to be reentered. The 8251A will be returned to the beginning to restart. Otherwise, the character streams transmission is activated. During the transmission process, you can read the status register to detect the states of the 8251A. If the 8251A enters its asynchronous mode, the next 1 or 2 bytes output will be the sync characters before the output of the control register, just as shown in **Figure 8-14**.

【Example 8-2】Suppose that the mode and control registers are at address E3H and the baud rate factor is 64. Please develop a program sequence which initializes the mode register and gives a command to enable the transmitter and begin an asynchronous transmission of 7-bit characters followed by an even-parity bit 1 stop bit and 16 baud rates factor.

The program sequence can be written as:

```
        MOV    AL,    01111010B
        OUT    0E3H,  AL
        MOV    AL,    00110011B
        OUT    0E3H,  AL
```

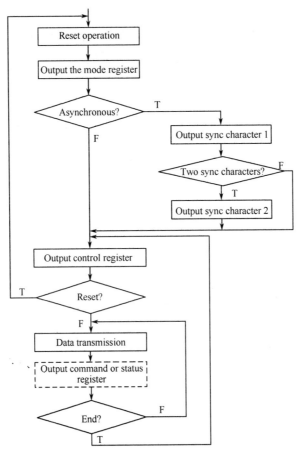

Figure 8-14 Flow chart of the disposition of output

【Example 8-3】 Assume that a 8251A works in synchronous mode, the mode and control registers are at address E3H. The characters consist of 7 data bits with even parity bit and two successive ASCII sync characters. Please develop a program sequence to initialize the 8251A.

The sequence can be written as:

    MOV AL, 00111000B
    OUT 0E3H, AL
    MOV AL, 16H
    OUT 0E3H, AL
    OUT 0E3H, AL
    MOV AL, 10010100B
    OUT 0E3H, AL

【Example 8-4】 Please give a program sequence which uses programmed I/O to output 256 characters from the 8251A in asynchronous mode with the baud rate factor 16. Assume that the mode and control registers are at address F1H. The characters are 7-bit followed by an e-ven-parity bit and 2 stop bits.

    MOV AL, 11111010B
    OUT 0F1H, AL

```
            MOV   AL,   37H         ; Enable transmitter and receiver
            OUT   0F1H, AL          ; And clear error bits
            MOV   BX,   0           ; Initialize index
            MOV   CX,   256         ; Put count in CX
    BEGIN:  IN    AL,   0F1H
            TEST  AL,   01H
            JZ    BEGIN
            MOV   AL,   [BX]
            OUT   0F0H, AL          ; Output character
            INC   BX
            LOOP  BEGIN             ; Continue outputting
```

According to the topic, the mode register is set as 11111010 B. The control register is set as 37H to ensure the transmitter and receiver to send or receive data respectively, and the error bits in control register is cleared. The loop continually tests the $T_X RDY$ bit until it is set 1.

[Example 8-5] **Figure 8-15** is a serial communication example of the 8251A. Please give a program sequence which uses programmed I/O to transmit 256 characters stored beginning at ARRY1 from an 8251A to another in asynchronous mode with the baud rate factor 64 and puts them in memory locations addressed at ARRY2. Assume that the address of the mode and control registers of the transmitter and receiver are at F1H and F3H respectively, and that of the data buffer for both are at F2H and F4H respectively. The characters are 8-bit followed by an even-parity bit and 1 stop bit.

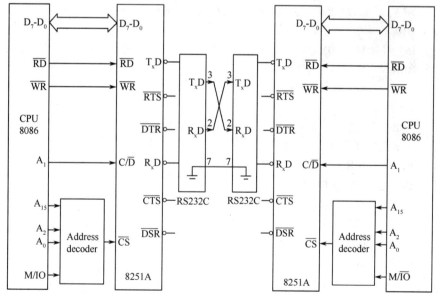

Figure 8-15   The diagram of a serial communication system

Initialization progarm for transmitter:

```
    STT:   MOV   DX,   0F1H
           MOV   AL,   0F1H
```

```
            OUT    DX,  AL
            MOV    AL,  11H
            OUT    DX,  AL
            MOV    DI,  OFFSET ARRY1    ;Initialize index
            MOV    CX,  256             ;Put count in CX
    NEXT:   MOV    DX,  0F1H
            IN     AL,  DX
            AND    AL,  01H
            JZ     NEXT
            MOV    DX,  0F2H
            MOV    AL,  [DI]
            OUT    DX,  AL
            INC    DI
            LOOP   NEXT
            HLT
```

Initialization program for receiver:

```
    SRR:    MOV    DX,  0F3H
            MOV    AL,  7FH
            OUT    DX,  AL
            MOV    AL,  04H
            OUT    DX,  AL
            MOV    DI,  OFFSET ARRY2    ;Initialize index
            MOV    CX,  256             ;Put count in CX
    COMT:   MOV    DX,  0F3H
            IN     AL,  DX
            ROR    AL,  1
            ROR    AL,  1
            JNC    COMT
            ROR    AL,  1
            ROR    AL,  1
            JC     SRR
            MOV    DX,  0F4H
            IN     AL,  DX
            MOV    [DI], AL
            INC    DI
            LOOP   COMT
            HLT
```

要点：

对 8251A 进行初始化编程必须先对系统进行复位，方式选择控制字必须紧跟在复位命令之后。如果定义 8251A 工作于异步方式，那么，必须紧跟操作命令控制字进行定义，然后才开始传送数据。在数据传送过程中，可重新定义操作命令字，或使用状态控制字读取 8251A 的状态，待数据传送结束，必须用操作命令控制字将 IR 位置"1"，向

8251A 传送内部复位命令后,8251A 才可重新接收方式选择命令字,改变工作方式完成其他传送任务。

8.5 Parallel Communication Interface

Parallel communication is accomplished by simultaneously transferring bits over separate lines. Comparing with the serial communication, it can transmit lines with a given maximum bit rate and higher information rates due to the use of several lines. Of course, it could lead to the costs increasement for the extra lines, especially when the distance increases.

A parallel interface may be designed for only outputting, only inputting, inputting and outputting over separate sets of lines, or performing I/O over a single set of bidirectional lines. It depends on the application you expect. For example, an interface would only output data when it is connected to a line printer and input data when it services a card reader. An interface that services both a scanner and a printer would require two sets of lines with one for input and another for output. Also, an interface could use only one set of bidirectional lines if the device does not input and output data simultaneously.

Figure 8-16 shows a representative parallel interface with handshaking control lines and separate input and output connections to an I/O device. In this case, an input would be carried out by first putting the data on the data bus and a 1 on the data-in ready line. The interface would respond by latching the data into the data-in buffer and putting a 1 on the data-in acknowledge line. Upon receiving the acknowledgment the device would drop the data and data-in ready signals. When the data are received by the interface, it would set a "ready" status bit and perhaps send out an interrupt request. After the CPU takes the data the interface would clear the "ready" status bit and put data lines in their high-impedance state. If an interrupt request were made, it could be handled in the usual manner.

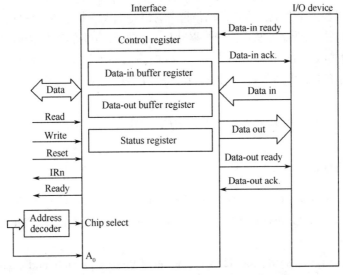

Figure 8-16 Representative parallel communication interfaces

For an output, the interface would set a "status" bit and, perhaps, make an interrupt request when the data-out buffer is available. After the CPU outputs the data, the interface would clear the output "ready" status bit, put the data on the data-out bus, and signal the I/O device over the data-out ready line. When the device is ready to accept the data, it would latch the data and then return a data-out acknowledgment. The interface would then drop its data-out ready signal and once again set the output "ready" status bit.

要点：

(1) 并行通信是指数据信息以字或字节为单位，各位同时进行传输的通信方式，优点是传输速度快，缺点是传输线多，所以适用于短距离（几米至几十米）、数据传输率高的场合。

从图 8-15 中看到，并行接口中有一个控制寄存器来接收 CPU 对它的控制命令，一个状态寄存器提供状态供 CPU 查询。为了实现输入输出，并行接口中还必定有相应的输入输出缓冲寄存器。

(2) 并行接口的输入过程。外部设备首先将数据送给接口，并使状态线"数据输入准备好"成为高电平。接口把数据接收到数据输入缓冲寄存器的同时，使"数据输入回答"线变为高电平，作为对外部输入设备的响应。外部设备接收到此信号，便撤除数据和"数据输入准备好"信号。数据到达接口中后，接口会在状态寄存器中设置"输入准备好"状态位，以便 CPU 对其进行查询，接口也可以在此时向 CPU 发一个中断请求。所以，CPU 既可以用软件查询方式，也可用中断方式来设法读取接口中的数据。CPU 从并行接口中读取数据之后，接口会自动清除状态寄存器中的"输入准备好"状态位，并且使数据总线处于高阻状态，此后，又可以开始下一个输入过程。

(3) 并行接口的输出过程。每当外部设备从接口取走一个数据后，接口就会将状态寄存器中的"输出准备好"状态位置"1"，以表示 CPU 当前可以往接口中输出数据，这个状态位可供 CPU 进行查询。此时，接口也可以向 CPU 发送一个中断请求。所以，CPU 既可以用软件查询方式，也可用中断方式设法往接口中输出一个数据。当 CPU 输出的数据到达接口的输出缓冲器后，接口会自动清除"输出准备好"状态位，并且将数据送往外部设备，同时，接口往外部设备发送一个"驱动信号"来启动外部设备接收数据。外部设备被启动后，开始接收数据，并往接口发一个"数据输出回答"信号。接口接收到此信号，便将状态寄存器中的"输出准备好"状态位重新置"1"，以便 CPU 输出下一个数据。

8.6 8255A Programmable Peripheral Interface

8.6.1 The Architecture of the 8255A

Intel's 8255A is a programmable peripheral interface chip. **Figure 8-17** is the interface diagram of the 8255A. **Figure 8-18** is the architecture of the 8255A and **Figure 8-19** is its pins assignment.

As shown in **Figure 8-18**, the 8255A consists of three separately addressable ports, denoted A, B, and C, the control units of group A and group B, data buffer, and read/write logical

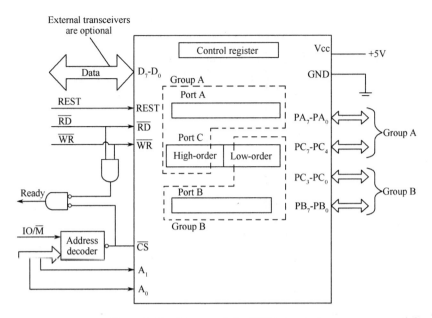

Figure 8-17 Diagram of the 8255A interface

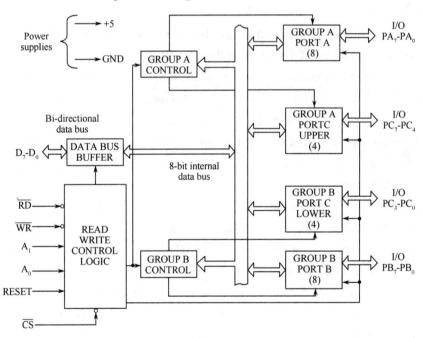

Figure 8-18 8255A block diagram

control unit.

1. The Addressable Ports: A, B and C

The ports A, B and C are 8-bit long. All of them can be programmed to input or output communicating with external device, although there are some differences in their functions.

Usually the port A and B are used as the input port or output independently, while the port C can be utilized as two 4 bit ports defined in the mode control register. The bits in the three

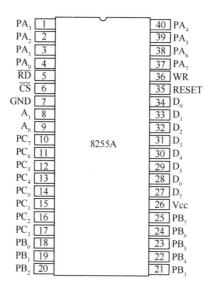

Figure 8-19   8255A Pin assignments

ports are attached to pins that may be connected to the I/O device. Three ports A, B and C can be divided into groups A and B, with group A consisting of the bits in port A and the four MSBs of port C and group B consisting of port B and the four LSBs of port C.

2. The Control Units of Group A and Group B

As said above, the A group consists of port A and upper 4 bits of port C (PC_4-PC_7). The port B and lower 4 bits of port C (PC_0-PC_3) form B group. Each group has its own control unit. These two control units can accept read / write command from the read / write control logic unit, and also receive the control character from the data bus to determine the read/write operation of group A and B.

3. Data Bus Buffer

Tri-state bidirectional 8-bit data buffer is a connected components between 8255A and CPU. CPU uses the IN and OUT instructions to read and write the data. The control and status information is transmitted through this buffer. The related pins are D_7-D_0.

4. Read/Write Logical Control Unit

This unit is responsible for the transceiving management of 8255A between internal and external device. It can not only receive the CPU commands but also send the operation command to other parts of 8255A. The control operation is accomplished by the 6 lines connected to the data bus buffer, I/O interface and the control units of A and B groups.

(1) $\overline{CS}$: chip select signal. Usually, it is obtained by the high address line decoding. An active low on this pin means this 8255A chip is selected.

(2) $\overline{RD}/\overline{WR}$: $\overline{RD}$ and $\overline{WR}$ signals. Both of them are active low.

(3) **RESET**: reset signal, high effective. When this pin is effective, it will remove all contents of internal registers (including control register), and set the port A, B and C as input mode.

(4) A_1 and A_0: ports selecting signals. The 8255A contains a control register and three separately addressable ports A, B and C, which of the four registers is being addressed is determined by the signals applied to the pins A_1 and A_0.

For example, $A_1 A_0 = 00$ makes port A available, $A_1 A_0 = 01$ is for port B, $A_1 A_0 = 10$ is for port C and $A_1 A_0 = 11$ activates control register.

Whether or not an 8255A is being accessed is determined by the signal on the $\overline{CS}$ pin and the direction of the access is according to the $\overline{RD}$ and $\overline{WR}$ signals. Therefore, the lowest port address assigned to an 8255A must be divisible by 4. A transfer description summary of the 8255A's is listed in **Table 8-4**.

Table 8-4　8255A transfer description

A_1	A_0	$\overline{RD}$	$\overline{WR}$	$\overline{CS}$	Transfer Description
0	0	0	1	0	Port A to data bus
0	1	0	1	0	Port B to data bus
1	0	0	1	0	Port C to data bus
0	0	1	0	0	Data bus to port A
0	1	1	0	0	Data bus to port B
1	0	1	0	0	Data bus to port C
1	1	1	0	0	Data bus to control register if $D_7 = 1$; if $D_7 = 0$ input from the data bus is treated as a Set/Reset instruction
×	×	×	×	1	D_7-D_0 go to high impedance state
1	1	0	1	0	illegal combination
×	×	1	1	0	D_7-D_0 go to high impedance state where 0 is low and 1 is high

要点：

由图 8-18 可见,8255A 由以下几部分组成：

（1）并行 I/O 端口 A、B、C。8255A 芯片内部包含 3 个 8 位端口：A 口,B 口,C 口,每一个端口都是 8 位的。端口 A 包含一个 8 位数据输出锁存/缓冲存储器和一个 8 位数据输入锁存器；端口 B 包含一个 8 位数据输入/输出、锁存/缓冲存储器和一个 8 位数据输入缓冲存储器；端口 C 包含一个输出锁存/缓冲存储器和一个输入缓冲存储器。它们都可以通过编程设定为输入或输出来与外设传送信息,但功能有些不同。

通常,A 口和 B 口作为独立的输入端口或输出端口,C 口则通过方式控制字被分成两个 4 位端口,它们分别作为 A 口和 B 口的输出控制线和状态输入线。

（2）A 组和 B 组控制电路(部件)。端口 A 和端口 C 高 4 位 $PC_7 \sim PC_4$ 构成 A 组,由 A 组控制部件实现控制功能。端口 B 和端口 C 低 4 位 $PC_3 \sim PC_0$ 构成 B 组,由 B 组控制部件实现控制功能。它们各有一个控制单元,可接收来自读/写控制部件的命令和 CPU 通过数据总线($D_7 \sim D_0$)送来的控制字,并根据它们来定义各个端口的操作方式。

（3）数据总线缓冲器。它是一个三态双向 8 位数据缓冲器,是 8255A 与 CPU 数据总线的连接部件。CPU 通过输入输出指令接收和发送的数据,控制字和状态信息都是通过该缓冲区传送的,因此,它是 CPU 和 8255A 之间交换信息的必经之路。$D_7 \sim D_0$ 是它的 8 根引脚。

(4) 读/写控制逻辑。这是 8255A 内部完成读/写控制功能的部件,负责管理 8255A 的内部和外部的传送过程。通过 6 条控制线完成对数据缓冲器、I/O 接口、A 组和 B 组控制部件的各种控制操作。

与 CPU 连接的 6 条控制线如下:

① $\overline{CS}$——片选信号。通常由高位地址线译码得到。$\overline{CS}$ 为低电平时,8255A 芯片被选中。

② $\overline{RD}$、$\overline{WR}$——读、写控制信号。由 CPU 输入,$\overline{RD}$ 有效,表示 CPU 读 8255A,应由 8255A 向 CPU 传送数据或状态信息。$\overline{WR}$ 有效,表示 CPU 写 8255A,应由 CPU 将控制字或数据写入 8255A。

③ RESET——复位信号,高电平有效。RESET 有效时,清除所有内部寄存器(包括控制寄存器),并把 A 口、B 口、C 口均设为输入方式。

④ A_1、A_0——端口选择信号。用来选择 3 个数据端口和 1 个控制字寄存器,即

$\overline{CS}$	A_1	A_0	选中端口
0	0	0	A 口
0	0	1	B 口
0	1	0	C 口
0	1	1	控制字寄存器(命令口)

A1、A0 常和系统地址线相连。A_1、A_0、$\overline{CS}$、$\overline{RD}$、$\overline{WR}$ 的组合,实现各种不同的操作功能,如表 8.4 所列。

8.6.2 The Function Description of the 8255A

The ports of 8255A can be defined as inputting or outputting and designated their communicating modes by programming. The 8255A has three kinds of basic communication modes such as Mode 0 (Basic input / output mode), Mode 1 (Strobed input/output mode) and Mode 2 (Bidirectional transmission mode). The modes are determined by the contents of the control register, which will be introduced later.

1. The format of the 8255A's Control Register

The control register is used to define the port working mode and to designate each port working for inputting or outputting. **Figure 8-20** is the format of 8255A control register. The working mode of each port is controlled by the bit associated with it.

(1) D_0: Specifies the lower half of port C working for inputting or outputting.

(2) D_1: Indicates the port B working for inputting or outputting.

(3) D_2: Specifies work mode selection of the group B.

(4) D_3: Denotes the upper half of port C serving for inputting or outputting.

(5) D_4: Designates the port A using for inputting or outputting.

(6) D_5 and D_6: Used together to specify the work mode of group A.

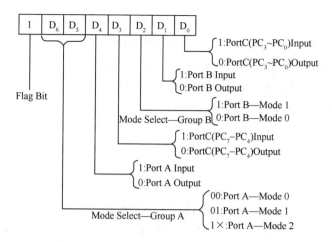

Figure 8-20 Format of 8255A control register

(7) D_7: Flag bit. It is always 1 in the control register.

The bits D_0, D_1, D_3 and D_4 specify the group A and B are used for input or output. If a bit is **0**, then the corresponding port or designated bits of port C is used for **output**, otherwise it is used for **input** when a bit is 1.

As shown in **Figure 8-20**, **port A** is associated with one of three modes, mode 0, mode 1, and mode 2, and **port B** is with one of the two modes, mode 0 and mode 1. **Port C** is in the mode 0 only when both A and B are designated in mode 0. **Port C** is usually used combining with port A and B to provide the control signal or state signal for port A and B.

2. The format of the 8255A's Set/Reset instruction for port C

Because the bits in port C are sometimes used as control bits, the 8255A designed so that they can be output to individually using a Set/Reset instruction.

The Set/Reset instruction of 8255A is used to set or clear the port C bit specified by the instruction. **Figure 8-21** is the format of Set and Reset Instruction. The set or reset operation of each bit of **port C** is controlled by the bit associated with it.

D_7	D_6	D_5	D_4	D_3	D_2	D_1	D_0
0	*	*	*	×	×	×	×

	D_3	D_2	D_1	
PC_0	0	0	0	
PC_1	0	0	1	0: Reset
PC_2	0	1	0	
PC_3	0	1	1	1: Set
PC_4	1	0	0	
PC_5	1	0	1	
PC_6	1	1	0	
PC_7	1	1	1	

Figure 8-21 Format of 8255A Set/Reset instruction for port C

(1) D_0: Specifies the bit operation of port C. It is used to indicate whether the bit of port C designated by D_3, D_2 and D_1 is to be set or cleared.

(2) $D_3 D_2 D_1$: Specify the bit of port C to be changed.

(3) $D_6D_5D_4$: These three bits are unused in the 8255A's Set and Reset Instruction.
(4) D_7: Flag bit. It is always 0 in the Set/Reset Instruction.

The Control Register and Set/Reset Instruction share the same port address. When the 8255A receives a command byte, it will examine the bit 7. If this bit is 1, the byte is treated as that from control register, otherwise, it is a Set/Reset instruction.

要点：

（1）8255A 中各端口可有 3 种工作方式：方式 0——基本输入/输出方式；方式 1——选通输入/输出方式；方式 2——双向传送方式。

端口 A 可处于 3 种工作方式（方式 0、方式 1 和方式 2），端口 B 只可处于两种工作方式（方式 0 和方式 1），端口 C 常常被分成高 4 位和低 4 位两部分，可分别用来传送数据或控制信息。用户可用软件来分别定义 3 个端口的工作方式，可使用的控制字有工作方式控制字和置位/复位控制字。

（2）控制字。

① 定义工作方式控制字：格式如图 8-20 所示。

通过工作方式控制字可将 3 个端口分别定义为 3 种不同方式的组合，当将端口 A 定义为方式 1 或方式 2 或将端口 B 定义为方式 1 时，端口 C 的某些位作为控制位使用，可通过置位/复位控制字来对控制端口 C 的各位分别进行置位/复位操作。

② 置位/复位控制字：只对端口 C 有效，其使用格式如图 8-21 所示。

8.6.3　The Communication Mode of 8255A

1. Mode 0

When a 8255A works in mode 0, its control register will be set as **Figure 8-22**.

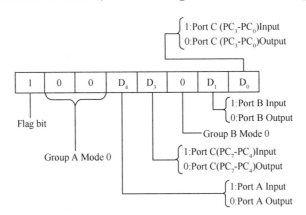

Figure 8-22　Format of 8255A control register in mode 0

If group A and group B are in mode 0, they can be divided into four sets such as port A, the upper 4 bits of port C, port B and the lower 4 bits of port C. Each set may be used for inputting or outputting, but not both. Bits D_4, D_3, D_1 and D_0 in the control register specify which sets are for input and which are for output. As shown in **Table 8-5**, there are 16 combinations of the four sets which are associated with these bits.

Table 8-5 Mode 0 port definition

#	A		B		Group A		Group B	
	D4	D3	D1	D0	Port A	Port C(Upper)	Port B	Port C(Lower)
0	0	0	0	0	Output	Output	Output	Output
1	0	0	0	1	Output	Output	Output	Input
2	0	0	1	0	Output	Output	Input	Output
3	0	0	1	1	Output	Output	Input	Input
4	0	1	0	0	Output	Input	Output	Output
5	0	1	0	1	Output	Input	Output	Input
6	0	1	1	0	Output	Input	Input	Output
7	0	1	1	1	Output	Input	Input	Input
8	1	0	0	0	Input	Output	Output	Output
9	1	0	0	1	Input	Output	Output	Input
10	1	0	1	0	Input	Output	Input	Output
11	1	0	1	1	Input	Output	Input	Input
12	1	1	0	0	Input	Input	Output	Output
13	1	1	0	1	Input	Input	Output	Input
14	1	1	1	0	Input	Input	Input	Output
15	1	1	1	1	Input	Input	Input	Input

2. Mode 1

When group A is in mode 1, the port A is used for input or output according to bit D_4 ($D_4 = 1$ indicates input), and the upper half of port C is used for handshaking and control signals. If group B is in mode 1, port B is applied to input or output according to bit D_1 of the control register ($D_1 = 1$ indicates input) with 3 bits of the lower half of port C as its handshaking and control signals.

(1) Inputting in mode 1.

Figure 8-23 is the diagram of both ports A and B working as inputting in mode 1.

As shown in **Figure 8-23**, the PC_5-PC_0 are assigned as control lines. The following are the symbols and definitions of these control lines.

Symbol	Definition	Port A	Port B
$\overline{STB}$	Strobe signal. A 0 applied to this pin causes PA_7-PA_0 or PB_7-PB_0 to be latched or "strobed" into port A or port B.	$\overline{STB}_A$-PC_4	$\overline{STB}_B$-PC_2
IBF	Indicates that the input buffer is full. It is 1 when a port (A or B) contains data that have not yet been input to the CPU. When a 0 is on this pin the device can input a new byte to the interface. It is actually an acknowledge signal to $\overline{STB}_A$.	IBF_A-PC_5	IBF_B-PC_1
INTR	Interrupt request signal sent to CPU. It is tied to one of the IR lines in the system bus	$INTR_A$-PC_3	$INTR_B$-PC_0
INTE	Interrupt enable signal.		

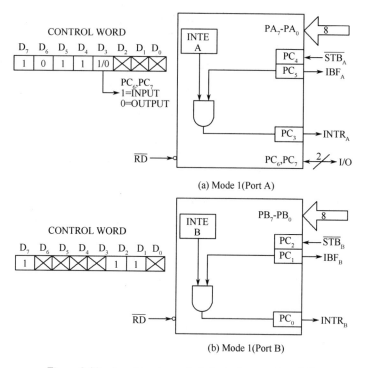

Figure 8-23　Inputting in mode 1 for both ports A and B

For Inputting in mode 1, the lines **PC$_6$** and **PC$_7$** may be used to output control signals to the device or input status from the device. If D_3 of the control register is 0, they are for outputting control signals; otherwise, they are for inputting status.

When the device inputs a data to 8255A, a strobe signal is issued on $\overline{STB}$ line. Then the data will be latched into the data-in register through the port A or port B. Meanwhile, $\overline{IBF}$ is set to 1 to indicate that the data-in buffer register is full. It is used not only to prevent device inputting new data, but also to offer a polling signal for the CPU. If the interruption operation is enabled, the interrupt request signal INTR will be send to CPU when the strobe signal $\overline{STB}$ is invalid. When the interrupt is responded, the CPU will send $\overline{RD}$ signal to read the data in. The active $\overline{RD}$ signal clears the interrupt request. At the moment that the $\overline{RD}$ signal became 1, the $\overline{IBF}$ goes low to indicate the data-in buffer register is empty and the device can input a new byte.

(2) Outputting in mode 1.

Figure 8-24 is the diagram of both ports A and B serving for outputting in mode 1.

For outputting in mode 1, the control lines for port A and B are listed as follows:

Symbol	Definition	Port A	Port B
$\overline{OBF}$	Indicates that the output buffer is full. It outputs A 0 to the device when port A or B is outputting new data to be taken by the device.	$\overline{OBF}_A$-PC$_7$	$\overline{OBF}_A$-PC$_1$

$\overline{ACK}$	Device puts a 0 on this pin when it is ready to accept data from Port A or B.	$\overline{ACK_A}$-PC_6	$\overline{ACK_B}$-PC_2
INTR	Interrupt request signal sent to CPU. It is tied to one of the IR lines in the system bus	$INTR_A$-PC_3	$INTR_B$-PC_0
INTE	Interrupt enable signal		

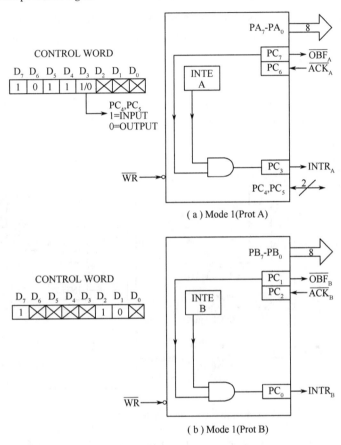

Figure 8-24 Outputting in mode 1 for ports A and B

For outputting in mode 1, the lines **PC₄** and **PC₅** serve the same purpose as described in inputting of mode 1 for PC_6, PC_7.

No matter inputting or outputting, in mode 1, PC_3 is denoted $INTR_A$ and is associated with group A and PC_0 becomes $INTR_B$ related to group B. Both of them are used as an interrupt request line. For inputting, INTR becomes 1 when new data are put in port A or B and is cleared when the CPU takes the data. For outputting, this pin is set to 1 when the contents of port A or B are taken by the device and is cleared when new data are sent from the CPU. The interrupt enable for group A is controlled by setting or clearing internal flags. Setting or clearing these flags is simulated by setting or clearing PC_4 when in input mode, and by PC_6 when in output mode, using a Set/Reset instruction. Similarly, the interrupt enable for group B is controlled by set/clear of PC_2 for both input and output.

The interrupt control is usually used in outputting in mode 1. The CPU transmits the data to 8255A and sends a $\overline{WR}$ signal during the ISR(interrupt service routine). The rising edge of $\overline{WR}$ not only clears the interrupt request INTR to indicate that the CPU is responding to the interrupt requested before, but also sets the $\overline{OBF}$ effectively low to notify the device to receive a data. As a device strobe signal, $\overline{OBF}$ enable the device to load the data in its data-in buffer. As soon as it receives the data, the device sends out an acknowledge signal $\overline{ACK}$ which sets the $\overline{OBF}$ to high to indicate that the data has been removed from the output buffer and makes interrupt request enable at its rising edge to ensure that a new interrupt request can be arrived at the CPU to begin a new data outputting.

3. Mode 2

This mode applies only to group A, although it also uses PC_3 for making interrupt requests. **Figure 8-25 is** the diagram of port A which works as bidirectional port in mode 2.

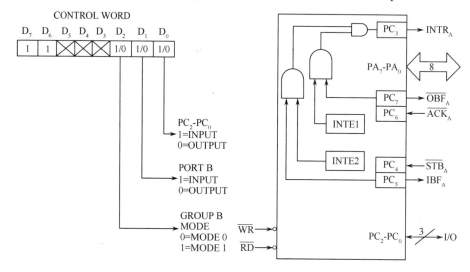

Figure 8-25 Port A in mode 2

In mode 2, port A is a bidirectional port and the four MSBs of port C are defined as follows:

Symbol	Definition	Port A
$\overline{STB}_A$	A 0 on this line causes the data on PA_7-PA_0 to be "strobed" into port A.	PC_4
IBF_A	Becomes 1 when port A is filled with new data from lines PA_7-PA_0 and is cleared when these data are taken by the CPU.	PC_5
$\overline{ACK}_A$	Indicates that the device is ready to accept data from PA_7-PA_0.	PC_6
$\overline{OBF}_A$	Becomes 0 when port A is filled with new data from the CPU and is set to 1 when data are taken by the device.	PC_7

While group A is in mode 2, group B may be in either mode 0 or mode 1. However, if group B is in mode 0, only PC_2-PC_0 can be used for input or output because group A has borrowed PC_3 to use as an interrupt request line. Normally, if group A is in mode 2, PC_2-PC_0 would be connected to control and status pins on the device attached to the port A lines. Port B may also be used for this purpose.

In all three modes port C reflects the signals on PC_7-PC_0 and can be read with an IN instruction.

要点:

8255A 中各端口可有三种基本的工作方式:

(1) 工作方式 0:这是 8255A 中各端口的基本输入/输出方式。它只完成简单的并行输入/输出操作,CPU 可以从指定端口输入信息,也可向指定端口输出信息。当 3 个端口均处于工作方式 0 时,共有 16 种工作方式的组合,如表 8-5 所列。

(2) 工作方式 1:被称作选通输入/输出方式,在此工作方式下,数据输入/输出操作要在选通信号的控制下完成。

工作方式 1 下的输入操作所需要使用的控制信号有:

① $\overline{STB}$——选通信号。由外部输入,低电平有效。

$\overline{STB}$有效时,将外部输入的数据锁存到所选端口的输入锁存器中。对 A 组来说,指定端口 C 的第 4 位(PC_4)接收向端口 A 输入的$\overline{STB}$信号。对 B 组来说,指定端口 C 的第 2 位(PC_2)接收向端口 B 输入的$\overline{STB}$信号。

② IBF——输入缓冲存储器满信号。向外部输出,高电平有效。

IBF 有效时,表示由输入设备输入的数据已占用该端口的输入锁存器,它实际是对$\overline{STB}$信号的应答信号,待 CPU 执行 IN 指令时,$\overline{RD}$有效,将输入数据读入 CPU,其后沿把 IBF 置零,表示输入缓冲器已空,外部设备可继续输入后续数据。对 A 组来说,指定端口 C 的第 5 位(PC_5)作为从端口 A 输出的 IBF 信号;对 B 组来说,指定端口的第 1 位(PC_1)作为从端口 B 输出的 IBF 信号。

③ INTR——中断请求信号。向 CPU 输出,高电平有效。

在 A 组和 B 组控制电路中分别设置一个内部中断触发器 INTEA 和 INTEB,前者由$\overline{STB_A}$(PC_4)控制置位,后者由$\overline{STB_B}$(PC_2)控制置位。

当任一组中的$\overline{STB}$有效,则把 IBF 置"1",表示当前缓冲器已满,并由$\overline{STB}$后沿置"1"各组的 INTE,于是输出 INTR 有效,向 CPU 发出中断请求信号。待 CPU 响应这一中断请求,可在中断服务程序中安排 IN 指令读取数据后置"0"于 IBF,外部设备才可继续输入后续数据。

工作方式 1 下的输出操作所需要的控制信号如下:

$\overline{OBF}$——输出缓冲存储器满信号。向外部输出,低电平有效。

有效时,表示 CPU 已将数据写入该端口正等待输出。当 CPU 执行 OUT 指令,$\overline{WR}$有效时,表示将数据锁存到数据输出缓冲寄存器,由$\overline{WR}$的上升沿将$\overline{OBF}$置为有效。对于 A 组,系统规定端口 C 的第 7 位用作从端口 A 输出的$\overline{OBF}$信号;对于 B 组,规定端口 C 的第 1 位(PC_1)用作从端口 B 输出的$\overline{OBF}$信号。

$\overline{ACK}$——外部应答信号。由外部输入,低电平有效。

$\overline{ACK}$有效,表示外部设备已收到由 8255A 输出的 8 位数据,实际上是对$\overline{OBF}$信号的应答信号。对于 A 组,指定端口 C 的第 6 位(PC_6)作为接收向端口 A 输入的$\overline{ACK}$信号;对于 B 组,指定端口 C 的第 2 位(PC_2)作为接收向端口 B 输入的$\overline{ACK}$信号。

INTR——中断请求信号。向 CPU 输出,高电平有效。

对于端口 A,内部中断触发器 INTEA 由 PC_6($\overline{ACK_A}$)置位,对于端口 B,INTEB 由 PC_2($\overline{ACK_B}$)置位。当$\overline{ACK}$有效时,$\overline{OBF}$被复位为高电平,并将相应端口的 INTE 置"1",于是 INTR 输出高电平,向 CPU 发出输出中断请求,待 CPU 响应该中断请求,可在中断服务程序中安排 OUT 指令继续输出后续字节。对于 A 组,指定端口 C 的第 3 位(PC_3)作为由端口 A 发出的 INTR 信号;对于 B 组,指定端口 C 的第 0 位(PC_0)作为由端口 B 发出的 IN-TR 信号。

采用工作方式 1 时,还允许将端口 A 和端口分别定义为输入和输出端口。

(3) 工作方式 2:被称作带选通的双向传送方式。

8255A 中只允许端口 A 处于工作方式 2,可用来在两台处理机之间实现双向并行通信。其控制信号由端口 C 提供,并可向 CPU 发出中断请求信号。

8.6.4 The Initialization and Programming of 8255A

【**Example 8-6**】Assume that a 8255A works in the mode 0 with the group A (port A and the four MSBs of port C) inputting and group B (port B and the four LSBs of port C) outputting and its ports addresses are 02E0H-02E3H. Please develop a program sequence to initializes the 8255A.

The sequences can be written as:

    MOV   DX,02E3H
    MOV   AL,83H
    OUT   DX,AL

【**Example 8-7**】As shown in **Figure 8-26**, the bits of group B in the 8255A are used as input which connected to 12 switches while the bits of group A are linked with 12 LEDs. A switch is corresponded to a LED. When a switch is turned on, the LED corresponding will shine. Otherwise, it will go out. Try to write program to achieve this function.

According to **Figure 8-26**, the ports address can be obtained as following:

A_{15}	A_{14}	A_{13}	A_{12}	A_{11}	A_{10}	A_9	A_8	A_7	A_6	A_5	A_4	A_3	A_2	A_1	A_0	Port	Hex
1	1	1	1	1	1	1	1	0	0	0	0	0	0	0	0	A	FF00H
1	1	1	1	1	1	1	1	0	0	0	0	0	0	0	1	B	FF01H
1	1	1	1	1	1	1	1	0	0	0	0	0	0	1	0	C	FF02H
1	1	1	1	1	1	1	1	0	0	0	0	0	0	1	1	CR	FF03H
								C	B	A	A_1	A_0					

    MOV   AL,   83H
    MOV   DX,   0FF03H   ;Initialize the 8255A
    OUT   DX,   AL
    MOV   DX,   0FF01H
    IN   AL,   DX   ;Read port B

```
DEC    DX                    ;Get the address of port A
OUT    DX,   AL              ;LED shines corresponding to the switch
MOV    DX,   0FF02H
IN     AL,   DX              ;Get data from port C
MOV    CL,   4
SHL    AL,   CL              ;Shift left to make lower 4-bit to upper 4-bit
OUT    DX,   AL              ;Outputting $PC_0$-$PC_3$ through $PC_4$-$PC_7$
MOV    AH,   4CH
INT    21H
```

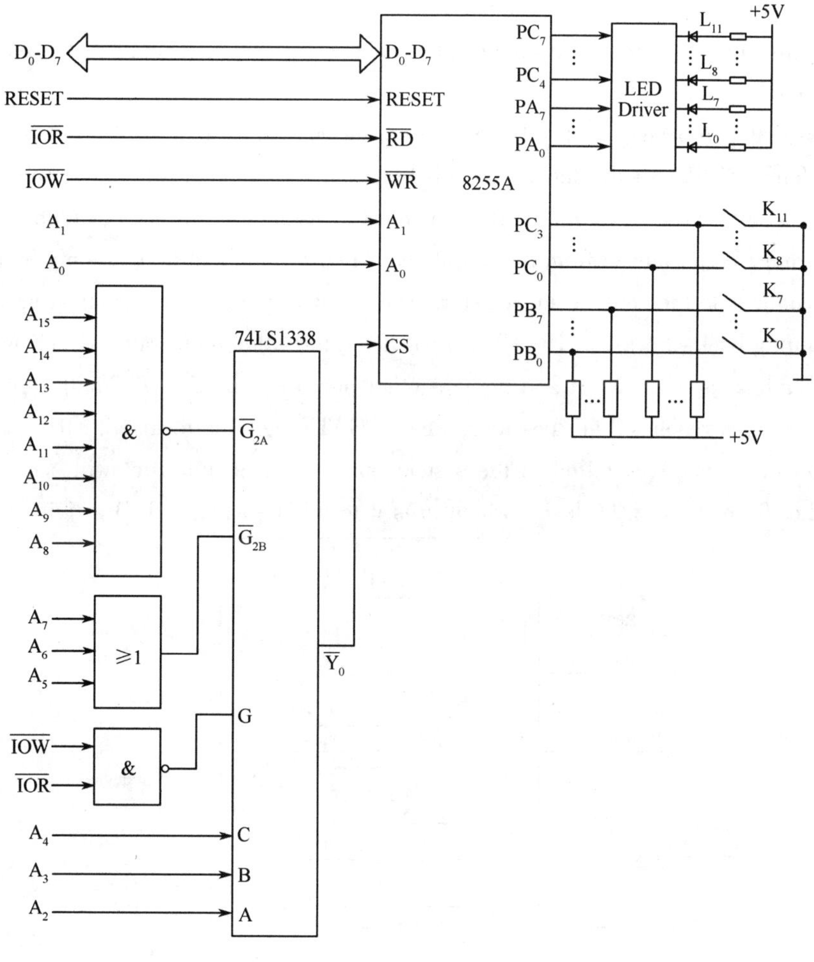

Figure 8-26 Circuit for Example 8-7

8.7 Programmable Timer and Event Counter

Quite often a device is needed to mark intervals of time for both the processor and external devices, count external events and make the count available to the processor, and provide external timing that can be programmed from the processor. Such a device is called a programmable interval time/event counter and some of its uses are to:

(1) Interrupt a time-sharing operating system at evenly spaced intervals so that it can switch programs.

(2) Output precisely timed signals with programmed periods to an I/O device (e.g. an A/D converter).

(3) Serve as a programmable baud rate generator.

(4) Measure time delays between external events.

(5) Count the number of times an event occurs in an external experiment and provide a means of inputting the count to the computer.

(6) Cause the processor to be interrupted after a programmed number of external events have occurred.

A representative design of an internal time/event counter is given in **Figure 8-27**. In this design the four registers on the left are accessible by the computer, with the upper two being output ports and the lower two input ports. The counter itself is not directly available to the processor, but must be initialized from the initial count register and can be read only by first transferring its contents to the counter out register. The counter operates by starting at an initial value and counting backward to 0. The CLK input determines the count rate, GATE is for enabling and disabling the CLK input and perhaps other purpose, and the OUT output becomes active when the count reaches 0 or, possibly, when a GATE signal is received. OUT may be connected to an interrupt request line in the system bus so that an interrupt will occur when the count reaches 0, or to an I/O device which uses it to initiate specific I/O activity.

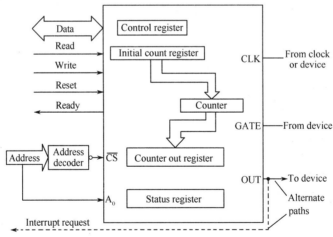

Figure 8-27 Typical interval timer/event counter

The operation is basically to enter a count in the initial count register, transfer the count to the counter, and cause the computer to count backward as pulses are applied to the CLK input. The current contents of the counter can be input at any time without disturbing the count by transferring them to the counter out register and then reading them. By buffering the count through the counter out register it doesn't have to be input to the processor immediately. The zero count indication would normally be applied to both the OUT pin and one of the bits in the status register. Thus either programmed I/O or interrupt I/O could be used to detect the zero count.

Notice that the control register includes the mode of operation. The mode determines exactly what happens when the count becomes 0 and/or a signal is applied to the gate input. Some possible actions are:

(1) The GATE input is used for enabling and disabling the CLK input.

(2) The GATE input may cause the counter to be reinitialized.

(3) The GATE input may stop the count and force OUT high.

(4) The count will give an OUT signal and stop when it reaches 0.

(5) The count will give an OUT signal and automatically be reinitialized from the Initial Count Register when the count reaches 0.

The modes could be defined by combinations of these possibilities.

As an example, consider the application of an interval timer to a time-sharing operating system. In this case a clock would be connected to the CLK input and OUT to an interrupt request line, possibly to a nonmaskable line. The GATE input would not be needed. When the system is brought up the initial count register would be filled with

$$\text{Initial count} = \text{Clock frequency} \times T$$

where T is the length of each time slice in seconds, and the mode would be set so that each time the count reaches 0 the contents of the initial register would be transferred to the counter and OUT would become active. Since OUT is used as an interrupt request, an interrupt routine for switching programs would be entered at the end of each period of T seconds.

8.8 Intel's 8253 Programmable Timer/Counter

In a digital computer system or a real time sampling or control system, a timing signal is usually needed. A device that generates such signal is called a programmable timer/ event counter. It is used to mark intervals of time for both the processor and external devices, count external events and make the count available to the processor, and provide external timing that can be programmed from the processor. Some of its uses are to:

(1) Interrupt a time-sharing operating system at evenly spaced intervals so that it can switch programs.

Output precisely timed signals with programmed periods to an I/O device (e.g. an A/D converter).

(2) Serve as a programmable baud rate generator.

(3) Measure time delays between external events.

Count the number of times an event occurs in an external experiment and provide a means of inputting the count to the computer.

(4) Cause the processor to be interrupted after a programmed number of external events have occurred.

According to the counting method, the programmable timer/event counter can be divided into addition timer/event counter and subtraction timer/event counter. In this section, we will take the Intel 8253, a subtraction timer/event counter, that was primarily designed for the Intel 8080/8085-processors but later used in x86-systems as an example to introduce its architecture, function and usage.

8.8.1 The Architecture of 8253

The 8253 is designed for use with the Intel™ Microcomputer systems. For a general purpose, as a multi-timing element, it can be treated as an array of I/O ports in the system software. A diagram of Intel's 8253 interval time/event counter is given in **Figure 8-28**. The 8253 consists of three identical counters(channels), each of which has CLK and GATE input and an OUT output. Each can be viewed as containing a Control and Status Register pair, a Counter Register (CR) for receiving the initial count, a Counter Element (CE) which performs the counting but is not directly accessible from the processor, and an Output Latch (OL) for latching the contents of the CE so that they can be read. The CR, CE, and OL are treated as pairs of 8-bit registers.

Figure 8-28 shows us the architecture of an 8253, which contains a Data Bus Buffer, the Read/Write Logic control circuit, 3 identical channels (counters), and the Control Register. It uses NMOS technology with a signal +5V supply and is packaged in a 24-pin plastic DIP. **Figure 8-29** is the 24 pins assignment of the 8253.

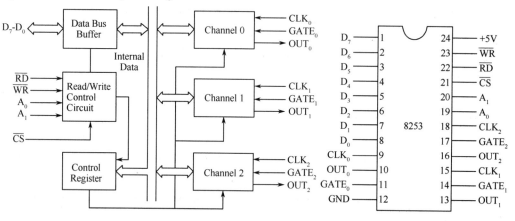

Figure 8-28 The 8253 internal block diagram   Figure 8-29 8253 pins

1. Data/Bus Buffer

The 3-state, bi-directional, 8-bit buffer is used to interface the 8253 to the system bus. Like a bridge, it connects to the microprocessor to transfer the data. The 8 input/output pins are usually labeled as $D_7 \cdots D_0$, where D_7 is the MSB. Data is transmitted or received by buffer with the execution if IN and OUT instructions. It has three basic functions:

(1) Programming the MODE of the 8253.

(2) Loading the CR.

(3) Reading the count values.

2. Read/Write Logic Control Circuit

The Read/Write Logic block receives the read/write control signal from CPU, and achieves controlling of internal components. The 8253 is enabled or disabled by $\overline{CS}$ so that no operation can occur to change the function unless the device has been selected. The 5 pins in the block are listed below.

(1) A_0, A_1: port select signal, as shown in **Table 8-6**. They are used to select one of the three counting channels or the control register. These inputs are normally connected to the address bus.

Table 8-6 The definition of A_0 and A_1

A_1	A_0	Port
0	0	Channel 0 (Count 0)
0	1	Channel 1 (Count 1)
1	0	Channel 2 (Count 2)
1	1	Control Register

(2) $\overline{CS}$: chip select signal. It is active low.

(3) $\overline{RD}$: read signal, low level effective. A "low" input on this pin informs the 8253 that the CPU is inputting the counters value selected by $A_0 A_1$.

(4) $\overline{WR}$: write signal, low level effective. The CPU loads the initial counting value into the selected channel, or writes the mode information into the control register.

The read/write operations of 8253 are shown in **Table 8-7**.

Table 8-7 8253's read/write operation

$\overline{CS}$	$\overline{RD}$	$\overline{WR}$	A_1	A_0	Operation
0	1	0	0	0	Write to channel 0
0	1	0	0	1	Write to channel 1
0	1	0	1	0	Write to channel 2
0	1	0	1	1	Write mode word to CR
0	0	1	0	0	Read from channel 0
0	0	1	0	1	Read from channel 1
0	0	1	1	0	Read from channel 2
0	0	1	1	1	No operation
1	×	×	×	×	Disable 3-state
0	1	1	×	×	No operation 3 state

Operation mode of the 8253 is changed by setting the above hardware signals. For example, to write a command to the Control Register, you need to set $\overline{CS}=0$, $\overline{RD}=1$, $\overline{WR}=0$, $A_1=A_0=1$.

3. Three Counting Channels

The 8253 has three 16-bit pre-settable subtraction counters, which are labeled as Counter (Channel) 0, Counter (Channel) 1 and Counter (Channel) 2 to count number in binary (counter range 0000H-FFFFH) or decimal (BCD) (counter range 0000-9999) in each channel. these three counters (channels) are identical in operation.

Each counting channel has 2 input pins-CLK (clock input) and GATE-and 1-pin, OUT, for data output. The CLK input determines the count rate, **GATE** is for enabling and disabling the CLK input and perhaps other purpose, and the OUT output becomes active when the count reaches 0 or, possibly, when a GATE signal is received. OUT may be connected to an interrupt request line in the system bus so that an interrupt will occur when the count reaches 0, or to an I/O device which uses it to initiate specific I/O activity.

As a counter, it operates by starting at an initial value and counting backward to 0. The operation is basically to enter a count in the initial count register, transfer the count to the counter, and cause the computer to count backward as pulses are applied to the CLK input. The current contents of the counter can be input at any time without disturbing the count by transferring them to the counter out register and then reading them. By buffering the count through the counter out register it doesn't have to be input to the processor immediately. The zero count indication would normally be applied to both the OUT pin and one of the bits in the status register. Thus either programmed I/O or interrupt I/O could be used to detect the zero count.

As a timer, the accurate clock pulse is input from pin CLK. The initial value of a counter is set according to the ratio of the timing length required to the clock cycle. Similar to the counter, the OUT output becomes active when the count reaches 0.

4. The Control Register

This register contains the programmed information which will be sent (by the microprocessor) to the device. It defines how the 8253 logically works. Each access to these ports takes about 1 μs.

To initialize the counters, the microprocessor must write a control word (CW) in this register. This can be done by setting proper values for the pins of the Read/Write Logic block and then by sending the control command to the Data/Bus Buffer block. The information stored in this register controls the operation MODE of each counter, selection of binary or BCD counting and the loading of each count register.

The control register contains 8 bits, labeled D_7-D_0. Each counter of 8253 is individually programmed by writing a control word to the control register ($A_0 A_1 = 11$). The control register format is defined as that in **Figure 8-30**.

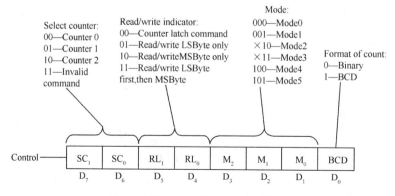

Figure 8-30 control register for 8253

Bit	Pins	Short Descriptions
$D_7 D_6$	SC_1, SC_0	*Select channel* *00 : Channel (Counter) 0* *01 : Channel (Counter) 1* *10 : Counter (Channel) 2* *11 : No used*
$D_5 D_4$	RL_1, RL_0	*Read/Write indicator, it indicates that the Read/Write operations* *00 : Counter latch command* *01 : Read/ write LSByte only* *10 : Read/ write MSByte only* *11 : Read/ write LSByte first, then MSByte*
$D_3 D_2 D_1$	$M_2 M_1 M_0$	*Mode selection- determine the mode 0 through 5* *For modes 2 and 3, the D_3 bit is ignored.*
D_0	BCD	*Specifies the counting format* *1 : BCD* *0 : Binary*

When setting the 8253, the microprocessor first sends a control word. The counting process will start after the 8253 has received these messages, and, in some cases, if it detects the rising from the GATE input signal.

Note that the Control Register can only be written into. No read operation of its contents is available.

As we discuss above, the 8253 is a component of the Intel Microcomputer systems and interfaces in the same manner as all other peripherals of the family. It is treated as an array of peripheral I/O ports. Three are counters and the forth is a control register for MODE programming. Basically, the select inputs A_0, A_1 connects to A_0, A_1 address bus signals of CPU. The $\overline{CS}$ can be derived directly from the address bus using a linear select method or connected to the output of a decoder.

要点：

（1）8253可编程计数器/定时器具有3个独立的计数器通道，最高计数速率为2.6MHz。3个独立通道分别称为计数器0,计数器1及计数器2。

（2）8253芯片的内部结构：

8253芯片由数据总线缓冲存储器、读/写控制电路、控制字寄存器及计数通道0～2组成,其内部结构流程图如图8-28所示。

① 数据总线缓冲存储器。由8位双向三态缓冲存储器构成,是CPU与8253之间交换信息的必经之路。

② 读/写控制电路。接收CPU送入的读/写控制信号,并完成对芯片内部各功能部件的控制功能。

$A_1 A_0$——端口选择信号,由CPU输入。8253内部有3个独立的通道和1个控制字寄存器。

$\overline{CS}$——片选信号。

$\overline{RD}$、$\overline{WR}$——读/写控制命令,由 CPU 输入,低电平有效。$\overline{RD}$有效时,CPU 读取由 A_1A_0 所选定的通道内计数器的内容。$\overline{WR}$有效时,CPU 将计数值写入各个通道的计数器中,或者是将方式控制字写入控制字寄存器中。

(3)计数通道 0~2。8253 内部包含 3 个功能完全相同的通道,每个通道内部设有一个 16 位计数器,可进行二进制或十进制(BCD 码)计数。

(4)控制字寄存器。控制字寄存器用来存放由 CPU 写入 8253 的方式选择控制字,由它来定义 8253 中各通道的工作方式,使用格式如图 8-3 所示。

8.8.2 The Operation Mode Definition of 8253

As shown in **Figure 8-30**, the bits D_3, D_2, and D_1 of the Control Word set the operating modes of the 8253. There are 6 modes in total. Given that n is the initial count, the modes are:

1. Mode 0(000): Interrupt on Terminal Count

Mode 0 is used for the generation of accurate time delay under software control. In this mode, GATE = 1 enables counting and GATE = 0 disables counting, and GATE has no effect on OUT. The counter will start counting from the initial COUNT value **n** loaded into it, down to 0. Counting rate is equal to the input clock frequency.

The OUT pin is set low after the mode set operation, and counting starts one clock cycle after the COUNT programmed. OUT goes low when there is an output to the control register and remains low until the terminal count reached, at which point OUT will go high and remain high until the selected counter is reloaded with the Control word or a new count is loaded. Mode 0 is primarily for event counting.

The Gate signal should remain active high for normal counting. If Gate goes low counting operation will be terminated and current count is latched till Gate pulse goes high again.

The timing diagram of Mode 0 is shown in **Figure 8-31**.

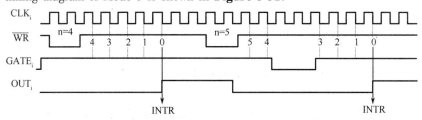

Figure 8-31 Timing diagram of mode 0

The main characteristics of mode 0 are listed as follows:

(1) A counting process is started up by the program only once when an initial counting value is loaded into the register. A new counting process will begin only when the initial counting number is replaced.

(2) The GATE is related to the counting operation. GATE = 1 enables counting and GATE = 0 disables counting.

(3) The down operation of a counter begins in the next CLK pulse that the initial counting value is loaded.

Mode 0's is used for event count and the OUT signal can be used as an interrupt request signal.

2. Mode 1 (001): **Programmable One Shot**

Mode 1 can only be started by the Hardware. Sometimes, it is called as Hardware Retriggerable one-shot. In this mode, 8253 can be used as mono-stable multi-vibrator. GATE input is used as trigger input. The timing diagram of mode 1 is shown in **Figure 8-32**.

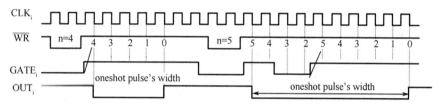

Figure 8-32 Timing diagram of mode 1

As shown in **Figure 8-32**, OUT will be initially high. When the initial value **n** is loaded, a 0-to-1 transition on GATE will lead the count n to be loaded, a 1-to-0 transition at OUT, and the count to begin. OUT will go low on the CLK pulse following a trigger to begin the one-shot pulse, and will go high when the count reaches 0, thus producing a negative-going OUT pulse which lasts **n** clock periods long. It remains high until the CLK pulse after the next trigger.

After writing the Control Word and initial count, the Counter is armed. A trigger results in loading the Counter and setting OUT low on the next CLK pulse, thus starting the one-shot pulse. An initial count of **n** will result in a one-shot pulse **n** CLK cycles in duration.

Characteristics of mode 1 are listed as follow:

(1) If a new count value is loaded while the output is low, it will not affect the duration of the one-shot pulse until the succeeding trigger. In that case, the Counter is loaded with the new count and the oneshot pulse continues until the new count expires.

(2) The one-shot is retriggerable, hence OUT will remain low for **n** CLK pulses after any trigger. The one-shot pulse can be repeated without rewriting the same count into the counter.

3. Mode 2 (X10): **Rate Generator**

Mode 2 can be started by the Software or Hardware. In this mode, the device acts as a divide-by-n counter, which is commonly used to generate a real-time clock interrupt. Its timing diagram is shown in **Figure 8-33**.

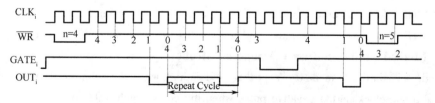

Figure 8-33 Timing diagram of mode 2

After loading CR with **n**, if the gate input goes high, the counter will start from the initial count. OUT goes from 1 to 0 when the count becomes 1 and remains low for one CLK pulse; then it returns to 1 and CR is reloaded automatically, thus giving a negative pulse at OUT after every n clock cycles. GATE = 1 enables the count and GATE = 0 disables the count. A 0-to-1 transition on GATE also causes the count to be reinitialized on the next clock pulse. This mode is used to provide a programmable periodic interval timer.

Like other modes, counting process will start the next clock cycle after COUNT is loaded. When this mode is set, the OUT will remain high until the count register is loaded. If the count register is reloaded among output pulses, the present period will not be affected, but the subsequent period will reflect the new value.

The time between the high pulses depends on the preset count in the counter's register, and is calculated using the following formula:

$$\text{counter} = \frac{f_{input}}{f_{output}}$$

Note that the values in the COUNT register range from n to 1; the register never reaches zero.

4. Mode 3 (X11): Square Wave Generator

Mode 3 can be started by the Software or Hardware. It is similar to mode 2 except that OUT goes low when half the initial count is reached and remains low until the count becomes 0. Hence the duty cycle is changed. As before, GATE enables and disables the count and a 0-to-1 transition on GATE reinitializes the count. When the counter reaches terminal count, the state of the output is changed and the counter is reloaded with the full count and the whole process is repeated. This mode may be used for baud rate generator. Its timing diagram is shown in **Figure 8-34**.

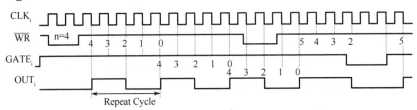

Figure 8-34 Timing diagram of mode 3

Suppose n is the number loaded into the counter (the COUNT message), the output will be:

(1) High for n/2 counts, and low for n/2 counts, if n is even.

(2) High for $(n-1)/2$ counts, and low for $(n-1)/2$ counts, if n is odd.

5. Mode 4 (100): Software Triggered Strobe

Mode 4 is similar to mode 0 except that OUT is high while the counting is taking place and produces a one-clock period negative pulse when the count reaches 0.

When the mode 4 is set, the OUT will be high. When the COUNT is loaded, the OUT will remain high until the counter reaches zero. On the terminal count, the OUT will goes low for 1 clock cycle (a strobe)-after that the OUT will become high again. The timing diagram of mode 4 is shown in **Figure 8-35**.

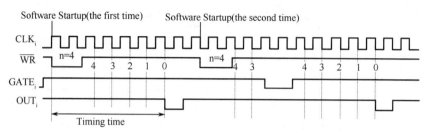

Figure 8-35 Timing diagram of mode 4

Characteristics of mode 4 are listed as follow:

(1) The initial counting value can only be used once in one counting process. A new counting process will begin only when the initial counting number is reloaded.

(2) GATE = 1 enables the count and GATE = 0 disables the count. In one counting duration, If the GATE goes low, the counting will be stop and will recount from the initial value when the GATE goes high.

(3) If the count register is reloaded during the counting, the new count will be loaded on the next CLK pulse.

6. Mode 5 (101): Hardware Triggered Strobe

In mode 5, after CR is loaded, a 0-to-1 transition on GATE will cause the counter to start counting during the next CLK pulse. The OUT will be high during the counting but will go low for one CLK period when the count becomes 0. The counter is retriggerable. The GATE can reinitialize counting at any time. The timing diagram of mode 5 is shown in **Figure 8-36**.

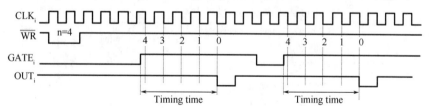

Figure 8-36 Timing diagram of mode 5

Mode 5 is very similar to mode 1. Both of them can only be stared by the Hardware. However, the counting process is triggered by the GATE input.

After receiving the Control Word and COUNT, the output will be set high. Once the device detects a rising edge on the GATE input, it will start counting. When the counter reaches terminal count, the output will go low for one clock cycle, after that it will become high again, to repeat the cycle on the next rising edge of GATE.

Table 8-8 is the summary of 8253's 6 modes and **Table 8-9** is the Gate pin operation summary.

Table 8-8 Summary of 8253's 6 modes

Mode	Description
Mode 0	realize timing or counting for the external event
Mode 1	produce negative oneshot, n times of the width of CLK
Mode 2	produce serial negative pulse, the same width with the cycle of CLK
Mode 3	produce continuous square wave
Mode 4 and Mode 5	have the same output waveform

Mode 0 and Mode 4 are started by Hardware, the initial counting value can be used only once. Mode 1 and Mode 5 are triggered by Software and can be reset initial counting value themselves when the counter reaches 0, but they need the rising edge of GATE to startup the new counting. Mode 2 and Mode 3 are strobed by Software/Hardware. They can reset initial counting value themselves and restart the count automatically.

Table 8-9 Pin operation summary

Working Mode	GATE = 0 & Trailing Edge	GATE Rising Edge	GATE = 1
Mode 0 (Interrupt on terminal count)	Disables counting	—	Enables counting
Mode1 (Programmable one shot)	—	(1) Initiates counting (2) Resets output after next clock	—
Mode 2 (Rate generator)	(1) Disables counting (2) Sets output immediately high	(1) Reloads counter (2) Initiates counting	Enables counting
Mode 3 (Square wave generator)	(1) Disables counting (2) Sets output immediately high	(1) Reloads counter (2) Initiates counting	Enables counting
Mode 4 (Software triggered strobe)	Disables counting	—	Enables counting
Mode 5 (Hardware triggered strobe)	—	Initiates counting	—

Notice that, for modes 0, 2, 3 and 4, GATE must be set to HIGH to enable counting. For mode 5, the rising edge of GATE starts the count.

要点：

8253 芯片的工作方式：

8253 中各通道均有 6 种可供选择的工作方式，以完成定时、计数或脉冲发生器等多种功能。

(1) 工作方式 0。工作方式 0 被称为计数结束中断方式，它的定时波形如图 8-31 所示。当通道被定义为方式 0 时，OUT$_i$ 输出为低电平；若门控信号 GATE 为高电平，当 CPU

向该通道写入计数值$\overline{WR}$信号有效时,OUT_i仍保持低电平,然后计数器开始减"1"计数,直到计数值为"0",此刻,OUT_i将输出由低电平向高电平跳变,可用它向CPU发出中断请求,OUT_i端输出的高电平一直维持到下次再写入计数值为止。

在工作方式0,门控信号GATE用来控制减"1"计数操作是否进行。当GATE=1时,允许减"1"计数;当GATE=0时,禁止减"1"计数;此时,计数值将保持GATE有效时的数值不变,待GATE重新有效后,减"1"计数继续进行。

（2）工作方式1。工作方式1被称作可编程单脉冲发生器,其定义波形如图8-32所示。进入这种工作方式,CPU装入计数值n后OUT_i输出高电平,无论当前GATE输入的是高电平还是低电平,都不开始减"1"计数,必须等到GATE由低电平向高电平跳变形成一个上升沿后,计数过程才会开始。与此同时,OUT_i输出由高电平向低电平跳变,形成了输出单脉冲的前沿,待计数值计到"0",OUT_i输出由低电平向高电平跳变,形成输出单脉冲的后沿,因此,由方式1所能输出单脉冲的宽度为CLK_i周期的n倍。

如果在减"1"计数过程中,GATE由高电平跳变为低电平,计数过程不影响,仍继续计数;但若重新遇到GATE的上升沿,则从初值开始重新计数,其效果会使输出的单脉冲加宽。

（3）工作方式2。工作方式2被称作速率波发生器,其定时波形如图8-33所示。进入方式2,OUT_i输出高电平,装入计数值n后,如果GATE为高电平,则立即开始计数,OUT_i保持为高电平不变;待计数值减到"1"和"0"之间,OUT_i将输出宽度为一个CLK_i周期的负脉冲。计数值为"0"时,计数初值n自动重新装入,实现循环计数,OUT_i将输出一定频率的负脉冲序列,其脉冲宽度固定为一个CLK_i周期,重复周期为CLK_i周期的n倍。

如果在减"1"计数过程中,GATE变为无效(输入0电平),则暂停减"1"计数,待GATE恢复有效后,从初值n开始重新计数。这样会改变输出脉冲的速率。

（4）工作方式3。工作方式3被称作方波发生器,其定时波型如图8-34所示。通道工作在方式3时,若计数值n为偶数,则可输出重复周期为n、占空比为1∶1的方波。

进入工作方式3,OUT_i输出低电平,装入计数值后,OUT_i立即跳变为高电平。如果当GATE为高电平,则立即开始减"1"计数,OUT_i保持为高电平,若n为偶数,则当计数值减到n/2时,OUT_i跳变为低电平,一直保持到计数值为"0",系统才自动重新置入计数值n,实现循环计数。这时OUT_i端输出的周期为$n \times CLK_i$周期,占空比为1∶1的方波序列;若n为奇数,则OUT_i端输出周期为$n \times CLK_i$周期,占空比为$((n+1)/2)/((n-1)/2)$的近似方波序列。

如果在操作过程中,GATE变为无效,则暂停减"1"计数过程,直到GATE再次有效,重新从初值n开始减"1"计数。

如果要改变输出方波的速率,则CPU可在任何时候重新装入新的计数初值n,并从下一个计数操作周期开始改变输出方波的速率。

（5）工作方式4。工作方式4被称作软件触发方式,其定时波形如图8-35所示。进入工作方式4,OUT_i输出高电平。装入计数值n后,如果GATE为高电平,则立即开始减"1"计数,直到计数值减到"0"为止,OUT_i输出宽度为一个CLK_i周期的负脉冲。由软件装入的计数值只有一次有效,如果要继续操作,必须重新置入计数初值n。如果在操作的过程中,GATE变为无效,则停止减"1"计数,到GATE再次有效时,减"1"计数从初值开始。

工作方式 4 可以完成定时功能,定时时间从装入计数值 n 开始,其定时时间 = n × CLK 周期,OUT$_i$ 输出负脉冲表示定时时间到。方式 4 也可完成计数功能,它要求计数的事件以脉冲的方式从 CLK$_i$ 输入,将计数次数作为计数初值装入后,由 CLK$_i$ 端输入的计数脉冲进行减"1"计数,直到计数值为"0",OUT$_i$ 端输出负脉冲表示计数次数到。当然,也可利用 OUT$_i$ 向 CFU 发出中断请求。因此,工作方式 4 与工作方式 0 很相似,只是方式 0 在 OUT$_i$ 端输出正阶跃信号、方式 4 在 OUT$_i$ 端输出负脉冲信号。

(6) 工作方式 5。工作方式 5 被称为硬件触发方式,其定时波形如图 8-36 所示。进入工作方式 5,OUT$_i$ 输出高电平,硬件触发信号由 GATE 端引入。因此,开始时 GATE 应输入为 0,装入计数初值 n 后,减"1"计数并不工作,一定要等到硬件触发信号由 GATE 端引入一个正阶跃信号,减"1"计数才会开始,待计数值计到"0",OUT$_i$ 将输出负脉冲,其宽度固定为一个 CLK$_i$ 周期,表示定时时间到或计数次数到。

这种工作方式下,当计数值计到"0"后,系统将自动重新装入计数值 n,但并不开始计数,只有当 GATE 端引入正跳沿时,减"1"计数才会开始,因此这是一种完全由 GATE 端引入的触发信号控制下的计数或定时功能。如果由 CLK$_i$ 输入的是一定频率的时钟脉冲,那么可完成定时功能,定时时间从 GATE 上升沿开始,到 OUT$_i$ 端输出负脉冲结束。如果从 CLK$_i$ 端输入的是要求计数的事件,则可完成计数功能,计数过程从 GATE 上升沿开始,到 OUT$_i$ 输出负脉冲结束。GATE 可由外部电路或控制现场产生,故硬件触发方式由此而得名。

8.8.3　Examples of 8253 Timer/Counter

Before you actually use the 8253 timer/counter, the system program must be developed for each counter of the 8253 with the mode and quantity desired. So you must write out the MODE control word and calculate the initial count for each counter at first. The program developing is quite flexible. Generally speaking, you can initialize each counter of the 8253 with its MODE control word respectively, then load the initial value to the selected counter register. Note that it must be loaded with the number of bytes programmed in the MODE control word. You can begin you program in any sequence of counter selection, eg., counter 0 does not have to be first and counter 3 last.

【Example 8-8】Suppose that 8253's ports addresses are 70H-73H, the counting channel 1 operates in mode 0 and the initial value is 130, please develop the instruction sequences to accomplish the initialization of the 8253.

```
MOV   AL, 71H        ;Control word 01110001, counting channel 1, BCD counting
OUT   73H, AL        ;Control word be written into control register
MOV   AL, 30H        ;Transfer the lower 8 bits first
OUT   71H, AL        ;Transfer the lower 8 bits to counting channel 1
MOV   AL, 01H        ;Transfer the higher 8 bits
OUT   71H, AL        ;Transfer the higher 8 bits to counting channel 1
```

【Example 8-9】Assume that 8253's ports addresses are 70H-73H, the counting channel 1 operates in mode 1 and the initial counting value is 20, please develop the instruction sequences to initialize the 8253 chip.

```
    MOV  AL, 52H        ;Control word 01010010, counting channel 1, BCD counting
    OUT  73H, AL        ;Control word be written into control register
    MOV  AL, 14H        ;Transfer the lower 8 bits first
    OUT  71H, AL        ;Transfer counting times to counting channel 1
```

[Example 8-10] Assume that 8253's ports addresses are 70H-73H and the counting channel 2 of 8253 is used as a frequency divider, please develop the instruction sequences to initialize the 8253 and accomplish the pulse frequency conversion from 1.19MHz to the frequency of 500Hz.

$$n(分频系数) = 1.19 \times 10^6 / 500 = 2380 = 094CH$$

```
    MOV  AL, 0B4H       ;Control word 10110100, counting channel 2 mode 2, BCD counting
    OUT  73H, AL        ;Control word be written into control register
    MOV  AL, 4CH
    OUT  72H, AL        ;Transfer the lower 8 bits to counting channel 2
    MOV  AL, 09H
    OUT  71H, AL        ;Transfer the higher 8 bits to counting channel 2
```

[Example 8-11] As shown in **Figure 8-37**, an 8086 system contains a 8253 chip. Please program to realize requirements listed as below:

(1) Use channel 0 to realize the external events counting and send an interrupt request to CPU after counting 100 times.

(2) Use channel 1 to produce square wave whose frequency is 1kHz.

(3) Use channel 2 as a standard clock.

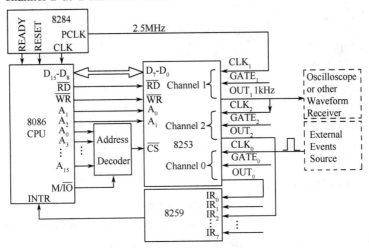

Figure 8-37 Connection diagram of the 8253 and the 8086 CPU

Analysis of the example: As request above, channel 0 is defined in mode 0 to realize the counting with the initial counting value 100 = 64H. The channel 1 is denoted in mode 3 for, output 1kHz square wave. Because the frequency of input CLK_1 is 2.5MHz, the initial counting value of channel 1 should be 2500 = 09C4H. Channel 2 is designated in mode 0 to accomplish the timing function which uses OUT_2 to send an interrupt request per second. Since its input clock frequency is 1kHz, the initial counting value of channel 2 should be set as 1000. Suppose the four ports of this 8253 are 1100H-1103H respectively. The program is given as below:

```
MOV DX, 1103H              ;Select working mode 0 for channel 0
MOV AL, 10H                ;Control word 00010000: channel 0, mode 0, only write lower 8
                            bits, Binary
OUT DX, AL
MOV DX, 1100H              ;Transfer counting value 100 to channel 0
MOV AL, 64H
OUT DX, AL
MOV DX, 1103H              ;Select working mode 3 for channel 1
MOV AL, 76H                ;Control word 01110110: channel 1, mode 3, lower 8 bits first, Bi-
                            nary
OUT DX, AL
MOV DX, 1101H              ;Transfer initial counting value (coefficient of dividing frequency)
                            2500 to channel 1
MOV AX, 09C4H
OUT DX, AL
MOV AL, AH
OUT DX, AL
MOV DX, 1103H              ;Select working mode 0 for channel 2
MOV AL, 0B1H               ;Control word 10110001: channel 2, mode 0, write 16 bits, BCD
                            counting
OUT DX, AL
MOV DX, 1102H              ;Transfer initial counting value 1000 to channel 2
MOV AX, 1000H
OUT DX, AL
MOV AL, AH
OUT DX, AL
MOV DX, 8259A even address port  ;Initialize 8259A
MOV AL, 13H                ;Define $ICW_1$
OUT DX, AL
MOV DX, 8259A odd address port
MOV AL, 50H                ;Define $ICW_2$, set the interrupt type code 50H
OUT DX, AL                 ;Interrupt type codes of Counting channel 0 and 2 are 50H and 52H
MOV AL, 03                 ;Define $ICW_4$, auto EOI mode
OUT DX, AL
MOV AL, 0FAH               ;Define $OCW_1$, open interrupt of $IR_0$ and $IR_2$
OUT DX, AL
STI
HLT
JMP HH
```

Tips

address decoder　地址译码器
addressable port　可寻址端口
alphanumerically　字母和数字混合编制
analogous　类似的，相似的，可比拟的
asynchronous serial communication　异步串行通信
baud rate　波特率（发报速率）
baud rate factor　异步传输时波特率因子
bidirectional lines　双向总线
break character　断码
built-in driver　内置驱动器
categorize　分类
chip select　片选
conduct　传输，传导
control register　控制寄存器
control register　操作命令控制字
counterpart　配对物
data bus buffer　数据总线缓冲器
data bus drivers and receivers　数据总线收发器
data-in ack　数据输入应答信号
data-in buffer register　数据输入缓存寄存器
data-in ready　数据输入准备好
data-in shift register　数据输入移位寄存器（串入并出）
data-out buffer register　数据输出缓存寄存器
data-out shift register　数据输出移位寄存器（并入串出）
DSR　数据装置准备好
DTR　数据中断准备好
echoed character　返回字符
EH　跟踪方式位（同步有效）
ER　清除错误标志位
examine　检测
external synchronous detect　外同步检测（内同步或外同步）
FE　帧校验错标志
flowchart　流程图
framing　帧校验
full duplex　全双工
half duplex　半双工

handshaking logic　联络信号逻辑
hardware retriggerable one-shot　可编程单脉冲发生器
hardware-triggered strobe　硬件触发方式
I/O interface　输入/输出接口
imbed　嵌入
interrupt acknowledgement　中断应答信号
interrupt on terminal count　计数结束中断方式
interrupt priority management　中断优先管理
interrupt type　中断类型码
IR　内部复位信号
latch　锁存
mode register　工作方式选择控制字
modulate/demodulate　调制/解调控制电路
number of data bits　定义数据字符的长度
number of stop bits　异步方式时停止位长度；
number of sync characters　同步字符数目（单同步或双同步）
OE　溢出错标志
one-way communication　单向通信
overrun　越界错误
paper tape punch　纸带打孔机
parallel　并行的
parallel communication interface　并行通信接口
parallel to serial　并行至串行转换
parity enable　奇偶校验允许
parity error　奇偶校验错误
parity type　奇偶校验方式
PE　奇偶错标志
periodic interval timer　速率波发生器
programmable communication interface　可编程通信接口
programmable timer and event counter　可编程定时/计数控制器
R/W control logic circuit　读/写控制逻辑电路
receive control circuit　接收控制电路
receiver　接收器
receiver buffer　接收缓冲器
receiver clock　接收时钟

reenter	再进入	status register	状态寄存器
reinitialize	重新初始化	status register	状态控制字
relay coil	继电器绕组线圈	strobe	锁存
RTS	请求发送信号	synchronous mode	同步模式
RxE	为1时允许接收	synchronous serial communication	同步串行通信
RxRDY	为0时接收器准备好	SYNDET	为1时达到同步
SBRK	为1发送短缺字符,为0正常工作	terminal	终端
serial communication interface	串行通信接口	time slice	时间片
serial input/parallel output	串行输入/并行输出	transmit buffer	发送缓冲器
serial to parallel	串行至并行转换	transmit clock	发送时钟
set/reset instruction	置位/复位控制	transmit control circuit	发送控制电路
shift register	移位寄存器	transmitter	发送器
single-board system	单板系统	TxE	为1时发送器空
software-triggered strobe	软件触发方式	TxEN	为1时允许发送
square-wave generator	方波发生器	TxRDY	为1时发送器准备好
status information	状态信息	verification	验证

Exercise

1. Assume the 7-bit ASCII code, even parity, and 2 stop bits and sketch the timing diagram of the letter C as it is transmitted over an asynchronous serial communication link.

2. If there is no dead time between characters, characters contains 7 information bits, there is a parity bit and 2 stop bits with each character, and the bit rate is 600 bits per second, what is the information transfer rate in characters per second? Compare this figure with the information transfer rate of a synchronous transmission having the same bit rate, a message format consisting of two synchronous transmissions having the same bit rate, a message format consisting of two sync characters followed by 200 seven-bit information characters, and no imbedded idle characters.

3. If the mode register of an 8251A contains 01111011, what will be the format of the transmitted character? In order for the receiver and transmitter baud rates to be 300 and 1200, respectively, what would the frequencies applied to RxC and TxC need to be?

4. Assuming that the content of the mode register of an 8251A are 00010100, determine the character and message formats of the 8251A's serial transmission.

5. Given that the sequence shown below is output with $A_0 = 1$ following a reset, describe the action taken as each output is received by the 8251A.

$$00010100$$
$$00010110$$
$$00010110$$
$$10110011$$
$$\cdots \qquad \text{with } A_0 = 0$$

01000000

6. Write a program sequence that will initialize an 8251A so that it will transmit asynchronous characters containing 7 information bits, 1 stop bit, and no parity bit. The baud rate factor is to be 64. The even address associated with the 8251A is to be 008A. Extend the sequence so that it outputs a command that activates Request to Send and Data Terminal Ready signals and enables the transmitter.

7. Extend the program sequence in Exercise 8 so that it will use programmed I/O to output the message

AN OVERRUN ERROR HAS OCCURRED

and then clear the overrun error bit.

8. Write an interrupt routine that will output the 20 bytes beginning at MESSAGE to the 8251A whose even address is 25C0. Each time the routine is entered it is to output 1 byte, and when it has output 20 bytes it is to put a 1 in FLAG. CNT is to be used to store the byte count.

9. Suppose that the beginning address of an 8255A is 0500, please write a program sequence that will:

(a) Put both groups A and B in mode 0 with ports A and C being input ports and port B being an output port.

(b) Put group A in mode 2 and group B in mode 1 with port B being an output.

(c) Put group A in mode 1 with port A being an input and PC_6 and PC_7 being outputs, and group B in mode 1 with port B being an input.

10. Assume an 8255A, whose beginning address is 0030, is such that group A is in mode 2 and group B is in mode 0 with PB_7-PB_0 as input and PC_2-PC_0 as outputs. Write a program that will output 100 bytes from an array beginning at OUT_ARRAY to the device connected to port A. Before outputting each byte the program must loop until PB_1, PB_2, and PB_3 are all active and after outputting the byte the program must set PC_1 to 1.

11. Using an 8255A, an 8282 latch, an 8286, and other logic, design a circuit for setting and reading a bank of eight relays. Assume that 8286 drivers are sufficient to drive the output circuits which supply the current to the relay coils. Specify the 8255A modes and all of the necessary control signals. The relay and their associated circuitry need not be included in the design.

Appendix A   8086/8088 Instruction Set Summary

The first two columns in Table A-1 give the mnemonic and a brief description of each instruction. The third column indicates the number of clock cycles needed to execute the instruction. For the 8086, four clock cycles need to be added for each word transfer with an odd address. For the 8088, four clock cycles need to be added for each word transfer. There are two numbers of clock cycles for the conditional branch and loop instructions; the one before the "/" is for when a branch is not taken and the other number is for when the branch is taken. The fourth column gives the number of bytes, or the range of numbers of bytes, for the instruction, and the fifth column indicates the possible flag settings using the following symbols:

–	Not affected
0	Cleared to 0
1	Set to 1
x	Set or cleared according to the result
u	Undefined
r	Restored from previously saved value
a(b)	a is clock cycles for byte operand, b is clock cycles for word operand
EA	effective address
rep	repeat prefix

Table A-1

Mnemonic	Description	Clock cycles	Number of bytes	Flags O	D	I	T	S	Z	A	P	C
AAA	ASCII adjust for addition	4	1	u	-	-	-	u	u	x	u	x
AAD	ASCII adjust for division	60	2	u	-	-	-	x	x	u	x	u
AAM	ASCII adjust for multiplication	83	2	u	-	-	-	x	x	u	x	u
AAS	ASCII adjust for subtraction	4	1	u	-	-	-	u	u	x	u	x
DAA	Decimal adjust for addition	4	1	u	-	-	-	x	x	x	x	x
DAS	Decimal adjust for subtraction	4	1	u	-	-	-	x	x	x	x	x
CALL	Call a procedure			-	-	-	-	-	-	-	-	-
	Intrasegment direct	19(23)	3									
	Intrasegment indirect through register	16(24)	2									
	Intrasegment indirect through memory	21(29) + EA	2-4									
	Intersegment direct	28(36)	5									
	Intersegment indirect	37(57) + EA	2-4									
CBW	Convert byte to word	2	1	-	-	-	-	-	-	-	-	-
CWD	Convert word to double word	5	1	-	-	-	-	-	-	-	-	-
CLC	Clear carry flag	2	1	-	-	-	-	-	-	-	-	0
CLD	Clear direction flag	2	1	-	0	-	-	-	-	-	-	-
CLI	Clear interrupt flag	2	1	-	-	0	-	-	-	-	-	-
CMC	Complement carry flag	2	1	-	-	-	-	-	-	-	-	x
STC	Set carry flag	2	1	-	-	-	-	-	-	-	-	1
STD	Set direction flag	2	1	-	1	-	-	-	-	-	-	-
STI	Set interrupt flag	2	1	-	-	1	-	-	-	-	-	-

(续)

Mnemonic	Description		Clock cycles	Number of bytes	Flags								
					O	D	I	T	S	Z	A	P	C
IN	Input from I/O port				-	-	-	-	-	-	-	-	-
		Fixed port	10(14)	2									
		Variable port	8(12)	1									
OUT	Output to I/O port				-	-	-	-	-	-	-	-	-
		Fixed port	10(14)	2									
		Variable port	8(12)	1									
CMP	Compare				x	-	-	-	x	x	x	x	x
	Register to register		3	2									
	Memory to register		9(13)+EA	2-4									
	Register to memory		9(13)+EA	2-4									
	Immediate to register		4	3-4									
	Immediate to memory		10(14)+EA	3-6									
	Immediate to accumulator		4	2-3									
CMPS	Compare string				x	-	-	-	x	x	x	x	x
CMPSB	Compare byte string/				x	-	-	-	x	x	x	x	x
CMPSW	Compare word string				x	-	-	-	x	x	x	x	x
	Not repeated		22(30)	1									
	Repeated		9+22(30)/rep	1									
NOP	No operation		3	1	-	-	-	-	-	-	-	-	-
HLT	Halt		2	1	-	-	-	-	-	-	-	-	-
INT	Interrupt				-	-	0	0	-	-	-	-	-
	Type = 3		52(72)	1									

(续)

Mnemonic	Description	Clock cycles	Number of bytes	Flags								
				O	D	I	T	S	Z	A	P	C
INT	Type-3	51(71)	2		r	r	r	r	r	r	r	r
IRET	Return from interrupt	24	1	r	r	r	r	r	r	r	r	r
INTO	Interrupt if overflow			-	-	0	0	-	-	-	-	-
	Interrupt is taken	53	1									
	Interrupt is not taken	4	1									
ADD	Addition			x	-	-	-	x	x	x	x	x
	Register to register	3	2									
	Memory to register	9(13) + EA	2-4									
	Register to memory	16(24) + EA	2-4									
	Immediate to register	4	3-4									
	Immediate to memory	17(25) + EA	3-6									
	Immediate to accumulator	4	2-3									
ADC	Add with carry			x	-	-	-	x	x	x	x	x
	Register to register	3	2									
	Memory to register	9(13) + EA	2-4									
	Register to memory	16(24) + EA	2-4									
	Immediate to register	4	3-4									
	Immediate to memory	17(25) + EA	3-6									
	Immediate to accumulator	4	2-3									
SUB	subtraction			x	-	-	-	x	x	x	x	x
	Register from register	3	2									
	Memory from register	9(13) + EA	2-4									

(续)

Mnemonic	Description	Clock cycles	Number of bytes	Flags								
				O	D	I	T	S	Z	A	P	C
SUB	Register from memory	16(24) + EA	2-4									
	Immediate from accumulator	4	2-3									
	Immediate from register	4	3-4									
	Immediate from memory	17(25) + EA	3-6									
SBB	Subtract with borrow			x	-	-	-	x	x	x	x	x
	Register from register	3	2									
	Memory from register	9(13) + EA	2-4									
	Register from memory	16(24) + EA	2-4									
	Immediate from accumulator	4	2-3									
	Immediate from register	4	3-4									
	Immediate from memory	17(25) + EA	3-6									
DEC	Decrement by 1			x	-	-	-	x	x	x	x	-
	16-bit register	2	1									
	8-bit register	3	2									
	Memory	15(23) + EA	2-4									
INC	Increment by 1			x	-	-	-	x	x	x	x	-
	16-bit register	2	1									
	8-bit register	3	2									
	Memory	15(23) + EA	2-4									
MUL	Unsigned multiplication			x	-	-	-	u	u	u	u	x
	8-bit register	70-77	2									
	16-bit register	118-133	2									

(续)

Mnemonic	Description	Clock cycles	Number of bytes	Flags								
				O	D	I	T	S	Z	A	P	C
MUL	8-bit memory	(76 – 83) + EA	2-4									
	16-bit memory	(124 – 139) + EA	2-4									
IMUL	Integer multiplication			x	-	-	-	u	u	u	u	x
	8-bit register	80 – 98	2									
	16-bit register	128 – 154	2									
	8-bit memory	(86 – 104) + EA	2-4									
	16-bit memory	(134 – 160) + EA	2-4									
DIV	Unsigned division			u	-	-	-	u	u	u	u	u
	8-bit register	80 – 90	2									
	16-bit register	144 – 162	2									
	8-bit memory	(86 – 96) + EA	2-4									
	16-bit memory	(150 – 168) + EA	2-4									
IDIV	Integer division			u	-	-	-	u	u	u	u	u
	8-bit register	101 – 112	2									
	16-bit register	165 – 184	2									
	8-bit memory	(107 – 118) + EA	2-4									
	16-bit memory	(171 – 190) + EA	2-4									
JA	Jump if above	16/4	2	-	-	-	-	-	-	-	-	-
JNBE	Jump if not below or equal	16/4	2	-	-	-	-	-	-	-	-	-
JAE	Jump if above or equal	16/4	2	-	-	-	-	-	-	-	-	-
JNB	Jump if not below	16/4	2	-	-	-	-	-	-	-	-	-
JNC	Jump if not carry	16/4	2	-	-	-	-	-	-	-	-	-

(续)

Mnemonic	Description	Clock cycles	Number of bytes	Flags							
				O	D	I	T	S	Z	A	P	C
JB	Jump if below	16/4	2	-	-	-	-	-	-	-	-	-
JNAE	Jump if not above or equal	16/4	2	-	-	-	-	-	-	-	-	-
JC	Jump if carry	16/4	2	-	-	-	-	-	-	-	-	-
JBE	Jump if below or equal	16/4	2	-	-	-	-	-	-	-	-	-
JNA	Jump if not above	16/4	2	-	-	-	-	-	-	-	-	-
JCXZ	Jump if CX is zero	18/6	2	-	-	-	-	-	-	-	-	-
JE	Jump if equal	16/4	2	-	-	-	-	-	-	-	-	-
JZ	Jump if zero	16/4	2	-	-	-	-	-	-	-	-	-
JG	Jump if greater	16/4	2	-	-	-	-	-	-	-	-	-
JNLE	Jump if not less or equal	16/4	2	-	-	-	-	-	-	-	-	-
JGE	Jump if greater or equal	16/4	2	-	-	-	-	-	-	-	-	-
JNL	Jump if not less	16/4	2	-	-	-	-	-	-	-	-	-
JL	Jump if less	16/4	2	-	-	-	-	-	-	-	-	-
JNGE	Jump if not greater or equal	16/4	2	-	-	-	-	-	-	-	-	-
JLE	Jump if less or equal	16/4	2	-	-	-	-	-	-	-	-	-
JNG	Jump if not greater	16/4	2	-	-	-	-	-	-	-	-	-
JMP	Jump	15	2	-	-	-	-	-	-	-	-	-
	Intrasegment direct short	15	3									
	Intrasegment direct	15	5									
	Intersegment direct	18 + EA	2-4									
	Intrasegment indirect through memory	11	2									
	Intrasegment indirect through register											

(续)

Mnemonic	Description	Clock cycles	Number of bytes	Flags								
				O	D	I	T	S	Z	A	P	C
JMP	ntersegment indirect	24 + EA	2-4	-	-	-	-	-	-	-	-	-
JNE	Jump if not equal	16/4	2	-	-	-	-	-	-	-	-	-
JNZ	Jump if not zero	16/4	2	-	-	-	-	-	-	-	-	-
JNO	Jump if not overflow	16/4	2	-	-	-	-	-	-	-	-	-
JNP	Jump if not parity	16/4	2	-	-	-	-	-	-	-	-	-
JPO	Jump if parity odd	16/4	2	-	-	-	-	-	-	-	-	-
JNS	Jump if not sign	16/4	2	-	-	-	-	-	-	-	-	-
JO	Jump if overflow	16/4	2	-	-	-	-	-	-	-	-	-
JP	Jump if parity	16/4	2	-	-	-	-	-	-	-	-	-
JPE	Jump if parity even	16/4	2	-	-	-	-	-	-	-	-	-
JS	Jump if sign	16/4	2	-	-	-	-	-	-	-	-	-
LEA	Load effective address	2 + EA	2-4	-	-	-	-	-	-	-	-	-
LDS	Load pointer using DS	16 + EA	2-4	-	-	-	-	-	-	-	-	-
LES	Load pointer using ES	16 + EA	2-4	-	-	-	-	-	-	-	-	-
LOCK	Lock bus	2	1	-	-	-	-	-	-	-	-	-
LODS	Load string			-	-	-	-	-	-	-	-	-
LODSB	Load byte string			-	-	-	-	-	-	-	-	-
LODSW	Load word string			-	-	-	-	-	-	-	-	-
	Not repeated	12(16)	1									
	Repeated	9 + 13(17)/rep	1									
LOOP	Loop	17/5	2	-	-	-	-	-	-	-	-	-
LOOPE	Loop it equal	18/6	2	-	-	-	-	-	-	-	-	-

315

(续)

Mnemonic	Description	Clock cycles	Number of bytes	Flags							
				O	D	I	T	S	Z	A	P	C
LOOPZ	Loop it zero	18/6	2	-	-	-	-	-	-	-	-	-
LOOPNE	Loop it not equal	19/5	2	-	-	-	-	-	-	-	-	-
LOOPNZ	Loop it not zero	19/5	2	-	-	-	-	-	-	-	-	-
MOV	Move			-	-	-	-	-	-	-	-	-
	Accumulator to memory	10(14)	3									
	Memory to accumulator	10(14)	3									
	Register to register	2	2									
	Memory to register	8(12)+EA	2-4									
	Register to memory	9(13)+EA	2-4									
	Immediate to register	4	2-3									
	Immediate to memory	10(14)+EA	3-6									
	Register to SS, DS or ES	2	2									
	Memory to SS, DS or ES	8(12)+EA	2-4									
	Segment register to register	2	2									
	Segment register to memory	9(13)+EA	2-4									
MOVS	Move string			-	-	-	-	-	-	-	-	-
MOVSB	Move byte string	18(26)	1									
MOVSW	Move word string			-	-	-	-	-	-	-	-	-
	Not repeated	18(26)	1									
	Repeated	9+17(25)/rep	1									
NEG	Negate			x	-	-	-	x	x	x	x	x
	Register	3	2									

(续)

Mnemonic	Description	Clock cycles	Number of bytes	O	D	I	T	S	Z	A	P	C
NEG	Memory	16(24) + EA	2-4	-	-	-	-	-	-	-	-	-
NOT	Logical NOT											
	Register	3	2									
	Memory	16(24) + EA	2-4									
OR	Logical OR			0	-	-	-	x	x	u	x	0
	Register to register	3	2									
	Memory to register	9(13) + EA	2-4									
	Register to memory	16(24) + EA	2-4									
	Immediate to accumulator	4	2-3									
	Immediate to register	4	3-4									
	Immediate to memory	17(25) + EA	3-6									
AND	Logical AND			0	-	-	-	x	x	u	x	0
	Register to register	3	2									
	Memory to register	9(13) + EA	2-4									
	Register to memory	16(24) + EA	2-4									
	Immediate to register	4	3-4									
	Immediate to memory	17(25) + EA	3-6									
	Immediate to accumulator	4	2-3									
XOR	Logical exclusive OR			0	-	-	-	x	x	u	x	0
	Register with register	3	2									
	Memory with register	9(13) + EA	2-4									
	Register with memory	16(24) + EA	2-4									

(续)

Mnemonic	Description	Clock cycles	Number of bytes	Flags								
				O	D	I	T	S	Z	A	P	C
XOR	Immediate with accumulator	4	2-3									
	Immediate with register	4	3-4									
	Immediate with memory	17(25) + EA	3-6									
POP	Pop word off stack	8	1	-	-	-	-	-	-	-	-	-
	Register	8	1									
	Segment register SS, DS, or ES	8	1									
	Memory	17 + EA	2-4									
POPF	Pop flags off stack	8	1	r	r	r	r	r	r	r	r	r
PUSH	Push word onto stack	10	1	-	-	-	-	-	-	-	-	-
	Register	10	1									
	Segment register	16 + EA	2-4									
	Memory											
PUSHF	Push flags onto stack	10	1	-	-	-	-	-	-	-	-	-
RCL	Rotate left through carry	2	2	x	-	-	-	-	-	-	-	x
	Register with single-shift	8 + 4/bit	2									
	Register with variable-shift	15(23) + EA	2-4									
	Memory with single-shift	20(28) + EA + 4/bit	2-4									
	Memory with variable-shift											
RCR	Rotate right through carry	2	2	x	-	-	-	-	-	-	-	x
	Register with single-shift	8 + 4/bit	2									
	Register with variable-shift	15(23) + EA	2-4									
	Memory with single-shift											

(续)

Mnemonic	Description	Clock cycles	Number of bytes	Flags							
				O	D	I	T	S	Z	A	P	C
RCR	Memory with variable-shift	20(28)+EA+4/bit	2-4	x	-	-	-	-	-	-	-	x
ROL	Rotate left											
	Register with single-shift	2	2									
	Register with variable-shift	8+4/bit	2									
	Memory with single-shift	15(23)+EA	2-4									
	Memory with variable-shift	20(28)+EA+4/bit	2-4									
ROR	Rotate right			x	-	-	-	-	-	-	-	x
	Register with single-shift	2	2									
	Register with variable-shift	8+4/bit	2									
	Memory with single-shift	15(23)+EA	2-4									
	Memory with variable-shift	20(28)+EA+4/bit	2-4									
REP	Repeat string operation	2	1	-	-	-	-	-	-	-	-	-
REPE	Repeat operation while equal	2	1	-	-	-	-	-	-	-	-	-
REPZ	Repeat operation while zero	2	1	-	-	-	-	-	-	-	-	-
REPNE	Repeat operation while not equal	2	1	-	-	-	-	-	-	-	-	-
REPNZ	Repeat operation while not zero	2	1	-	-	-	-	-	-	-	-	-
RET	Return from procedure			-	-	-	-	-	-	-	-	-
	Intrasegment	8	1									
	Intrasegment with constant	12	3									
	Intersegment	18	1									
	Intersegment with constant	17	3									

Mnemonic	Description	Clock cycles	Number of bytes	Flags								
				O	D	I	T	S	Z	A	P	C
SAHF	Store AH into flags	4	1	-	-	-	-	r	r	r	r	r
LAHF	Load AH from flags	4	1	-	-	-	-	-	-	-	-	-
SAL	Shift arithmetic left			x	-	-	-	x	x	u	x	x
SHL	Shift logical left			x	-	-	-	x	x	u	x	x
	Register with single-shift	2	2									
	Register with variable-shift	8+4/bit	2									
	Memory with single-shift	15(23)+EA	2-4									
	Memory with variable-shift	20(28)+EA+4/bit	2-4									
SAR	Shift arithmetic right			x	-	-	-	x	x	u	x	x
	Register with single-shift	2	2									
	Register with variable-shift	8+4/bit	2									
	Memory with single-shift	15(23)+EA	2-4									
	Memory with variable-shift	20(28)+EA+4/bit	2-4									
SHR	Shift logical right			x	-	-	-	x	x	u	x	x
	Register with single-shift	2	2									
	Register with variable-shift	8+4/bit	2									
	Memory with single-shift	15(23)+EA	2-4									
	Memory with variable-shift	20(28)+EA+4/bit	2-4									
SCAS	Scan string			x	-	-	-	x	x	x	x	x
SCASB	Scan byte string			x	-	-	-	x	x	x	x	x
SCASW	Scan word string			x	-	-	-	x	x	x	x	x

(续)

(续)

Mnemonic	Description	Clock cycles	Number of bytes	Flags								
				O	D	I	T	S	Z	A	P	C
STOS	Not repeated	15(19)	1	-	-	-	-	-	-	-	-	-
	Repeated	9+15(19)/rep	1	-	-	-	-	-	-	-	-	-
STOSB	Store string											
	Store byte string											
STOSW	Store word string											
	Not repeated	11(15)	1	-	-	-	-	-	-	-	-	-
	Repeated	9+10(14)/rep	1	-	-	-	-	-	-	-	-	-
TEST	Test											
	Register with register	3	2	0	-	-	-	x	x	u	x	0
	Memory with register	9(13)+EA	2-4									
	Immediate with accumulator	4	2-3									
	Immediate with register	5	3-4									
	Immediate with memory	11+EA	3-6									
WAIT	Wait while TEST pin not assert	3+5n	1	-	-	-	-	-	-	-	-	-
XCHG	Exchange											
	Register with accumulator	3	1	-	-	-	-	-	-	-	-	-
	Register with memory	17(25)+EA	2-4									
	Register with register	4	2									
XLAT	Translate	11	1	-	-	-	-	-	-	-	-	-

Appendix B Vocabulary and Terms

（NONE） 空缺,表示本段不与任何段连接
20-bit address adder 20位地址加法器
6-byte instruction queue 6字节指令队列
abacus 算盘
abbreviation 缩写
absorption rules 合并规则
acknowledgement 应答
ADC（add with carry flag） 带进位的加法指令
address bus 地址总线
address connection 地址线连接方式
address decoder 地址译码器
addressable port 可寻址端口
addressing mode 寻址方式
aforementioned 上述的,前述的
albeit 虽然,即使
algorithm 运算法则
align 排列,定位方式
alignment 调整
alignment directive 定位伪指令
alphanumerically 字母和数字混合编制
Altair 8800 1957年4月,MITS发布第一个通用型Altair 8800,是世界上第一台微型计算机
ambiguity 多义性
ampersand &(= and)的记号名称
analogous 类似的,相似的,可比拟的
AND operation 与运算
anemic address space 有限的地址空间
append 添加
appendix 附录
arbitrary 任意的
arithmetic 算术
arithmetic & logic unit（ALU） 算术逻辑单元
ASCII 美国信息交换标准码
assembler 汇编程序
assembly instruction 汇编指令
assembly language 汇编语言

associative 结合律
ASSUME 段寄存器说明伪指令
asterisk 星号"＊"
asynchronous serial communication 异步串行通信
AT paragraph-address 表示本段定位在表达式值指定的段地址处
audible bell 振铃
auto-decrement 自动减少
auto-increment 自动增加
auxiliary 辅助的,补助的
auxiliary carry flag（AF） 辅助进位标志
Babylonians 古巴比伦人
base pointer register 基数指针寄存器
base relative-plus-index addressing 相对的基址加变址寻址
base-plus-index addressing 基址加变址寻址
baud rate 波特率(发报速率)
baud rate factor 异步传输时波特率因子
BCD 二进制编码的十进制
bidirectional lines 双向总线
binary 二进制的
Binary number 二进制数
bipolar 双极的,有两极的
BIOS（basic input/output system） 基本输入输出系统
Boolean algebra 布尔代数
Boolean logic 布尔逻辑(布尔代数的基础是一整套定义了基本函数的逻辑运算,包括与、或、非等运算)
Borland Turbo Borland公司开发的系列语言软件
break character 断码
buffer 缓冲器
built-in driver 内置驱动器
bus 总线
bus cycle 总线周期
bus grant 总线授予

bus interface unit (BIU)　总线接口部件
bus request　总线请求
byte　字节(8位)
BYTE　起始地址为任意值
byte operand　字节型操作数
cable　电缆
cache　高速缓冲存储器
CALL & RET　子程序调用与返回指令
call procedure　子程序调用
carriage return　回车
carry flag (CF)　进位标志
categorize　分类
central processing unit　中央处理器(CPU)
char　字符
charge injection　电荷注入
charge distribution　电荷分布
chip select　片选
chore　零碎工作
chunk of data　数据块
circuitry　电路，线路
class　类别名
CLI (clear interrupt)　清除中断标志
clock frequency　时钟频率
clock period　时钟周期
CMOS (Complementary Metal Oxide Semiconductor)　互补金属氧化物半导体
CMPS (string comparison)　串比较指令
coercion operator　强制转换操作符
colon　冒号
combine　连接方式
comment　注释
COMMON　表示本段与同名、同类别的段共用同一段起始地址，即同名同段相重叠，段的长度是最长段的长度
commutative　交换律，可交换的
compatibility　兼容性
compatible　兼容的
compiler　编译器
conditional jump　条件跳转
conditional-jumping/decision　有条件跳转/决策
conduct　传输，传导
confront　使面临
compatible　兼容的

comprise　包括；由……组成
consecutive　连续的
console　控制台
contiguous　相邻的
control bus　控制总线
control connection　控制线连接方式
control matrix (CM)　控制矩阵
control register　控制寄存器，操作命令控制字
control unit (CU)　控制部件
copyright violation　侵犯版权
counterpart　配对物
cross-reference　交叉引用
CS (code segment) register　指令段寄存器
cyclic　循环的
DAA (decimal adjust for addition)　压缩BCD码的加法调整指令
DAS (decimal adjust for subtraction)　压缩BCD码的减法调整指令
data bus　数据总线
data bus buffer　数据总线缓冲器
data bus drivers and receivers　数据总线收发器
data connection　数据线连接方式
data definition　数据定义
data item　数据项
data link escape　删除
data locality　数据局部性
data-in ACK　数据输入应答信号
data-in buffer register　数据输入缓存寄存器
data-in ready　数据输入准备好
data-in shift register　数据输入移位寄存器(串入并出)
data-out buffer register　数据输出缓存寄存器
data-out shift register　数据输出移位寄存器(并入串出)
debug　调试
debugger　调试器
decimal　十进制的
decoder　解码器
default　缺省值，默认值
denominator　分母
denote　指示，表示
descriptive　描述，说明
descriptor　描述符

destination operand 目的操作数
detest 憎恶
diagrammatically 用图表表示
dictate 命令
direct addressing 直接寻址
direct memory access(DMA) 直接存储器存取方式
direction flag (DF) 方向标志
directive 伪指令
diskette 磁盘
displacement 偏移量
distributive 分配律
dividend 被除数
divisor 除数
DOS (disc operate system) 磁盘操作系统
double-precision 双精度数
DS (data segment) register 数据段寄存器
DSR 数据装置准备好
DTR 数据中断准备好
D-trigger D 触发器
dummy 虚的
duplication operator 重复操作符
duty cycle 占空比
echo 回送
echoed character 返回字符
effective address 有效地址
EH 跟踪方式位(同步有效)
EI (enable interrupt)/DI (disable interrupt) 中断允许/中断禁止
eject 弹出
electrical disturbance 电路中断
encoding instruction 编译指令
end of transmit 传输结束
ENIAC (Electronic Numerical Integrator And Computer) 电子数字积分计算机,世界上第一台电子计算机,于1946年2月15日在美国诞生
enquiry 查询
entity 实体
equivalent 相等的
ER 清除错误标志位
erase 擦除
erroneous 错误的,不正确的
error diagnostics 错误诊断信息

ES (extra segment) register 附加段寄存器
examine 检测
executable module 可执行模块
execution unit (EU) 执行部件
expansion 扩展
exponent 指数
exponential value 指数值
external bus 外部总线
external interrupt 外部中断
external synchronous detect 外同步检测(内同步或外同步)
extrapolate 推断;推知
fabricated 制作好的,已制作完成的
facilitate 使容易,促进
falling edge 下降沿
far call 远程调用
FE 帧校验错标志
feasible 可行的
flag manipulation instruction 标志位操作指令
flash memory 闪存
flags register 标志寄存器
flip-flop 触发器
floating point numbers 浮点数(实数)
flow control 程序流程控制
flowchart 流程图
form feed 换页
fraction 分数
framing 帧校验
full duplex 全双工
gate 门电路
gauge 测量,测定
general-purpose register 通用寄存器
generic 同类的
half duplex 半双工
handshaking logic 联络信号逻辑
handshaking signal 握手信号(同步交换信号)
hardware interrupt 硬件中断
hardware retriggerable one-shot 可编程单脉冲发生器
hardware-triggered strobe 硬件触发方式
hex number 十六进制数
hexadecimal 十六进制的
high order byte 高字节

high-impedance　高阻态,高阻抗
high-impedance state　高阻状态
HLDA　对 HOLD 信号的应答信号
HMOS　高密度金属氧化物半导体工艺(一种单片机封装工艺)
hundreds position　百位
hyphen　连字号
I/O interface　输入/输出接口
I/O-mapped　I/O 映射方式
identical equation　恒等式
idle state　空闲状态
imbed　嵌入
immediate addressing　立即数寻址
immediate operand　立即数
inactive　非激活的,闲置的
inactive low　非当前(非活动态)低电平
incentive　动机,诱因,刺激
increment　增量
index register　变址寄存器
ingenious　有独创性的,精制的,具有创造才能
initialize　初始化
instruction　指令
instruction decoder (ID)　指令译码器
instruction queue　指令队列
instruction register (IR)　指令寄存器
INT function　中断功能
INT instruction　中断指令
integrated　综合的,完整的
integrated circuit　集成电路(IC)
internal bus　内部总线
internal interrupt　内部中断
interrupt acknowledgement　中断应答信号
interrupt flag (IF)　中断标志
interrupt handler　中断处理
interrupt line　中断线
interrupt on terminal count　计数结束中断方式
interrupt priority management　中断优先管理
interrupt type　中断类型码
interrupt vector　中断向量
interrupt vector table　中断向量表
interruption　中断
inter-segment jumps　段间跳转
INTR (maskable interrupt request)　可屏蔽中断请求
intransigent　非妥协性的,不妥协的人
intra-segment jumps　段内跳转
intrinsically　固有地
inverted　反相
invoke　调用
IP (instruction pointer) register　指令指针寄存器
IR　内部复位信号
IRET (interrupt return)　中断返回
IRP directive　重复定义语句 1
IRPC directive　重复定义语句 2
iteration　重复操作
Kilobyte (KB)　千字节(1024 字节)
label　标号
LAHF (load AH with flag)　读取标志指令
latch　锁存器
LEA (load effective address)　取有效地址指令
line feed　换行,移行
linker　连接程序
local directive　局域符号定义语句
logic circuit　逻辑电路
loop　循环
looping construct　循环结构
looping/iteration　循环/迭代
low order byte　低字节
lowercase　小写
machine control instruction　控制指令
machine language　机器码,目标码
macro definition　宏定义
macro expansion　宏扩展
macro invocation　宏调用
macro processing　宏处理
magnetic disk　磁盘
mantissa　尾数
manufacturer　制造商
MASK operator　返回记录字段位或使用情况
master cascade enable (MCE)　主控级联允许信号
maximum mode　最大工作模式
Megabyte (MB)　兆字节(1024KB = 1048576 字节)
MEMORY　表示本段在连接时定位在所有段之上,即高地址处
Memory address register (MAR)　存储地址寄存器

memory cell 存储单元
Memory data register (MDR) 存储器数据寄存器
memory location 存储单元
memory-mapped 内存映射方式
merge 合并
microprocessor 微处理器
microsecond 微秒(10^{-6}s)
mild controversy 轻微的争论
minimum mode 最小工作模式
miscellaneous instruction 其他指令
mnemonic 助记符
mode register 工作方式选择控制字
modulate/demodulate 调制/解调控制电路
module 模块
Morgan's theorem 摩根定律
MOS(Metal Oxide Semiconductor) 金属氧化物半导体
Most Significant Bit (MSB) 最高有效位
multiplex 多元的
multiplexing 多路传输
multiplication 乘法
NA (no auxiliary carry)/AC (auxiliary carry) 无辅助进位/有辅助进位
nanosecond 纳秒(10^{-9}s)
NC (no carry)/CY (carry) 无进位/有进位
near call 近程调用
negate 求反
negative-going edge 下降沿,负沿
negative pulse 负脉冲
nibble 半字节(4位)
NMI (non-maskable interrupt) 非屏蔽中断
NMOS(N-channel Metal Oxide Semiconductor) N沟道金属氧化物半导体
non-volatile 非易失性
non-volatile memory 永久性存储器,非易失存储器
NOT operation 非运算
number of data bits 定义数据字符的长度
number of stop bits 异步方式时停止位长度;
number of sync characters 同步字符数目(单同步或双同步)
numbering system 数制系统
numerator 分子

numeric coprocessor 浮点运算处理器(数值协同处理器)
NV (no overflow)/OV (overflow) 未溢出/溢出
NXOR operation 异或非运算
NZ (nonzero)/ZR (zero) 非零/为零
object program 目标程序
obsolete 废弃的
obviate 排除
octal 八进制的
odd address 奇地址
OE 溢出错标志
offset address 偏移地址
omission 省略
one's complement 反码
one-way communication 单向通信
op-code 操作码
operand 操作数
OR operation 或运算
ORG 定位伪指令
oscillator 振荡器
overflow flag (OF) 溢出标志
overhaul 改进,革新
overhead 在头上的,高架的
overlap 重叠
overlay 覆盖
overrun 越界错误
Packed BCD 压缩的BCD码
PAGE 起始地址的低8位为0,也称为页边界
paper tape punch 纸带打孔机
PARA 起始地址的低4位为0,也称为段边界,为默认定位方式
paragraph 节(128位)
parallel 并行的
parallel communication interface 并行通信接口
parallel to serial 并行至串行转换
parameter 参数
parentheses 圆括号
parity 奇偶校验
parity enable 奇偶校验允许
parity error 奇偶校验错误
parity flag (PF) 奇偶标志
parity type 奇偶校验方式
partition 分割,划分

Pascal statement　Pascal 语句
PCI　周边元件扩展接口
PE　奇偶错标志
periodic interval timer　速率波发生器
periodic rate　周期率
peripheral data enable(PDEN)　外设数据允许信号
peripheral device　外围设备
permissible　可允许的
permutation　排列
pertain　与……相关;属于;适用(于)
physical address　物理地址
pin　管脚
PL(plus sign)/NG(negative)　正数/负数
plug　塞子,插头,插销
PO(parity odd)/PE(parity even)　奇校验/偶校验
portend　预示
position notation　位置记数法
power　幂
pre-assign　预赋值
predecessor　先驱
primitives　基本类型
program counter(PC)　程序计数器
program flow via simple decision　程序流程控制
programmable communication interface　可编程通信接口
programmable timer and event counter　可编程定时/计数控制器
PROM　可编程的只读存储器
prompt　提示符
proportional　与……成比例的
proverbial　众所周知的
pseudo-C array declaration　C 语言数组声明(语句)
PSW　程序状态标志寄存器
PUBLIC　告诉连接程序与本段其他同名同类别的段连接起来,公用一个段的起始地址,形成一个物理段
quadword　四倍字长(64 位)
quote　引用,引证,提供
quotient　商
R/W control logic circuit　读/写控制逻辑电路

radix　基
RAM(Random Access Memory)　随机存储器
real Numbers　实数
reboot　重新启动
receipt　接收
receive control circuit　接收控制电路
receiver　接收器
receiver buffer　接收缓冲器
receiver clock　接收时钟
reciprocal　倒数
RECORD　记录伪指令定义
reenter　再进入
register addressing　寄存器寻址
register indirect addressing　寄存器间接寻址
register relative addressing　寄存器相对寻址
reinitialize　重新初始化
relay coil　继电器绕组线圈
relinquish　放弃
remainder　余数
repeat prefix　重复前缀
repetition directive　重复定义语句
reside　驻留
retain　保持
return address　返回地址
ring counter(RC)　环形计数器
rising edge　上升沿
ROM(Read Only Memory)　只读存储器
RTL(Resistor Transistor Logic)　电阻晶体管逻辑(电路)
RTS　请求发送信号
RxE　为 1 时允许接收
RxRDY　为 0 时接收器准备好
sacrifice　牺牲
saga　传奇
SAHF　设置标志指令
SBB(subtract with carry flag)　带借位的减法指令
SBRK　为 1 发送短缺字符,为 0 正常工作
SCAS(string scan)　串扫描指令
segment　段
segment address　段地址
segment definition　段定义
segment override prefix　设置段前缀

327

segment register　段寄存器
selection connection　选通线连接方式
semicolon　分号(即";")
semiconductor　半导体
serial communication interface　串行通信接口
serial input/parallel output　串行输入/并行输出
serial to parallel　串行至并行转换
set/reset instruction　置位/复位控制
shift register　移位寄存器
shuffle　运载,传输
sign flag (SF)　符号标志
signed integer　有符号整数
simultaneously　同时发生的,同时存在的
single-board system　单板系统
single-chip　单片的
single-precision　单精度数
skip over　跳过,略去
socket　插孔,插座
solder　焊,焊接
software interrupt　软件中断
software-triggered strobe　软件触发方式
sophistication　复杂性
source operand　源操作数
source program　源程序
space　空格
square brackets　方括号
square-wave generator　方波发生器
SS (stack segment) register　堆栈段寄存器
STACK　堆栈段
stack pointer　堆栈指针
stack pointer register　堆栈指针寄存器
state-of-the-art　艺术级的
status information　状态信息
status line　状态线
status register　状态寄存器
status register　状态控制字
statute of "self-locking"　自锁状态
STI (set interrupt)　设置中断标志
storage allocation　存储单元分配
strap　跨接
string comparison instruction　串比较指令
string instruction　串操作指令
strobe　选通信号,选通
STRUCTURE　结构伪指令定义
sub-procedure calls　子程序调用
subroutines　子程序
subscript　下标,脚注
suffix　后缀
superscalar　超标量体系结构
symbolic instruction　符号代码,符号指令
symmetric　对称的
synchronize　同步
synchronous mode　同步模式
synchronous serial communication　同步串行通信
SYNDET　为1时达到同步
syntax　句法
system clock　系统时钟
tens position　十位
terminal　终端
termination　终止
textual substitution　原文代替
the least significant byte　低字节
the most significant byte　高字节
three-state driver　三态驱动器
throughput　产量
time slice　时间片
trace command　跟踪指令
trailing edge of clock pulse　时钟脉冲下降沿
transceiver　收发器(总线驱动器)
transmit buffer　发送缓冲器
transmit clock　发送时钟
transmit control circuit　发送控制电路
transmitter　发送器
trap flag (TF)　跟踪标志
tri-state gate　三态门
TTL (transistor-transistor logic)　晶体管-晶体管逻辑
two's complement　补码
TxE　为1时发送器空
TxEN　为1时允许发送
TxRDY　为1时发送器准备好
type coercion operator　类型强制操作数
ultraviolet　紫外的
units position　个位
Un-packed BCD　非压缩的BCD码
unsigned integer　无符号整数

UP（up direction）/DN（down direction） 增量方向/减量方向
uppercase 大写
usher 引领
vaguely 含糊地
variant 变体（型）
variation 变化,变动
verification 验证
vice versa 反之亦然
volatile 易失性
volatile memory 非永久性存储器,易失存储器
Von Neumann architecture（VNA） 冯·诺依曼体系结构
wait state 等待状态
weight 权
whole number 整数
WIDTH operator 返回记录或字段所占的位数
word 字（16位）
WORD 起始地址的最低位为0,也称为字边界
word operand 字型操作数
workhorse 重负荷机器
XCHG（exchange instruction） 交换指令
XLAT（translate instruction） 换码指令
XOR operation 异或运算
zero flag（ZF） 零标志

Reference

[1] Brey Barry B. The Intel Microprocessors Architecture, Programming, and Interfacing. NJ: Prentice Hall, Inc. ,2003.

[2] Peter Abel. IBM® PC 汇编语言与程序设计. 北京:清华大学出版社,1998.

[3] Liu Y, Gibson G A, Liu Y, et al. Microcomputer systems: the 8086/8088 family architecture, programming and design. Second edition. Englewood Cliffs NJ: Prentice Hall, Inc. ,1984.

[4] 郑学坚,朱定华. 微型计算机原理及应用. 4版. 北京:清华大学出版社,2013.

[5] 冯博琴,吴宁. 微型计算机原理与接口技术. 3版. 北京:清华大学出版社,2011.

[6] 8253/8253-5 Programmable Interval Timer November 1986 Order Number:231306-001.

[7] 沈美明,温冬婵. IBM-PC 汇编语言程序设计. 2版. 北京:清华大学出版社,2012.

[8] 戴梅萼,史嘉权. 微型计算机技术及应用. 4版. 北京:清华大学出版社,2008.

[9] 张菊鹏. 计算机硬件技术基础. 2版. 北京:清华大学出版社,2000.

[10] 尹建华. 微型计算机原理与接口技术. 2版. 北京:高等教育出版社,2008.

[11] 王克义. 计算机硬件技术基础. 北京:清华大学出版社,2014.

[12] 李继灿. 计算机硬件技术基础. 3版. 北京:清华大学出版社,2015.